LIVERPOOL JMU LIBRARY
3 1111 01329 0208

VOLUME FOUR HUNDRED AND EIGHTY-EIGHT

METHODS IN ENZYMOLOGY

Biothermodynamics, Part C

METHODS IN ENZYMOLOGY

Editors-in-Chief

JOHN N. ABELSON AND MELVIN I. SIMON

Division of Biology
California Institute of Technology
Pasadena, California

Founding Editors

SIDNEY P. COLOWICK AND NATHAN O. KAPLAN

VOLUME FOUR HUNDRED AND EIGHTY-EIGHT

METHODS IN ENZYMOLOGY

Biothermodynamics, Part C

EDITED BY

MICHAEL L. JOHNSON
University of Virginia Health Science Center
Department of Pharmacology
Charlottesville, Virginia, USA

JO M. HOLT AND GARY K. ACKERS
Emeritus, Department of Biochemistry and Molecular Biophysics
Washington University School of Medicine
St. Louis, Missouri, USA

AMSTERDAM • BOSTON • HEIDELBERG • LONDON
NEW YORK • OXFORD • PARIS • SAN DIEGO
SAN FRANCISCO • SINGAPORE • SYDNEY • TOKYO
Academic Press is an imprint of Elsevier

Academic Press is an imprint of Elsevier
525 B Street, Suite 1900, San Diego, CA 92101-4495, USA
30 Corporate Drive, Suite 400, Burlington, MA 01803, USA
32 Jamestown Road, London NW1 7BY, UK

First edition 2011

Copyright © 2011, Elsevier Inc. All Rights Reserved.

No part of this publication may be reproduced, stored in a retrieval system or transmitted in any form or by any means electronic, mechanical, photocopying, recording or otherwise without the prior written permission of the publisher

Permissions may be sought directly from Elsevier's Science & Technology Rights Department in Oxford, UK: phone (+44) (0) 1865 843830; fax (+44) (0) 1865 853333; email: permissions@elsevier.com. Alternatively you can submit your request online by visiting the Elsevier web site at http://elsevier.com/locate/permissions, and selecting *Obtaining permission to use Elsevier material*

Notice
No responsibility is assumed by the publisher for any injury and/or damage to persons or property as a matter of products liability, negligence or otherwise, or from any use or operation of any methods, products, instructions or ideas contained in the material herein. Because of rapid advances in the medical sciences, in particular, independent verification of diagnoses and drug dosages should be made

For information on all Academic Press publications
visit our website at elsevierdirect.com

ISBN: 978-0-12-381268-1
ISSN: 0076-6879

Printed and bound in United States of America
11 12 13 10 9 8 7 6 5 4 3 2 1

Working together to grow
libraries in developing countries

www.elsevier.com | www.bookaid.org | www.sabre.org

ELSEVIER BOOK AID International Sabre Foundation

Contents

Contributors

Andrew J. Andrews
Department of Biochemistry and Molecular Biology, Colorado State University, Fort Collins, Colorado, USA

Stefan T. Arold
Department of Biochemistry and Molecular Biology, University of Texas MD Anderson Cancer Center, Houston, Texas, USA

Zeljko Bajzer
Department of Biochemistry and Molecular Biology, Mayo Clinic College of Medicine, Rochester, Minnesota, USA

Nicole A. Becker
Department of Biochemistry and Molecular Biology, Mayo Clinic College of Medicine, Rochester, Minnesota, USA

Dorothy Beckett
Department of Chemistry and Biochemistry, Center for Biological Structure and Organization, University of Maryland, Maryland, USA

Wlodzimierz Bujalowski
Department of Biochemistry and Molecular Biology, and Department of Obstetrics and Gynecology, The Sealy Center for Structural Biology, Sealy Center for Cancer Cell Biology, The University of Texas Medical Branch at Galveston, Galveston, Texas, USA

James L. Cole
Department of Molecular and Cell Biology, and Department of Chemistry, University of Connecticut, Storrs, Connecticut, USA

Deborah G. Conrady
Department of Molecular Genetics, Biochemistry & Microbiology, University of Cincinnati College of Medicine, Cincinnati, Ohio, USA

Ernesto J. Fuentes
Department of Biochemistry, Roy J. and Lucille A. Carver College of Medicine, University of Iowa, and Holden Comprehensive Cancer Center, Iowa City, Iowa, USA

Petr Herman
Institute of Physics, Charles University, Ke Karlovu, Prague, Czech Republic

Andrew B. Herr
Department of Molecular Genetics, Biochemistry & Microbiology, University of Cincinnati College of Medicine, Cincinnati, Ohio, USA

Nicholas G. Housden
Department of Biology (Area 10), University of York, York, United Kingdom

Maria J. Jezewska
Department of Biochemistry and Molecular Biology, The Sealy Center for Structural Biology, Sealy Center for Cancer Cell Biology, The University of Texas Medical Branch at Galveston, Galveston, Texas, USA

Jason D. Kahn
Department of Chemistry and Biochemistry, University of Maryland, College Park, Maryland, USA

Colin Kleanthous
Department of Biology (Area 10), University of York, York, United Kingdom

John E. Ladbury
Department of Biochemistry and Molecular Biology, University of Texas MD Anderson Cancer Center, Houston, Texas, USA

Katherine Launer-Felty
Department of Molecular and Cell Biology, University of Connecticut, Storrs, Connecticut, USA

J. Ching Lee
Department of Biochemistry and Molecular Biology, The University of Texas Medical Branch at Galveston, Galveston, Texas, USA

Vince J. LiCata
Department of Biological Sciences, Louisiana State University, Baton Rouge, Louisiana, USA

Chin-Chi Liu
Department of Biological Sciences, Louisiana State University, Baton Rouge, Louisiana, USA

Aaron L. Lucius
Department of Chemistry, The University of Alabama at Birmingham, Birmingham, Alabama, USA

Karolin Luger
Department of Biochemistry and Molecular Biology, Colorado State University, Fort Collins, Colorado, and Howard Hughes Medical Institute, USA

L. James Maher III
Department of Biochemistry and Molecular Biology, Mayo Clinic College of Medicine, Rochester, Minnesota, USA

Justin M. Miller
Department of Chemistry, The University of Alabama at Birmingham, Birmingham, Alabama, USA

Justin P. Peters
Department of Biochemistry and Molecular Biology, Mayo Clinic College of Medicine, Rochester, Minnesota, USA

Burki Rajendar
Department of Chemistry, The University of Alabama at Birmingham, Birmingham, Alabama, USA

Emily M. Rueter
Department of Biochemistry and Molecular Biology, Mayo Clinic College of Medicine, Rochester, Minnesota, USA

Tyson R. Shepherd
Department of Biochemistry, Roy J. and Lucille A. Carver College of Medicine, University of Iowa, Iowa City, Iowa, USA

C. Jason Wong
Department of Molecular and Cell Biology, University of Connecticut, Storrs, Connecticut, USA

PREFACE

This volume is the continuation in a series of Methods in Enzymology volumes which promotes thermodynamics as an important tool for the study of biological systems. One of the many examples of biological thermodynamics is the cooperative binding of oxygen by hemoglobin.

Hemoglobin is the quintessential example of a ligand-binding protein. Most biochemistry textbooks explain that the hemoglobin tetramer exists in two structural states, a low-affinity structure without oxygen bound and a high-affinity structure with oxygen bound. This is the classic two-state allosteric model as presented by Monod, Wyman, and Changeux (1965, *J. Mol. Biol.*, **12**, 88–118) and extended by Ackers and Johnson (1981, *J. Mol. Biol.* **147**, 559–582). Unfortunately, this model tells us nothing about the specific molecular interactions that are altered by the binding of oxygen, which force the hemoglobin to shift to the alternative structural state.

Investigating hemoglobin at this level is analogous to viewing only the first and last act of a tragedy by William Shakespeare (e.g., King Lear or Macbeth). The first act being a celebration and in the last act, the stage is full of dead bodies. While only the first and last acts are provocative, the truly wonderful part of a Shakespearean play is the character interactions in the intervening steps between the initial and final states, which force the last act to follow from the first act.

Thermodynamics provides a conceptual and mathematical framework, that is, a "logic tool," which allows the investigation of the specific molecular interactions and the concomitant energetics such as those that are altered by the binding of oxygen, which force the hemoglobin to shift to the alternative structural and/or association states.

MICHAEL L JOHNSON, JO HOLT AND GARY K ACKERS

Methods in Enzymology

Volume I. Preparation and Assay of Enzymes
Edited by Sidney P. Colowick and Nathan O. Kaplan

Volume II. Preparation and Assay of Enzymes
Edited by Sidney P. Colowick and Nathan O. Kaplan

Volume III. Preparation and Assay of Substrates
Edited by Sidney P. Colowick and Nathan O. Kaplan

Volume IV. Special Techniques for the Enzymologist
Edited by Sidney P. Colowick and Nathan O. Kaplan

Volume V. Preparation and Assay of Enzymes
Edited by Sidney P. Colowick and Nathan O. Kaplan

Volume VI. Preparation and Assay of Enzymes *(Continued)*
Preparation and Assay of Substrates
Special Techniques
Edited by Sidney P. Colowick and Nathan O. Kaplan

Volume VII. Cumulative Subject Index
Edited by Sidney P. Colowick and Nathan O. Kaplan

Volume VIII. Complex Carbohydrates
Edited by Elizabeth F. Neufeld and Victor Ginsburg

Volume IX. Carbohydrate Metabolism
Edited by Willis A. Wood

Volume X. Oxidation and Phosphorylation
Edited by Ronald W. Estabrook and Maynard E. Pullman

Volume XI. Enzyme Structure
Edited by C. H. W. Hirs

Volume XII. Nucleic Acids (Parts A and B)
Edited by Lawrence Grossman and Kivie Moldave

Volume XIII. Citric Acid Cycle
Edited by J. M. Lowenstein

Volume XIV. Lipids
Edited by J. M. Lowenstein

Volume XV. Steroids and Terpenoids
Edited by Raymond B. Clayton

Volume XVI. Fast Reactions
Edited by Kenneth Kustin

Volume XVII. Metabolism of Amino Acids and Amines (Parts A and B)
Edited by Herbert Tabor and Celia White Tabor

Volume XVIII. Vitamins and Coenzymes (Parts A, B, and C)
Edited by Donald B. McCormick and Lemuel D. Wright

Volume XIX. Proteolytic Enzymes
Edited by Gertrude E. Perlmann and Laszlo Lorand

Volume XX. Nucleic Acids and Protein Synthesis (Part C)
Edited by Kivie Moldave and Lawrence Grossman

Volume XXI. Nucleic Acids (Part D)
Edited by Lawrence Grossman and Kivie Moldave

Volume XXII. Enzyme Purification and Related Techniques
Edited by William B. Jakoby

Volume XXIII. Photosynthesis (Part A)
Edited by Anthony San Pietro

Volume XXIV. Photosynthesis and Nitrogen Fixation (Part B)
Edited by Anthony San Pietro

Volume XXV. Enzyme Structure (Part B)
Edited by C. H. W. Hirs and Serge N. Timasheff

Volume XXVI. Enzyme Structure (Part C)
Edited by C. H. W. Hirs and Serge N. Timasheff

Volume XXVII. Enzyme Structure (Part D)
Edited by C. H. W. Hirs and Serge N. Timasheff

Volume XXVIII. Complex Carbohydrates (Part B)
Edited by Victor Ginsburg

Volume XXIX. Nucleic Acids and Protein Synthesis (Part E)
Edited by Lawrence Grossman and Kivie Moldave

Volume XXX. Nucleic Acids and Protein Synthesis (Part F)
Edited by Kivie Moldave and Lawrence Grossman

Volume XXXI. Biomembranes (Part A)
Edited by Sidney Fleischer and Lester Packer

Volume XXXII. Biomembranes (Part B)
Edited by Sidney Fleischer and Lester Packer

Volume XXXIII. Cumulative Subject Index Volumes I-XXX
Edited by Martha G. Dennis and Edward A. Dennis

Volume XXXIV. Affinity Techniques (Enzyme Purification: Part B)
Edited by William B. Jakoby and Meir Wilchek

VOLUME 123. Vitamins and Coenzymes (Part H)
Edited by FRANK CHYTIL AND DONALD B. MCCORMICK

VOLUME 124. Hormone Action (Part J: Neuroendocrine Peptides)
Edited by P. MICHAEL CONN

VOLUME 125. Biomembranes (Part M: Transport in Bacteria, Mitochondria, and Chloroplasts: General Approaches and Transport Systems)
Edited by SIDNEY FLEISCHER AND BECCA FLEISCHER

VOLUME 126. Biomembranes (Part N: Transport in Bacteria, Mitochondria, and Chloroplasts: Protonmotive Force)
Edited by SIDNEY FLEISCHER AND BECCA FLEISCHER

VOLUME 127. Biomembranes (Part O: Protons and Water: Structure and Translocation)
Edited by LESTER PACKER

VOLUME 128. Plasma Lipoproteins (Part A: Preparation, Structure, and Molecular Biology)
Edited by JERE P. SEGREST AND JOHN J. ALBERS

VOLUME 129. Plasma Lipoproteins (Part B: Characterization, Cell Biology, and Metabolism)
Edited by JOHN J. ALBERS AND JERE P. SEGREST

VOLUME 130. Enzyme Structure (Part K)
Edited by C. H. W. HIRS AND SERGE N. TIMASHEFF

VOLUME 131. Enzyme Structure (Part L)
Edited by C. H. W. HIRS AND SERGE N. TIMASHEFF

VOLUME 132. Immunochemical Techniques (Part J: Phagocytosis and Cell-Mediated Cytotoxicity)
Edited by GIOVANNI DI SABATO AND JOHANNES EVERSE

VOLUME 133. Bioluminescence and Chemiluminescence (Part B)
Edited by MARLENE DELUCA AND WILLIAM D. MCELROY

VOLUME 134. Structural and Contractile Proteins (Part C: The Contractile Apparatus and the Cytoskeleton)
Edited by RICHARD B. VALLEE

VOLUME 135. Immobilized Enzymes and Cells (Part B)
Edited by KLAUS MOSBACH

VOLUME 136. Immobilized Enzymes and Cells (Part C)
Edited by KLAUS MOSBACH

VOLUME 137. Immobilized Enzymes and Cells (Part D)
Edited by KLAUS MOSBACH

VOLUME 138. Complex Carbohydrates (Part E)
Edited by VICTOR GINSBURG

VOLUME 139. Cellular Regulators (Part A: Calcium- and Calmodulin-Binding Proteins)
Edited by ANTHONY R. MEANS AND P. MICHAEL CONN

VOLUME 140. Cumulative Subject Index Volumes 102–119, 121–134

VOLUME 141. Cellular Regulators (Part B: Calcium and Lipids)
Edited by P. MICHAEL CONN AND ANTHONY R. MEANS

VOLUME 142. Metabolism of Aromatic Amino Acids and Amines
Edited by SEYMOUR KAUFMAN

VOLUME 143. Sulfur and Sulfur Amino Acids
Edited by WILLIAM B. JAKOBY AND OWEN GRIFFITH

VOLUME 144. Structural and Contractile Proteins (Part D: Extracellular Matrix)
Edited by LEON W. CUNNINGHAM

VOLUME 145. Structural and Contractile Proteins (Part E: Extracellular Matrix)
Edited by LEON W. CUNNINGHAM

VOLUME 146. Peptide Growth Factors (Part A)
Edited by DAVID BARNES AND DAVID A. SIRBASKU

VOLUME 147. Peptide Growth Factors (Part B)
Edited by DAVID BARNES AND DAVID A. SIRBASKU

VOLUME 148. Plant Cell Membranes
Edited by LESTER PACKER AND ROLAND DOUCE

VOLUME 149. Drug and Enzyme Targeting (Part B)
Edited by RALPH GREEN AND KENNETH J. WIDDER

VOLUME 150. Immunochemical Techniques (Part K: *In Vitro* Models of B and T Cell Functions and Lymphoid Cell Receptors)
Edited by GIOVANNI DI SABATO

VOLUME 151. Molecular Genetics of Mammalian Cells
Edited by MICHAEL M. GOTTESMAN

VOLUME 152. Guide to Molecular Cloning Techniques
Edited by SHELBY L. BERGER AND ALAN R. KIMMEL

VOLUME 153. Recombinant DNA (Part D)
Edited by RAY WU AND LAWRENCE GROSSMAN

VOLUME 154. Recombinant DNA (Part E)
Edited by RAY WU AND LAWRENCE GROSSMAN

VOLUME 155. Recombinant DNA (Part F)
Edited by RAY WU

VOLUME 156. Biomembranes (Part P: ATP-Driven Pumps and Related Transport: The Na, K-Pump)
Edited by SIDNEY FLEISCHER AND BECCA FLEISCHER

VOLUME 190. Retinoids (Part B: Cell Differentiation and Clinical Applications)
Edited by LESTER PACKER

VOLUME 191. Biomembranes (Part V: Cellular and Subcellular Transport: Epithelial Cells)
Edited by SIDNEY FLEISCHER AND BECCA FLEISCHER

VOLUME 192. Biomembranes (Part W: Cellular and Subcellular Transport: Epithelial Cells)
Edited by SIDNEY FLEISCHER AND BECCA FLEISCHER

VOLUME 193. Mass Spectrometry
Edited by JAMES A. MCCLOSKEY

VOLUME 194. Guide to Yeast Genetics and Molecular Biology
Edited by CHRISTINE GUTHRIE AND GERALD R. FINK

VOLUME 195. Adenylyl Cyclase, G Proteins, and Guanylyl Cyclase
Edited by ROGER A. JOHNSON AND JACKIE D. CORBIN

VOLUME 196. Molecular Motors and the Cytoskeleton
Edited by RICHARD B. VALLEE

VOLUME 197. Phospholipases
Edited by EDWARD A. DENNIS

VOLUME 198. Peptide Growth Factors (Part C)
Edited by DAVID BARNES, J. P. MATHER, AND GORDON H. SATO

VOLUME 199. Cumulative Subject Index Volumes 168–174, 176–194

VOLUME 200. Protein Phosphorylation (Part A: Protein Kinases: Assays, Purification, Antibodies, Functional Analysis, Cloning, and Expression)
Edited by TONY HUNTER AND BARTHOLOMEW M. SEFTON

VOLUME 201. Protein Phosphorylation (Part B: Analysis of Protein Phosphorylation, Protein Kinase Inhibitors, and Protein Phosphatases)
Edited by TONY HUNTER AND BARTHOLOMEW M. SEFTON

VOLUME 202. Molecular Design and Modeling: Concepts and Applications (Part A: Proteins, Peptides, and Enzymes)
Edited by JOHN J. LANGONE

VOLUME 203. Molecular Design and Modeling: Concepts and Applications (Part B: Antibodies and Antigens, Nucleic Acids, Polysaccharides, and Drugs)
Edited by JOHN J. LANGONE

VOLUME 204. Bacterial Genetic Systems
Edited by JEFFREY H. MILLER

VOLUME 205. Metallobiochemistry (Part B: Metallothionein and Related Molecules)
Edited by JAMES F. RIORDAN AND BERT L. VALLEE

Volume 206. Cytochrome P450
Edited by Michael R. Waterman and Eric F. Johnson

Volume 207. Ion Channels
Edited by Bernardo Rudy and Linda E. Iverson

Volume 208. Protein–DNA Interactions
Edited by Robert T. Sauer

Volume 209. Phospholipid Biosynthesis
Edited by Edward A. Dennis and Dennis E. Vance

Volume 210. Numerical Computer Methods
Edited by Ludwig Brand and Michael L. Johnson

Volume 211. DNA Structures (Part A: Synthesis and Physical Analysis of DNA)
Edited by David M. J. Lilley and James E. Dahlberg

Volume 212. DNA Structures (Part B: Chemical and Electrophoretic Analysis of DNA)
Edited by David M. J. Lilley and James E. Dahlberg

Volume 213. Carotenoids (Part A: Chemistry, Separation, Quantitation, and Antioxidation)
Edited by Lester Packer

Volume 214. Carotenoids (Part B: Metabolism, Genetics, and Biosynthesis)
Edited by Lester Packer

Volume 215. Platelets: Receptors, Adhesion, Secretion (Part B)
Edited by Jacek J. Hawiger

Volume 216. Recombinant DNA (Part G)
Edited by Ray Wu

Volume 217. Recombinant DNA (Part H)
Edited by Ray Wu

Volume 218. Recombinant DNA (Part I)
Edited by Ray Wu

Volume 219. Reconstitution of Intracellular Transport
Edited by James E. Rothman

Volume 220. Membrane Fusion Techniques (Part A)
Edited by Nejat Düzgüneş

Volume 221. Membrane Fusion Techniques (Part B)
Edited by Nejat Düzgüneş

Volume 222. Proteolytic Enzymes in Coagulation, Fibrinolysis, and Complement Activation (Part A: Mammalian Blood Coagulation Factors and Inhibitors)
Edited by Laszlo Lorand and Kenneth G. Mann

Volume 223. Proteolytic Enzymes in Coagulation, Fibrinolysis, and Complement Activation (Part B: Complement Activation, Fibrinolysis, and Nonmammalian Blood Coagulation Factors)
Edited by Laszlo Lorand and Kenneth G. Mann

Volume 224. Molecular Evolution: Producing the Biochemical Data
Edited by Elizabeth Anne Zimmer, Thomas J. White, Rebecca L. Cann, and Allan C. Wilson

Volume 225. Guide to Techniques in Mouse Development
Edited by Paul M. Wassarman and Melvin L. DePamphilis

Volume 226. Metallobiochemistry (Part C: Spectroscopic and Physical Methods for Probing Metal Ion Environments in Metalloenzymes and Metalloproteins)
Edited by James F. Riordan and Bert L. Vallee

Volume 227. Metallobiochemistry (Part D: Physical and Spectroscopic Methods for Probing Metal Ion Environments in Metalloproteins)
Edited by James F. Riordan and Bert L. Vallee

Volume 228. Aqueous Two-Phase Systems
Edited by Harry Walter and Göte Johansson

Volume 229. Cumulative Subject Index Volumes 195–198, 200–227

Volume 230. Guide to Techniques in Glycobiology
Edited by William J. Lennarz and Gerald W. Hart

Volume 231. Hemoglobins (Part B: Biochemical and Analytical Methods)
Edited by Johannes Everse, Kim D. Vandegriff, and Robert M. Winslow

Volume 232. Hemoglobins (Part C: Biophysical Methods)
Edited by Johannes Everse, Kim D. Vandegriff, and Robert M. Winslow

Volume 233. Oxygen Radicals in Biological Systems (Part C)
Edited by Lester Packer

Volume 234. Oxygen Radicals in Biological Systems (Part D)
Edited by Lester Packer

Volume 235. Bacterial Pathogenesis (Part A: Identification and Regulation of Virulence Factors)
Edited by Virginia L. Clark and Patrik M. Bavoil

Volume 236. Bacterial Pathogenesis (Part B: Integration of Pathogenic Bacteria with Host Cells)
Edited by Virginia L. Clark and Patrik M. Bavoil

Volume 237. Heterotrimeric G Proteins
Edited by Ravi Iyengar

Volume 238. Heterotrimeric G-Protein Effectors
Edited by Ravi Iyengar

VOLUME 239. Nuclear Magnetic Resonance (Part C)
Edited by THOMAS L. JAMES AND NORMAN J. OPPENHEIMER

VOLUME 240. Numerical Computer Methods (Part B)
Edited by MICHAEL L. JOHNSON AND LUDWIG BRAND

VOLUME 241. Retroviral Proteases
Edited by LAWRENCE C. KUO AND JULES A. SHAFER

VOLUME 242. Neoglycoconjugates (Part A)
Edited by Y. C. LEE AND REIKO T. LEE

VOLUME 243. Inorganic Microbial Sulfur Metabolism
Edited by HARRY D. PECK, JR., AND JEAN LEGALL

VOLUME 244. Proteolytic Enzymes: Serine and Cysteine Peptidases
Edited by ALAN J. BARRETT

VOLUME 245. Extracellular Matrix Components
Edited by E. RUOSLAHTI AND E. ENGVALL

VOLUME 246. Biochemical Spectroscopy
Edited by KENNETH SAUER

VOLUME 247. Neoglycoconjugates (Part B: Biomedical Applications)
Edited by Y. C. LEE AND REIKO T. LEE

VOLUME 248. Proteolytic Enzymes: Aspartic and Metallo Peptidases
Edited by ALAN J. BARRETT

VOLUME 249. Enzyme Kinetics and Mechanism (Part D: Developments in Enzyme Dynamics)
Edited by DANIEL L. PURICH

VOLUME 250. Lipid Modifications of Proteins
Edited by PATRICK J. CASEY AND JANICE E. BUSS

VOLUME 251. Biothiols (Part A: Monothiols and Dithiols, Protein Thiols, and Thiyl Radicals)
Edited by LESTER PACKER

VOLUME 252. Biothiols (Part B: Glutathione and Thioredoxin; Thiols in Signal Transduction and Gene Regulation)
Edited by LESTER PACKER

VOLUME 253. Adhesion of Microbial Pathogens
Edited by RON J. DOYLE AND ITZHAK OFEK

VOLUME 254. Oncogene Techniques
Edited by PETER K. VOGT AND INDER M. VERMA

VOLUME 255. Small GTPases and Their Regulators (Part A: Ras Family)
Edited by W. E. BALCH, CHANNING J. DER, AND ALAN HALL

Volume 274. RNA Polymerase and Associated Factors (Part B)
Edited by Sankar Adhya

Volume 275. Viral Polymerases and Related Proteins
Edited by Lawrence C. Kuo, David B. Olsen, and Steven S. Carroll

Volume 276. Macromolecular Crystallography (Part A)
Edited by Charles W. Carter, Jr., and Robert M. Sweet

Volume 277. Macromolecular Crystallography (Part B)
Edited by Charles W. Carter, Jr., and Robert M. Sweet

Volume 278. Fluorescence Spectroscopy
Edited by Ludwig Brand and Michael L. Johnson

Volume 279. Vitamins and Coenzymes (Part I)
Edited by Donald B. McCormick, John W. Suttie, and Conrad Wagner

Volume 280. Vitamins and Coenzymes (Part J)
Edited by Donald B. McCormick, John W. Suttie, and Conrad Wagner

Volume 281. Vitamins and Coenzymes (Part K)
Edited by Donald B. McCormick, John W. Suttie, and Conrad Wagner

Volume 282. Vitamins and Coenzymes (Part L)
Edited by Donald B. McCormick, John W. Suttie, and Conrad Wagner

Volume 283. Cell Cycle Control
Edited by William G. Dunphy

Volume 284. Lipases (Part A: Biotechnology)
Edited by Byron Rubin and Edward A. Dennis

Volume 285. Cumulative Subject Index Volumes 263, 264, 266–284, 286–289

Volume 286. Lipases (Part B: Enzyme Characterization and Utilization)
Edited by Byron Rubin and Edward A. Dennis

Volume 287. Chemokines
Edited by Richard Horuk

Volume 288. Chemokine Receptors
Edited by Richard Horuk

Volume 289. Solid Phase Peptide Synthesis
Edited by Gregg B. Fields

Volume 290. Molecular Chaperones
Edited by George H. Lorimer and Thomas Baldwin

Volume 291. Caged Compounds
Edited by Gerard Marriott

Volume 292. ABC Transporters: Biochemical, Cellular, and Molecular Aspects
Edited by Suresh V. Ambudkar and Michael M. Gottesman

VOLUME 345. G Protein Pathways (Part C: Effector Mechanisms)
Edited by RAVI IYENGAR AND JOHN D. HILDEBRANDT

VOLUME 346. Gene Therapy Methods
Edited by M. IAN PHILLIPS

VOLUME 347. Protein Sensors and Reactive Oxygen Species (Part A: Selenoproteins and Thioredoxin)
Edited by HELMUT SIES AND LESTER PACKER

VOLUME 348. Protein Sensors and Reactive Oxygen Species (Part B: Thiol Enzymes and Proteins)
Edited by HELMUT SIES AND LESTER PACKER

VOLUME 349. Superoxide Dismutase
Edited by LESTER PACKER

VOLUME 350. Guide to Yeast Genetics and Molecular and Cell Biology (Part B)
Edited by CHRISTINE GUTHRIE AND GERALD R. FINK

VOLUME 351. Guide to Yeast Genetics and Molecular and Cell Biology (Part C)
Edited by CHRISTINE GUTHRIE AND GERALD R. FINK

VOLUME 352. Redox Cell Biology and Genetics (Part A)
Edited by CHANDAN K. SEN AND LESTER PACKER

VOLUME 353. Redox Cell Biology and Genetics (Part B)
Edited by CHANDAN K. SEN AND LESTER PACKER

VOLUME 354. Enzyme Kinetics and Mechanisms (Part F: Detection and Characterization of Enzyme Reaction Intermediates)
Edited by DANIEL L. PURICH

VOLUME 355. Cumulative Subject Index Volumes 321–354

VOLUME 356. Laser Capture Microscopy and Microdissection
Edited by P. MICHAEL CONN

VOLUME 357. Cytochrome P450, Part C
Edited by ERIC F. JOHNSON AND MICHAEL R. WATERMAN

VOLUME 358. Bacterial Pathogenesis (Part C: Identification, Regulation, and Function of Virulence Factors)
Edited by VIRGINIA L. CLARK AND PATRIK M. BAVOIL

VOLUME 359. Nitric Oxide (Part D)
Edited by ENRIQUE CADENAS AND LESTER PACKER

VOLUME 360. Biophotonics (Part A)
Edited by GERARD MARRIOTT AND IAN PARKER

VOLUME 361. Biophotonics (Part B)
Edited by GERARD MARRIOTT AND IAN PARKER

Volume 362. Recognition of Carbohydrates in Biological Systems (Part A)
Edited by Yuan C. Lee and Reiko T. Lee

Volume 363. Recognition of Carbohydrates in Biological Systems (Part B)
Edited by Yuan C. Lee and Reiko T. Lee

Volume 364. Nuclear Receptors
Edited by David W. Russell and David J. Mangelsdorf

Volume 365. Differentiation of Embryonic Stem Cells
Edited by Paul M. Wassauman and Gordon M. Keller

Volume 366. Protein Phosphatases
Edited by Susanne Klumpp and Josef Krieglstein

Volume 367. Liposomes (Part A)
Edited by Nejat Düzgüneş

Volume 368. Macromolecular Crystallography (Part C)
Edited by Charles W. Carter, Jr., and Robert M. Sweet

Volume 369. Combinational Chemistry (Part B)
Edited by Guillermo A. Morales and Barry A. Bunin

Volume 370. RNA Polymerases and Associated Factors (Part C)
Edited by Sankar L. Adhya and Susan Garges

Volume 371. RNA Polymerases and Associated Factors (Part D)
Edited by Sankar L. Adhya and Susan Garges

Volume 372. Liposomes (Part B)
Edited by Nejat Düzgüneş

Volume 373. Liposomes (Part C)
Edited by Nejat Düzgüneş

Volume 374. Macromolecular Crystallography (Part D)
Edited by Charles W. Carter, Jr., and Robert W. Sweet

Volume 375. Chromatin and Chromatin Remodeling Enzymes (Part A)
Edited by C. David Allis and Carl Wu

Volume 376. Chromatin and Chromatin Remodeling Enzymes (Part B)
Edited by C. David Allis and Carl Wu

Volume 377. Chromatin and Chromatin Remodeling Enzymes (Part C)
Edited by C. David Allis and Carl Wu

Volume 378. Quinones and Quinone Enzymes (Part A)
Edited by Helmut Sies and Lester Packer

Volume 379. Energetics of Biological Macromolecules (Part D)
Edited by Jo M. Holt, Michael L. Johnson, and Gary K. Ackers

Volume 380. Energetics of Biological Macromolecules (Part E)
Edited by Jo M. Holt, Michael L. Johnson, and Gary K. Ackers

Volume 417. Functional Glycomics
Edited by Minoru Fukuda

Volume 418. Embryonic Stem Cells
Edited by Irina Klimanskaya and Robert Lanza

Volume 419. Adult Stem Cells
Edited by Irina Klimanskaya and Robert Lanza

Volume 420. Stem Cell Tools and Other Experimental Protocols
Edited by Irina Klimanskaya and Robert Lanza

Volume 421. Advanced Bacterial Genetics: Use of Transposons and Phage for Genomic Engineering
Edited by Kelly T. Hughes

Volume 422. Two-Component Signaling Systems, Part A
Edited by Melvin I. Simon, Brian R. Crane, and Alexandrine Crane

Volume 423. Two-Component Signaling Systems, Part B
Edited by Melvin I. Simon, Brian R. Crane, and Alexandrine Crane

Volume 424. RNA Editing
Edited by Jonatha M. Gott

Volume 425. RNA Modification
Edited by Jonatha M. Gott

Volume 426. Integrins
Edited by David Cheresh

Volume 427. MicroRNA Methods
Edited by John J. Rossi

Volume 428. Osmosensing and Osmosignaling
Edited by Helmut Sies and Dieter Haussinger

Volume 429. Translation Initiation: Extract Systems and Molecular Genetics
Edited by Jon Lorsch

Volume 430. Translation Initiation: Reconstituted Systems and Biophysical Methods
Edited by Jon Lorsch

Volume 431. Translation Initiation: Cell Biology, High-Throughput and Chemical-Based Approaches
Edited by Jon Lorsch

Volume 432. Lipidomics and Bioactive Lipids: Mass-Spectrometry–Based Lipid Analysis
Edited by H. Alex Brown

VOLUME 433. Lipidomics and Bioactive Lipids: Specialized Analytical Methods and Lipids in Disease
Edited by H. ALEX BROWN

VOLUME 434. Lipidomics and Bioactive Lipids: Lipids and Cell Signaling
Edited by H. ALEX BROWN

VOLUME 435. Oxygen Biology and Hypoxia
Edited by HELMUT SIES AND BERNHARD BRÜNE

VOLUME 436. Globins and Other Nitric Oxide-Reactive Protiens (Part A)
Edited by ROBERT K. POOLE

VOLUME 437. Globins and Other Nitric Oxide-Reactive Protiens (Part B)
Edited by ROBERT K. POOLE

VOLUME 438. Small GTPases in Disease (Part A)
Edited by WILLIAM E. BALCH, CHANNING J. DER, AND ALAN HALL

VOLUME 439. Small GTPases in Disease (Part B)
Edited by WILLIAM E. BALCH, CHANNING J. DER, AND ALAN HALL

VOLUME 440. Nitric Oxide, Part F Oxidative and Nitrosative Stress in Redox Regulation of Cell Signaling
Edited by ENRIQUE CADENAS AND LESTER PACKER

VOLUME 441. Nitric Oxide, Part G Oxidative and Nitrosative Stress in Redox Regulation of Cell Signaling
Edited by ENRIQUE CADENAS AND LESTER PACKER

VOLUME 442. Programmed Cell Death, General Principles for Studying Cell Death (Part A)
Edited by ROYA KHOSRAVI-FAR, ZAHRA ZAKERI, RICHARD A. LOCKSHIN, AND MAURO PIACENTINI

VOLUME 443. Angiogenesis: *In Vitro* Systems
Edited by DAVID A. CHERESH

VOLUME 444. Angiogenesis: *In Vivo* Systems (Part A)
Edited by DAVID A. CHERESH

VOLUME 445. Angiogenesis: *In Vivo* Systems (Part B)
Edited by DAVID A. CHERESH

VOLUME 446. Programmed Cell Death, The Biology and Therapeutic Implications of Cell Death (Part B)
Edited by ROYA KHOSRAVI-FAR, ZAHRA ZAKERI, RICHARD A. LOCKSHIN, AND MAURO PIACENTINI

VOLUME 447. RNA Turnover in Bacteria, Archaea and Organelles
Edited by LYNNE E. MAQUAT AND CECILIA M. ARRAIANO

VOLUME 448. RNA Turnover in Eukaryotes: Nucleases, Pathways and Analysis of mRNA Decay
Edited by LYNNE E. MAQUAT AND MEGERDITCH KILEDJIAN

VOLUME 449. RNA Turnover in Eukaryotes: Analysis of Specialized and Quality Control RNA Decay Pathways
Edited by LYNNE E. MAQUAT AND MEGERDITCH KILEDJIAN

VOLUME 450. Fluorescence Spectroscopy
Edited by LUDWIG BRAND AND MICHAEL L. JOHNSON

VOLUME 451. Autophagy: Lower Eukaryotes and Non-Mammalian Systems (Part A)
Edited by DANIEL J. KLIONSKY

VOLUME 452. Autophagy in Mammalian Systems (Part B)
Edited by DANIEL J. KLIONSKY

VOLUME 453. Autophagy in Disease and Clinical Applications (Part C)
Edited by DANIEL J. KLIONSKY

VOLUME 454. Computer Methods (Part A)
Edited by MICHAEL L. JOHNSON AND LUDWIG BRAND

VOLUME 455. Biothermodynamics (Part A)
Edited by MICHAEL L. JOHNSON, JO M. HOLT, AND GARY K. ACKERS (RETIRED)

VOLUME 456. Mitochondrial Function, Part A: Mitochondrial Electron Transport Complexes and Reactive Oxygen Species
Edited by WILLIAM S. ALLISON AND IMMO E. SCHEFFLER

VOLUME 457. Mitochondrial Function, Part B: Mitochondrial Protein Kinases, Protein Phosphatases and Mitochondrial Diseases
Edited by WILLIAM S. ALLISON AND ANNE N. MURPHY

VOLUME 458. Complex Enzymes in Microbial Natural Product Biosynthesis, Part A: Overview Articles and Peptides
Edited by DAVID A. HOPWOOD

VOLUME 459. Complex Enzymes in Microbial Natural Product Biosynthesis, Part B: Polyketides, Aminocoumarins and Carbohydrates
Edited by DAVID A. HOPWOOD

VOLUME 460. Chemokines, Part A
Edited by TRACY M. HANDEL AND DAMON J. HAMEL

VOLUME 461. Chemokines, Part B
Edited by TRACY M. HANDEL AND DAMON J. HAMEL

VOLUME 462. Non-Natural Amino Acids
Edited by TOM W. MUIR AND JOHN N. ABELSON

VOLUME 463. Guide to Protein Purification, 2nd Edition
Edited by RICHARD R. BURGESS AND MURRAY P. DEUTSCHER

Volume 479. Functional Glycomics
Edited by Minoru Fukuda

Volume 480. Glycobiology
Edited by Minoru Fukuda

Volume 481. Cryo-EM, Part A: Sample Preparation and Data Collection
Edited by Grant J. Jensen

Volume 482. Cryo-EM, Part B: 3-D Reconstruction
Edited by Grant J. Jensen

Volume 483. Cryo-EM, Part C: Analyses, Interpretation, and Case Studies
Edited by Grant J. Jensen

Volume 484. Constitutive Activity in Receptors and Other Proteins, Part A
Edited by P. Michael Conn

Volume 485. Constitutive Activity in Receptors and Other Proteins, Part B
Edited by P. Michael Conn

Volume 486. Research on Nitrification and Related Processes, Part A
Edited by Martin G. Klotz

Volume 487. Computer Methods, Part C
Edited by Michael L. Johnson and Ludwig Brand

Volume 488. Biothermodynamics, Part C
Edited by Michael L. Johnson, Jo M. Holt and Gary K. Ackers

CHAPTER ONE

Measurement and Analysis of Equilibrium Binding Titrations: A Beginner's Guide

Dorothy Beckett

Contents

Abstract

Binding events are central to biology. Simple binding of a substrate to an enzyme initiates catalysis. Formation of protein:protein complexes is integral to signal transduction. Binding of multiple proteins to the ribosomal ribonucleic acid (rRNA) results in ribosome assembly. Consequently, elucidation of mechanisms of biological processes requires binding measurements. Such measurements reveal, among other things, the relevant concentrations required for binding partners to form a complex and are indispensible to understanding the relationship between structure and biological function. This article is intended to serve as a primer for biologists who are contemplating performing binding studies. The focus is on practical aspects of design and analysis of binding measurements for a simple process. The information that one can extract from such measurements is also addressed. Theoretical background on binding for both simple and complex systems can be found in many textbooks and monographs including those by Hammes [Hammes, G. G. (2000). Thermodynamics and Kinetics for the Biological Sciences. Wiley, New York, NY], Weber [Weber, G. (1992).

Department of Chemistry and Biochemistry, Center for Biological Structure and Organization, University of Maryland, Maryland, USA

Methods in Enzymology, Volume 488
ISSN 0076-6879, DOI: 10.1016/B978-0-12-381268-1.00001-X
© 2011 Elsevier Inc.
All rights reserved.

Protein Interactions. Chapman and Hall, New York, NY], and Wyman and Gill [Wyman, J. and Gill, S. J. (1990). Binding and Linkage. University Science Books, Mill Valley, CA]. While the first reference is excellent for beginners, the latter two, in addition to discussion of simple binding, contain theoretical background for complex binding processes.

1. Material Requirements for Binding Measurements

Binding component purity is of greatest importance to the success of any binding measurement. First, any impurity can potentially interfere with the binding measurement. Second, if the components are impure it is impossible to know concentrations. This second point cannot be overemphasized because implicit in the design and analysis of any binding measurement is an assumption of accurate and precise knowledge of concentration.

The buffer conditions used for binding measurements must be standardized. Solution variables including pH, temperature, and salt concentration influence binding in unpredictable ways. Consequently, buffer conditions must be well defined and carefully controlled.

Finally, it is extremely important, when one is working with proteins, to have information about oligomeric state. Is the protein a monomer or an oligomer? Moreover, if a protein forms an oligomer, is the oligomer extremely stable or does it readily dissociate into a mixture of species? In the absence of this knowledge, it is impossible to analyze any binding data. Methods to determine oligomeric state include size exclusion chromatography, analytical ultracentrifugation, and light scattering. An assumption in the binding measurements described in this chapter is that the receptor protein of interest is monomeric.

2. Monitoring a Binding Reaction

Many possible methods for monitoring binding reactions exist. Optical methods such as fluorescence or circular dichroism are among the most convenient. Methods that monitor changes in molecular weight or size include size exclusion chromatography and analytical ultracentrifugation. Labeling of one of the binding partners coupled with a method to separate the bound partner from the free, such as polyacrylamide gel electrophoresis or filtration, provides yet another avenue for monitoring a binding process. Identification of the appropriate method for a binding system can be one of the most challenging aspects of performing binding measurements. Detailed descriptions of the application of a few methods are provided below.

Spectroscopy is one of the most convenient methods for monitoring binding. For example, many proteins, providing that their sequences contain tryptophan or tyrosine residues, are characterized by intrinsic fluorescence signals (See Lakowicz, 1983, for a comprehensive introduction to fluorescence spectroscopy). Since the intensity of the protein's fluorescence spectrum as well as its maximum wavelength is sensitive to the environments of the fluorescent side chains, ligand binding frequently results in a fluorescence change. An example of this phenomenon is shown in Fig. 1.1 in which the intrinsic spectra of a protein are shown in the absence and in the presence of different concentrations of a ligand. In this case, the protein is the biotin protein ligase (BPL) from *Pyrococcus horikoshii* and the ligand is biotin. Subtraction of the integrated area of the ligand-bound spectrum from the ligand-free reveals that biotin binding results in quenching of 30% of the original fluorescence signal. This difference provides a handle for monitoring the binding by measuring the integrated intrinsic fluorescence spectrum or the spectrum intensity as a function of added ligand concentration. The fluorescence spectrum starts at the intensity characteristic of the unliganded protein, and in the binding experiment, as it is converted to the ligand-bound form by addition of the ligand, the protein's spectrum decreases until it is fully saturated with ligand. Before embarking on measuring binding curves, it is extremely important to establish that a final intensity or saturation point can be reached at some ligand concentration.

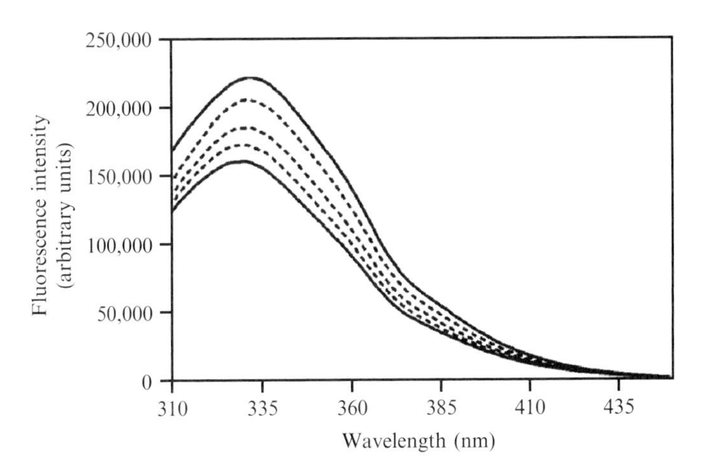

Figure 1.1 Representative fluorescence spectra of the protein *Pyrococcus horikoshii* Biotin Protein Ligase obtained in the presence and absence of the ligand biotin. The spectra shown as solid lines are those obtained in the absence (most intense) and presence (least intense) of saturating biotin. The dashed lines are spectra obtained at biotin concentrations that are subsaturating. The excitation wavelength is 295 nm and emission was monitored from 310 to 440 nm. Data were acquired using an ISS PC-1 fluorimeter.

In other words, there must be conditions of ligand concentration at which addition of more ligand results in no further change in the fluorescence spectrum. The regime of ligand concentration in which the spectral intensity does change provides information about the strength of the protein–ligand interaction.

The electrophoretic mobility shift assay (EMSA) method that relies on separation of the bound and free components in a binding reaction is very popular for measurement of protein:nucleic acid binding reactions. Please refer to Hellman and Fried (2007) for a recent review of the method. The success of the method depends on the distinct electrophoretic mobilities in a native acrylamide gel of the free nucleic acid molecule and the molecule when it is bound to a protein. In the absence of binding protein, all of the nucleic acid is characterized by high mobility in a gel. As protein is added to the nucleic acid, binding occurs, and a fraction of the nucleic acid is now characterized by the lower mobility that characterizes the bound state. As the protein concentration is further increased, the fraction of nucleic acid that is shifted to the species with lower electrophoretic mobility increases until all the nucleic acid migrates as the bound species. The protein concentration regime in which fractions of both the bound and free species exist provides information about the strength of the interaction. Although this method is relatively simple, it does require setting up multiple binding reactions, all containing an identical amount of nucleic acid but each containing a different protein concentration. Once the system reaches equilibrium, the reactions must be loaded onto a gel, subjected to electrophoresis to separate bound from free nucleic acid and the resulting gel is imaged to quantitate the amount of free and bound nucleic acid in each lane. Thus, the complexity of this measurement is significantly greater than a simple spectroscopic measurement in which all measurements are performed in one pot or cuvette.

Other methods for monitoring binding include Isothermal Titration Calorimetry, fluorescence anisotropy, Förster resonance energy transfer, and Surface Plasmon Resonance. Each has associated strengths and pitfalls. With respect to measuring binding affinities, all that is important is that the method provides a reliable signal for monitoring the conversion of one or the other binding partner from fully free to fully bound.

In application of any method to measuring an equilibrium binding process, it is important to demonstrate that at each combination of components in the binding reaction equilibrium has been obtained. If the system is at equilibrium, no time-dependence should be observed in the measured signal. The practical approach for testing that a binding process has reached equilibrium is to measure the time-dependence. In a titration monitored using fluorescence, the intensities can be measured at different times following ligand addition to establish the time required to reach equilibrium. Likewise, in binding measurements monitored using EMSA reaction mixtures can be subjected to electrophoresis at different times following mixing to determine the equilibration time. Similar controls are a necessary part of developing any equilibrium binding assay.

3. The Binding Equation and Its Relationship to Binding Measurements

In combining partners in a binding measurement, a major goal is to determine the equilibrium constant or how tightly two partners bind. Continuing with the theme of using fluorescence to monitor the binding, one of the partners, for example, the protein is placed in a cuvette and the fluorescence is monitored upon each addition of a binding partner or ligand to the protein solution. The chemical expression that describes the binding process is

$$\mathrm{P} + \mathrm{L} \rightleftharpoons \mathrm{PL} \tag{1.1}$$

As the ligand concentration increases, the reaction is driven to the right or toward complex formation. The equilibrium dissociation complex, K_{D}, provides a measure of the strength of an interaction, and the expression that relates the equilibrium dissociation constant to the concentrations of the species in the binding reaction is

$$K_{\mathrm{D}} = \frac{[\mathrm{P}][\mathrm{L}]}{[\mathrm{PL}]} \tag{1.2}$$

For both the protein, P, and ligand, L, the concentrations signify free concentrations, not total. The equilibrium constant can also be expressed in terms of complex formation and is then referred to as an equilibrium association constant, K_{A} and

$$K_{\mathrm{A}} = \frac{[\mathrm{PL}]}{[\mathrm{P}][\mathrm{L}]} \tag{1.3}$$

The association and dissociation constants are the inverse of one another. It is advantageous to express binding in terms of the dissociation constant because the number then provides an idea of the concentrations that are relevant to a binding process. For example, an equilibrium dissociation constant of 1×10^{-6} M indicates that concentrations in the micromolar range are relevant for a binding reaction. The Gibbs free energy of the interaction can be obtained from the equilibrium dissociation constant using the expression, $\Delta G^{\mathrm{o}} = RT\ln K_{\mathrm{D}}$.

In performing a binding titration, the concentration of the added ligand is increased from zero to some maximal value and the most probable state of the protein changes from free, P, to the bound, PL. The state of the protein in the binding experiment can be expressed as fractional saturation or $\bar{Y}$, which is:

$$\bar{Y} = \frac{[\mathrm{PL}]}{[\mathrm{P}]_\mathrm{T}} \tag{1.4}$$

where $[\mathrm{P}]_\mathrm{T}$ is the total protein concentration used for the measurement. By combining Eqs. (1.2) and (1.4), the fractional saturation can also be expressed in terms of the equilibrium dissociation constant and the free ligand concentration in the reaction mixture.

$$\begin{aligned}
K_\mathrm{D} &= \frac{[\mathrm{P}][\mathrm{L}]}{[\mathrm{PL}]}\\
[\mathrm{PL}] &= [\mathrm{P}][\mathrm{L}]/K_\mathrm{D}\\
\bar{Y} &= \frac{[\mathrm{PL}]}{[\mathrm{P}]_\mathrm{T}}\\
[\mathrm{P}]_\mathrm{T} &= [\mathrm{P}] + [\mathrm{PL}]\\
\bar{Y} &= \frac{[\mathrm{P}][\mathrm{L}]/K_\mathrm{D}}{[\mathrm{P}] + [\mathrm{P}][\mathrm{L}]/K_\mathrm{D}}\\
\bar{Y} &= \frac{[\mathrm{L}]}{K_\mathrm{D} + [\mathrm{L}]}
\end{aligned} \tag{1.5}$$

However, in an actual binding measurement, the *total*, not the free, protein and ligand concentrations are the known variables. Therefore, an expression that relates the fractional saturation to the total ligand and protein concentrations added to the reaction and the equilibrium constant proves more useful for analyzing the data and is derived as follows:

$$\begin{aligned}
K_\mathrm{D} &= \frac{[\mathrm{P}][\mathrm{L}]}{[\mathrm{PL}]}\\
[\mathrm{P}] &= [\mathrm{P}]_\mathrm{T} - [\mathrm{PL}]\\
[\mathrm{L}] &= [\mathrm{L}]_\mathrm{T} - [\mathrm{PL}]\\
K_\mathrm{D} &= \frac{([\mathrm{P}]_\mathrm{T} - [\mathrm{PL}])([\mathrm{L}]_\mathrm{T} - [\mathrm{PL}])}{[\mathrm{PL}]}\\
K_\mathrm{D}[\mathrm{PL}] &= [\mathrm{P}]_\mathrm{T}[\mathrm{L}]_\mathrm{T} - ([\mathrm{P}]_\mathrm{T} + [\mathrm{L}]_\mathrm{T})[\mathrm{PL}] + [\mathrm{PL}]^2\\
[\mathrm{PL}]^2 &- ([\mathrm{P}]_\mathrm{T} + [\mathrm{L}]_\mathrm{T} + K_\mathrm{D})[\mathrm{PL}] + [\mathrm{P}]_\mathrm{T}[\mathrm{L}]_\mathrm{T} = 0\\
[\mathrm{PL}] &= \frac{([\mathrm{P}]_\mathrm{T} + [\mathrm{L}]_\mathrm{T} + K_\mathrm{D}) - \left((-([\mathrm{P}]_\mathrm{T} + [\mathrm{L}]_\mathrm{T} + K_\mathrm{D}))^2 - 4[\mathrm{P}]_\mathrm{T}[\mathrm{L}]_\mathrm{T}\right)^{1/2}}{2}
\end{aligned} \tag{1.6}$$

and the fractional saturation can then be expressed as:

$$\bar{Y} = \left(\frac{([\mathrm{P}]_\mathrm{T} + [\mathrm{L}]_\mathrm{T} + K_\mathrm{D}) - \left((-([\mathrm{P}]_\mathrm{T} + [\mathrm{L}]_\mathrm{T} + K_\mathrm{D}))^2 - 4[\mathrm{P}]_\mathrm{T}[\mathrm{L}]_\mathrm{T} \right)^{1/2}}{2} \right) \Big/ [\mathrm{P}]_\mathrm{T} \quad (1.7)$$

In any binding measurement the experimentalist controls the total protein and ligand concentrations and, therefore, the only unknown in this equation is the equilibrium dissociation constant or K_D.

Equation (1.7) is still not useful for directly determining the equilibrium constant for a reaction directly from the measured dependence of a binding signal on the concentration of added binding partner. In the example of monitoring fluorescence, the integrated intensity of the fluorescence spectrum can be measured as a function of the added ligand concentration. The expression in Eq. (1.7) does not allow relating the measured fluorescence intensities to the equilibrium constant and binding partner concentrations. This is accomplished first relating the measured signal to fractional saturation. Using the fluorescence titration as an example, the fluorescence signal associated with free protein is designated F_FREE and the signal associated with the ligand-bound protein is F_BOUND. Each measured fluorescence signal in a binding titration is characterized by a value F, which changes as the titration proceeds. Using these designations for the fluorescence signals, the expression for fractional saturation in terms of the measurements is:

$$\bar{Y} = \frac{F_\mathrm{FREE} - F}{F_\mathrm{FREE} - F_\mathrm{BOUND}} \quad (1.8)$$

At low ligand concentrations, F is similar in magnitude to F_FREE and at ligand concentrations close to those required for saturation of the protein F is similar in magnitude to F_BOUND. Therefore, as expected, the value of $\bar{Y}$ ranges from 0 to 1.

Combining Eqs. (1.7) and (1.8) leads to an equation that relates the measured fluorescence signal F to the signals for the bound and free forms, the total ligand and protein concentrations, and the equilibrium dissociation constant:

$$\frac{F_\mathrm{FREE} - F}{F_\mathrm{FREE} - F_\mathrm{BOUND}} = \left(\frac{([\mathrm{P}]_\mathrm{T} + [\mathrm{L}]_\mathrm{T} + K_D) - \left((-([\mathrm{P}]_\mathrm{T} + [\mathrm{L}]_\mathrm{T} + K_D))^2 - 4[\mathrm{P}]_\mathrm{T}[\mathrm{L}]_\mathrm{T} \right)^{1/2}}{2} \right) \Big/ [\mathrm{P}]_\mathrm{T}$$

$$F_\mathrm{FREE} - F = (F_\mathrm{FREE} - F_\mathrm{BOUND}) \left(\frac{([\mathrm{P}]_\mathrm{T} + [\mathrm{L}]_\mathrm{T} + K_D) - \left((-([\mathrm{P}]_\mathrm{T} + [\mathrm{L}]_\mathrm{T} + K_\mathrm{D}))^2 - 4[\mathrm{P}]_\mathrm{T}[\mathrm{L}]_\mathrm{T} \right)^{1/2}}{2} \right) \Big/ [\mathrm{P}]_\mathrm{T}$$

$$F = F_\mathrm{FREE} - (F_\mathrm{FREE} - F_\mathrm{BOUND}) \left(\frac{([\mathrm{P}]_\mathrm{T} + [\mathrm{L}]_\mathrm{T} + K_D) - \left((-([\mathrm{P}]_\mathrm{T} + [\mathrm{L}]_\mathrm{T} + K_\mathrm{D}))^2 - 4[\mathrm{P}]_\mathrm{T}[\mathrm{L}]_\mathrm{T} \right)^{1/2}}{2} \right) \Big/ [\mathrm{P}]_\mathrm{T} \quad (1.9)$$

This equation can be used directly to analyze the data acquired in a titration experiment, or F versus $[L]_T$, to obtain the best estimates of the values for the fluorescence signals of the free and bound proteins and the equilibrium dissociation constant.

4. Plotting and Analysis of Binding Data

Once a reliable method to monitor binding is in hand titrations can be performed. However, the proper design of the titration experiment is critical to obtaining the desired information about affinity. Simulation of data can be very helpful in this design. In addition to appropriate design, binding data must be properly analyzed to obtain the equilibrium constant for the binding reaction. Finally, the choice of method to display the data in a plot can be critical with respect to the insight that can be obtained into the binding process.

Prior to introducing examples of binding data, tools for both analyzing and simulating data will be briefly discussed. Data analysis provides a means to get best estimates of the parameters that describe a binding process. As indicated in Eq. (1.8),

$$F = F_{\mathrm{FREE}} - (F_{\mathrm{FREE}} - F_{\mathrm{BOUND}})\left(\frac{([\mathrm{P}]_\mathrm{T} + [\mathrm{L}]_\mathrm{T} + K_\mathrm{D}) - \left((-([\mathrm{P}]_\mathrm{T} + [\mathrm{L}]_\mathrm{T} + K_\mathrm{D}))^2 - 4[\mathrm{P}]_\mathrm{T}[\mathrm{L}]_\mathrm{T}\right)^{1/2}}{2}\right)\Big/[\mathrm{P}]_\mathrm{T}$$

For a simple 1:1 binding process, the measured fluorescence versus the total ligand concentration added to a protein in a titration is analyzed to find the three unknowns, namely, the fluorescence signal for the free protein, F_{FREE}, the signal for the bound protein, F_{BOUND}, and the equilibrium constant K_D. These unknowns are referred to as parameters. In nonlinear least squares regression, the data are tested against a binding model, such as the one shown above, to obtain the parameters that have the greatest likelihood of describing the binding reaction. Again, in the case of monitoring the binding using the intrinsic protein fluorescence as the signal these parameters include the fluorescence signals for the free and bound protein, the endpoints of the titration curve, and the equilibrium dissociation constant. Many commercial programs now exist for performing nonlinear least squares analysis. The program used to generate and analyze the data shown in this chapter is GraphPad Prism. Ideally, any program used for this purpose should allow the user to enter an equation of his/her choice so that any binding model can be tested. These same programs frequently allow the user to simulate what the data should look like assuming a particular model and values. The ability to simulate data helps to streamline experimental design. Excellent introductions to nonlinear least squares parameter estimation are provided in Johnson (1992) and Motulsky and Ransnas (1987). The practical information about use of analysis programs is typically provided with the specific program. Support materials provided with GraphPad Prism are superb.

An example of titration data that would be obtained from monitoring a binding process is provided in Fig. 1.2. These data were simulated using Eq. (1.9) and the maximal fluorescence, F_{FREE}, was assumed to be 40,000 and the minimal, F_{BOUND}, was assumed to be 33,600. The equilibrium dissociation constant was set at 5×10^{-7} *M* and the total protein concentration was set at 5×10^{-7} *M*. The ligand concentration ranged from 1×10^{-8} to 1×10^{-5} *M*. In simulating the data, the error in each measured fluorescence values was assumed to be 3% of the total difference between the intensities of the free and bound forms and distributed in a Gaussian shape around the best-fit value. This characteristic distribution is an assumption, although not a requirement, of most common nonlinear regression analyses. The titration data exhibit a hyperbolic dependence on ligand concentration, which is expected of any binding curve for a simple 1:1 interaction. The line shown with the data is the best-fit curve that is generated using Eq. (1.9) and the parameters obtained from the nonlinear least squares analysis of the binding data. By inspection, the data are well described by the line. This conclusion is based on the fact that the data points are randomly distributed

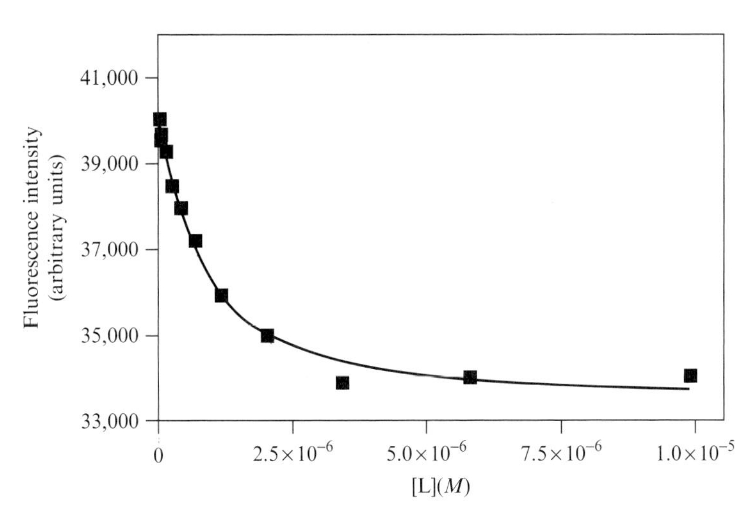

Figure 1.2 Fluorescence titration for a hypothetical protein with a ligand. Please see the text for details of the data simulation. The data points were simulated using Eq. (1.8) in Prism 4 (Prism). The solid line the best-fit curve that is generated using the best-fit parameters obtained from nonlinear regression of the data, again using Eq. (1.8).

Table 1.1 Simulation and analysis of binding from a titration monitored by fluorescence

Parameter/variable	Simulated	Best-fit values
$[P]_T$ (M)	5×10^{-7}	– (held constant)
K_D (M)	5×10^{-7}	$5.4(4.6/6.3) \times 10^{-7}$
F_{FREE}	40,000	40,007 (39,904/40,110)
F_{BOUND}	33,500	33,385 (33,180/33,590)

about the best-fit curve. The parameters obtained from the analysis, Table 1.1, while not exactly those that were used to simulate the data, are close. The errors or uncertainties shown with the best-fit parameters provide limits on the parameter values. These errors are the inevitable consequence of the uncertainties associated with the measured fluorescence values. However, in simulating and analyzing the data, the ligand and protein concentrations are both assumed to be known with infinite accuracy and precision. This assumption also holds, with some exceptions, in real experiments.

An alternative format for presenting titration data is fractional saturation versus either the ligand concentration or the logarithm of the ligand concentration (Fig. 1.3). The transformation is achieved using Eq. (1.7):

$$\bar{Y} = \frac{F_{FREE} - F}{F_{FREE} - F_{BOUND}}$$

with the best-fit values of F_{FREE} and F_{BOUND} obtained from the nonlinear least square analysis of the data and the measured value of F obtained in the titrations as input. As expected, the magnitude of Fractional Saturation ranges from 0 to 1. In addition, inspection of the figure in which Fractional Saturation is plotted versus the log[biotin] reveals that ~2 log units of ligand concentration are required to change the fractional saturation from ~0.1 to 0.9. Please see Weber (1992) for a detailed discussion of this point. This "span" of the binding curve is characteristic of all simple binding processes. If the span is more narrow or wider than 2 log units, either the binding model is more complicated than a simple 1:1 process or the experimental design is not correct for the system.

5. Protein Concentration Is Important: Equilibrium Versus Stoichiometric Conditions

A common example of inappropriate experimental design is the use of a total receptor protein concentration that is too high to obtain an equilibrium constant for a particular binding system. In this case, the conditions are

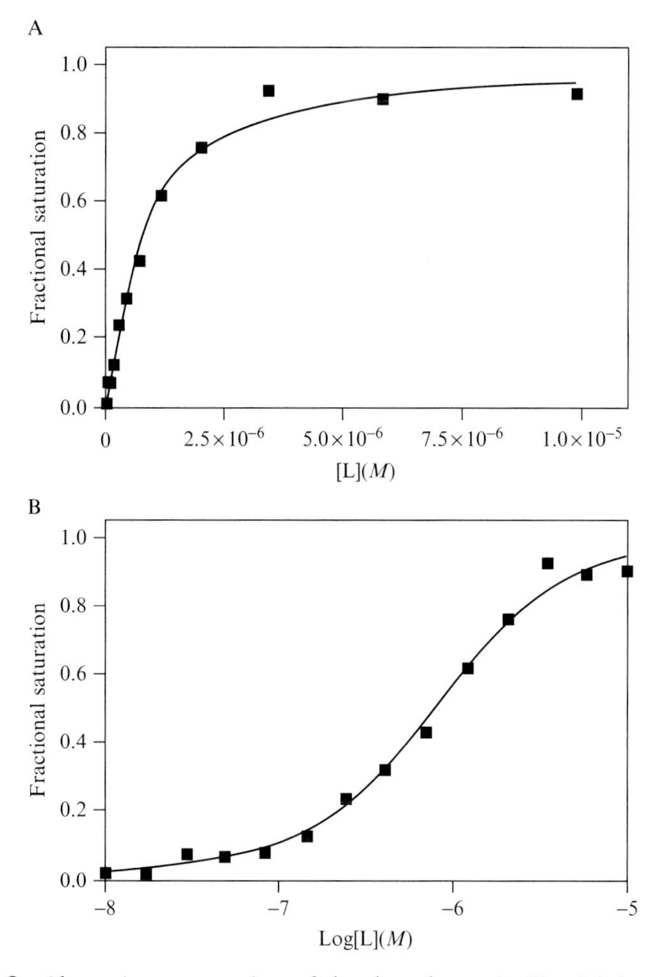

Figure 1.3 Alternative presentation of the data shown in Fig. 1.2 in which the "measured" fluorescence intensities, F, are normalized to values between 0 and 1, or $\bar{Y}$, using the best-fit values of F_{FREE} and F_{BOUND}. The two plotting methods are (A) $\bar{Y}$ versus [L] or the direct plot and (B) $\bar{Y}$ versus log [L], the semilogarithmic plot. Solid lines are the best-fit curves generated using the parameters obtained from nonlinear regression of the data.

appropriate for determining the stoichiometry of the complex but not the equilibrium constant governing its formation. An explicit assumption of Eq. (1.7), which is derived using Eq. (1.2), is that all three of the equilibrium species including free protein, P, free ligand, L, and the protein ligand complex, PL are present in the transition region of the binding measurement. However, in binding titrations, these conditions are achieved only when the protein concentration is in the appropriate range relative to the magnitude of the equilibrium constant. The ideal conditions are those in which the protein concentration is at a value similar to that of the equilibrium dissociation constant. The equation that relates fractional saturation to the receptor protein and the ligand concentrations,

$$\bar{Y} = \left(\frac{([\mathrm{P}]_\mathrm{T} + [\mathrm{L}]_\mathrm{T} + K_\mathrm{D}) - \left((-([\mathrm{P}]_\mathrm{T} + [\mathrm{L}]_\mathrm{T} + K_\mathrm{D}))^2 - 4[\mathrm{P}]_\mathrm{T}[\mathrm{L}]_\mathrm{T} \right)^{1/2}}{2} \right) \Big/ [\mathrm{P}]_\mathrm{T}$$

can be used to simulate the binding curves that will be obtained at different total concentrations of the receptor, $[\mathrm{P}]_\mathrm{T}$. In these simulations, a 1:1 binding model is assumed and the "experiments" are performed at total receptor concentrations ranging from 5×10^{-8} to 5×10^{-5} (Fig. 1.4). The equilibrium dissociation constant for the binding reaction is the same as that used for the titration described above, or 5×10^{-7} *M*. Titration curves obtained using these different total protein concentrations differ in appearance. In plots in which fractional saturation is plotted versus ligand concentration, the curves obtained at lower total protein concentration show more curvature and with increasing total protein concentration the curve shape approaches two intersecting lines. The binding conditions for this series of curves shift from "equilibrium" to "stoichiometric." In equilibrium conditions, as expected, all three species including the free protein, the free ligand, and the protein ligand complex are present in the transition region between fractional saturation values of 0 and 1. By contrast, in stoichiometric conditions, performed at high protein concentration, only the free protein and the complex are present in the transition region, which corresponds to the initial linear portion of the curve. Once the protein is saturated, only the free ligand and the complex are present. Stoichiometric conditions are those in which total protein concentration is much greater than the value of the equilibrium dissociation constant or $[\mathrm{P}]_\mathrm{T} \gg K_\mathrm{D}$. In practice, this means a protein concentration at least 100-fold greater than the equilibrium dissociation constant.

Titrations performed under stoichiometric conditions can be useful for obtaining information about stoichiometry of a complex or fractional activities of the components used in a binding reaction. At truly stoichiometric concentrations, the intersection of the two lines of the titration corresponds to the stoichiometry of the complex formed between the protein and ligand, provided that concentrations are accurately known and that both the protein and ligand components of the binding experiment are 100% active. If the concentrations are, indeed, well established and the stoichiometry is known from some other method, such as high-resolution structural data, a titration performed under stoichiometric conditions provides information about the fractional activity of one of the binding partners.

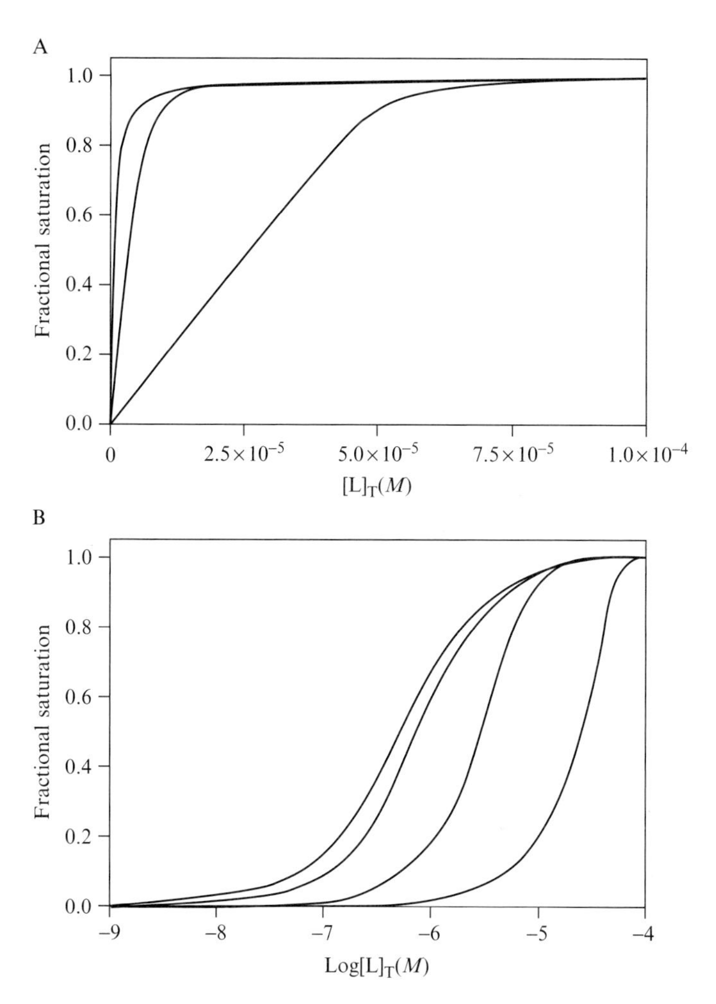

Figure 1.4 Effect of the total protein concentration, $[P]_T$, on the titration curves obtained for a simple binding process. The curves are shown as (A) direct plots and (B) semilogarithmic plots. The equilibrium dissociation constant for the binding process is 5×10^{-7} *M* and the $[P]_T/K_D$ values are 0.1, 1.0, 10, and 100 for the curves from left to right in each plot. In the direct plots, the curves obtained at ratios of 0.1 and 1.0 overlap.

Equilibrium constants that are obtained from titration data performed under stoichiometric conditions are not accurate and, in the extreme, have very large errors associated with them. The reference by Weber (1992) provides a helpful discussion of this issue. There is also a good presentation of stoichiometric versus equilibrium binding conditions provided in Chapter 14 of van Holde *et al.* (2006).

6. When Are Total and Free Ligand Concentrations Equal?

The following equation, particularly using canned programs, is often employed for analyzing binding data:

$$\bar{Y} = \frac{[\mathrm{L}]}{K_{\mathrm{D}} + [\mathrm{L}]}$$

in which, instead of the free ligand concentration, [L] is the total ligand concentration in the reaction mixture. However, the relationship between fractional saturation and ligand concentration is derived from the relationship between the equilibrium constant and the free ligand and free protein concentration and the complex concentration (Eq. (1.2)). In many binding titrations, the free ligand concentration differs significantly from the total because some of the ligand is bound up with protein. There are, however, circumstances in which it is valid to approximate the free with the total ligand concentration. These correspond to measurements in which the "receptor" protein concentration is much less than the equilibrium dissociation constant governing the reaction. Measurements of protein binding to nucleic acids are frequently adequately analyzed using this approximation. This is because in these titrations the nucleic acid is often labeled with ^{32}P phosphate, which allows detection at picomolar concentrations. Equilibrium constants governing protein binding to nucleic acids can be in the nanomolar range of concentration and, thus, in the titration mixture the concentration of the complex is always negligible compared to the total protein concentration. Before using the simplifying assumption that $[\mathrm{L}]_{\mathrm{FREE}}=[\mathrm{L}]_{\mathrm{TOTAL}}$, it is important to know the relative values of the receptor concentration and the equilibrium constant governing a binding reaction.

7. Deviations from Simple Binding

This chapter has thus far focused on measuring and analyzing data from simple 1:1 binding process. However, biology is complex and the binding processes that contribute to it are similarly complex. Two examples of potential complexity in binding curves are shown in Fig. 1.5. The first panel (Fig. 1.5A) provides data in which the span of the curve is less than 1.9 log units in ligand concentration. This curve was generated assuming cooperative binding of two ligands to two identical sites. The equilibrium

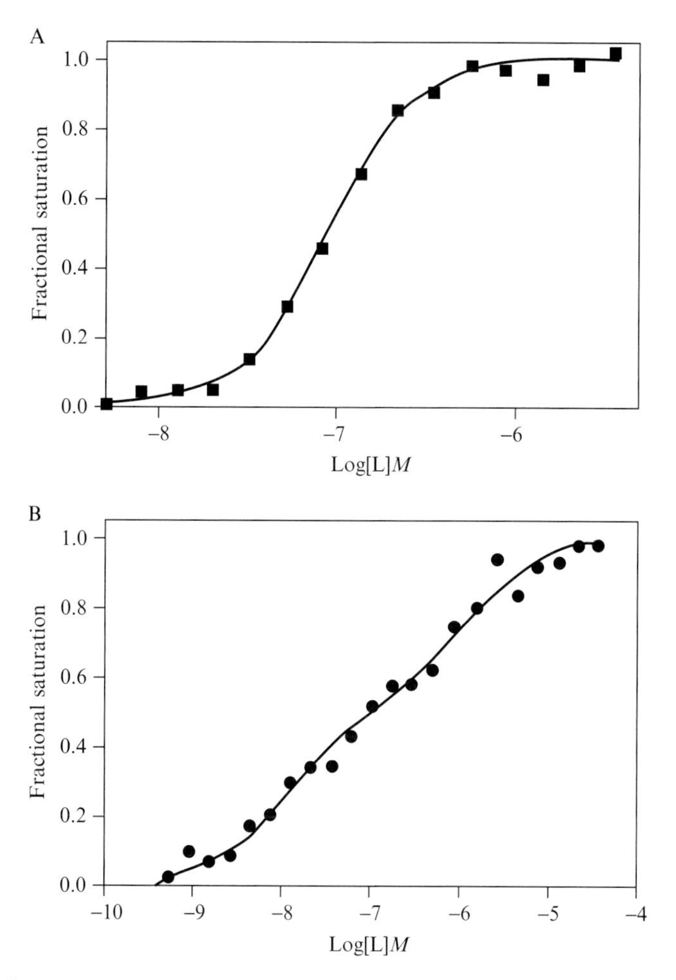

Figure 1.5 Titrations that indicate deviations from simple binding. (A) The curve is steeper than expected for simple 1:1 binding. (B) The curve is broader than expected for simple 1:1 binding. The solid lines are the best-fit curves obtained from analysis of the data using the models indicated in the text.

constant governing the interaction of each ligand with each site in the absence of binding to the other site, or the intrinsic equilibrium dissociation constant, was set at 1×10^{-6} M and the cooperativity constant for the interaction between the two ligand binding simultaneously to the two sites was set at 100. Although this model accurately describes the data, there are other possible models that could describe it equally well. One possibility is that protein oligomerization is occurring simultaneously with ligand binding. In this case, the curve would be steep because of the coupling of oligomerization with binding in the system. This ambiguity in data interpretation emphasizes the importance of measuring the oligomerization properties of a protein prior to performing binding studies. A second possible deviation from simple binding is illustrated in Fig. 1.5B. In this case, the span of the curve is much broader than the 1.9 log units expected of simple binding. The data in the curve were generated assuming that two ligands are binding to two sites independently but with distinct equilibrium dissociation constants of 1×10^{-8} and 1×10^{-6} *M*. An alternative model of negatively cooperative binding in which binding of the first ligand causes the second ligand to bind less tightly would describe the data equally well. Distinguishing which model is valid requires additional information about the system. For example, if the protein is a dimer of identical subunits the broad curve is probably due to negative cooperativity in ligand binding to the two sites.

Other complications may arise in binding data and, by using a combination of common sense and additional experimental measurements, these complications can be interpreted to reveal the properties of the system under study.

REFERENCES

Hammes, G. G. (2000). *Thermodynamics and Kinetics for the Biological Sciences*. Wiley, New York, NY.

Hellman, L. M., and Fried, M. G. (2007). Electrophoretic mobility shift assay (EMSA) for detecting protein–nucleic acid interactions. *Nat. Protoc.* **2**(8), 1849–1861.

Johnson, M. L. (1992). Why, when, and how biochemists should use least squares. *Anal. Biochem.* **206**(2), 215–225.

Lakowicz, J. R. (1983). *Principles of Fluorescence Spectroscopy*. Plenum Press, New York and London.

Motulsky, H. J., and Ransnas, L. A. (1987). Fitting curves to data using nonlinear regression: A practical and nonmathematical review. *FASEB J.* **1**(5), 365–374.

van Holde, K. E., Johnson, W. C., and Ho, P. S. (2006). *Principle of Physical Biochemistry*. Pearson/Prentice Hall, Upper Saddle River, NJ.

Weber, G. (1992). *Protein Interactions*. Chapman and Hall, New York, NY.

Wyman, J., and Gill, S. J. (1990). *Binding and Linkage*. University Science Books, Mill Valley, CA.

CHAPTER TWO

Macromolecular Competition Titration Method: Accessing Thermodynamics of the Unmodified Macromolecule–Ligand Interactions Through Spectroscopic Titrations of Fluorescent Analogs

Wlodzimierz Bujalowski*,† *and* Maria J. Jezewska*

Contents

* Department of Biochemistry and Molecular Biology, The Sealy Center for Structural Biology, Sealy Center for Cancer Cell Biology, The University of Texas Medical Branch at Galveston, Galveston, Texas, USA
† Department of Obstetrics and Gynecology, The Sealy Center for Structural Biology, Sealy Center for Cancer Cell Biology, The University of Texas Medical Branch at Galveston, Galveston, Texas, USA

Methods in Enzymology, Volume 488
ISSN 0076-6879, DOI: 10.1016/B978-0-12-381268-1.00002-1
© 2011 Elsevier Inc.
All rights reserved.

Abstract

Analysis of thermodynamically rigorous binding isotherms provides fundamental information about the energetics of the ligand–macromolecule interactions and often an invaluable insight about the structure of the formed complexes. The Macromolecular Competition Titration (MCT) method enables one to quantitatively obtain interaction parameters of protein–nucleic acid interactions, which may not be available by other methods, particularly for the unmodified long polymer lattices and specific nucleic acid substrates, if the binding is not accompanied by adequate spectroscopic signal changes. The method can be applied using different fluorescent nucleic acids or fluorophores, although the etheno-derivatives of nucleic acid are especially suitable as they are relatively easy to prepare, have significant blue fluorescence, their excitation band lies far from the protein absorption spectrum, and the modification eliminates the possibility of base pairing with other nucleic acids. The MCT method is not limited to the specific size of the reference nucleic acid. Particularly, a simple analysis of the competition titration experiments is described in which the fluorescent, short fragment of nucleic acid, spanning the exact site-size of the protein–nucleic acid complex, and binding with only a 1:1 stoichiometry to the protein, is used as a reference macromolecule. Although the MCT method is predominantly discussed as applied to studying protein–nucleic acid interactions, it can generally be applied to any ligand–macromolecule system by monitoring the association reaction using the spectroscopic signal originating from the reference macromolecule in the presence of the competing macromolecule, whose interaction parameters with the ligand are to be determined.

1. Introduction

Macromolecule–ligand interactions play a fundamental role in metabolic activities of enzymes, structural proteins, and nucleic acids (Boschelli, 1982; Brenowitz *et al.*, 1986; Bujalowski and Jezewska, 1995; Bujalowski and Klonowska, 1993, 1994a,b; Bujalowski and Lohman, 1986, 1987; Bujalowski *et al.*, 1989, 1994; Chabbert *et al.*, 1987; Crothers, 1968; deHaseth *et al.*, 1977; Draper and von Hippel, 1978; Epstein, 1978; Fernando and Royer, 1992; Fried and Crothers, 1981; Garner and Revzin, 1981; Heyduk and Lee, 1990; Kowalczykowski *et al.*, 1981, 1986; Lohman and Bujalowski, 1991; Lohman and Ferrari, 1994; McGhee and von Hippel, 1974; McSwiggen *et al.*, 1988; Menetski and Kowalczykowski, 1985; Porschke and Rauh, 1983;

Revzin and von Hippel, 1977; Scatchard, 1949). The interactions can achieve various levels of complexity. For instance, the binding process may be very specific, limited to a single type of ligand molecule. Specificity may encompass a group of ligand molecules of homologous structures or the interactions may include several ligand molecules, and the binding process may be characterized by cooperativity. In any case, the first step in elucidating the intrinsic energetics of the association, the forces which drive the interactions, the functionally relevant conformational changes of the participating species is the determination of the equilibrium binding isotherm over a range of the total average degree of binding, $\Sigma\Theta_i$ (number of moles of the ligand bound per one mole of the macromolecule). Equilibrium binding isotherms, for ligand binding to a macromolecule, represent the relationship between $\Sigma\Theta_i$ and the free ligand concentration. A true thermodynamic isotherm reflects only this relationship and does not depend upon any particular binding models or any particular assumptions. The extraction of physically meaningful interaction parameters that characterize the examined complex is only possible when such a relationship is available and is achieved by comparing the experimental isotherms to theoretical predictions that incorporate known molecular aspects of the examined systems (Brenowitz *et al.*, 1986; Bujalowski and Klonowska, 1993, 1994a,b; Bujalowski and Lohman, 1986, 1987; Bujalowski *et al.*, 1989, 1994; Fried and Crothers, 1981; Garner and Revzin, 1981; Heyduk and Lee, 1990; Lohman and Bujalowski, 1991; Lohman and Ferrari, 1994; Revzin and von Hippel, 1977; Scatchard, 1949).

In our discussion, we will concentrate on the thermodynamic analyses of the ligand–macromolecule binding, including protein–DNA interactions, using spectroscopic approaches (Bujalowski and Klonowska, 1993, 1994a,b; Bujalowski *et al.*, 1994; Lohman and Bujalowski, 1991). Spectroscopic methods are widely used in studying the most complex ligand–nucleic acid associations (Boschelli, 1982; Bujalowski and Klonowska, 1993, 1994a,b; Bujalowski and Lohman, 1986, 1987; Bujalowski *et al.*, 1994; Chabbert *et al.*, 1987; Draper and von Hippel, 1978; Fernando and Royer, 1992; Kowalczykowski *et al.*, 1986; Lohman and Ferrari, 1994; McSwiggen *et al.*, 1988; Menetski and Kowalczykowski, 1985; Porschke and Rauh, 1983). This is because the spectroscopic methods are relatively easy to apply, do not require large quantities of materials, allow one to perform solution experiments without perturbing the studied equilibrium, and can be applied at very low, submicromolar concentrations of interacting macromolecules, thus, eliminating any correction for nonideality necessary when high concentrations are required, for example, in transport or chromatographic methods. Moreover, the observed changes of the spectroscopic parameters provide additional information about the structural changes accompanying the examined interactions. We will discuss the quantitative analyses as applied to the use of the fluorescence intensity, which is the most widely applied spectroscopic technique in energetics and dynamics studies of the ligand–macromolecule interactions in solution (Bujalowski and Klonowska, 1993, 1994a,b; Bujalowski and Lohman, 1987; Bujalowski *et al.*, 1994; Lohman and Bujalowski, 1991). Nevertheless, the derived relationships are general and applicable to any physico-chemical signal used to monitor the association (Bujalowski, 2006; Jezewska and Bujalowski, 1996; Lohman and Bujalowski, 1991; Szymanski *et al.*, 2010). The primary limitation of spectroscopic methods for studying many ligand–macromolecule interactions is the lack of a sufficient spectroscopic signal change accompanying the complex formation. Lack of adequate changes in spectroscopic properties of the interacting molecules hinders the rigorous quantitative analyses of such systems. It can be overcome by introducing a fluorescence label into the ligand or the macromolecule. Nevertheless, besides the nonspecific nature of labeling a protein, or a polymer nucleic acid, which limits these approaches to some particular applications, for example, use of short nucleic acids, labeling SH groups of the protein, there is always a problem as to what extent the fluorescent marker perturbs the system, or reflects the interactions between the unmodified components.

The Macromolecular Competition Titration (MCT) method is particularly suited to analyze macromolecule–ligand interactions where the formation of the complexes is not accompanied by any adequate spectroscopic signal change (Bujalowski, 2006; Bujalowski and Jezewska, 2009; Jezewska and Bujalowski, 1996; Szymanski *et al.*, 2010). The method is based on quantitative titrations of the reference fluorescent macromolecule/ligand with a macromolecule/ligand, in the presence of a competing, nonfluorescent macromolecule/ligand, whose interaction parameters are to be determined. The approach allows the determination of the total average degree of binding and the free macromolecule/ligand concentrations, over a large degree of binding range, unavailable by other methods, and construction of a model-independent, thermodynamic binding isotherm. In the case of protein binding to a long, one-dimensional nucleic acid lattice, a more convenient parameter than $\Sigma\Theta_i$ is the total average binding density $\Sigma\nu_i$, (moles of ligand bound per mole of bases or base pairs; Bujalowski, 2006; Jezewska and Bujalowski, 1996). Using the MCT method, the determination of $\Sigma\Theta_i$, or $\Sigma\nu_i$, is independent of the *a priori* knowledge of the binding characteristics of the macromolecule/ligand to the reference ligand/macromolecule.

2. A Single Titration Curve: Some Simple Considerations of Possible Pitfalls

Fluorescence titration, or any titration method based on observing a physico-chemical signal, is an indirect method of determining the binding isotherm. This is because the association of the ligand with the macromolecule is examined by monitoring changes in the parameter of the system (here, the emission intensity) accompanying the formation of the complex, not the total average degree of binding or the binding density (Bujalowski, 2006; Bujalowski and Jezewska, 2000, 2009; Bujalowski and Klonowska, 1993, 1994a,b; Bujalowski and Lohman, 1987; Bujalowski *et al.*, 1994; Jezewska *et al.*, 1996, 1998a,b, 2001, 2003, 2006; Lohman and Bujalowski, 1991; Rajendran *et al.*, 1998, 2001; Szymanski *et al.*, 2010). These changes must then be correlated with $\Sigma\Theta_i$, or $\Sigma\nu_i$, or with the fractional saturation of the macromolecule, and the concentration of the free ligand. Nevertheless, the functional relationship between the observed spectroscopic signal and the total average degree of binding or the total average binding density is never *a priori* known. The only exceptions from this rule are systems where, at saturation, a single ligand molecule binds to the macromolecule. But then, the maximum stoichiometry of the system is already known. In much more general cases, where there is a possibility that multiple ligand molecules can participate in the binding process, the maximum stoichiometry is one of the thermodynamic/structural parameters that must be determined. Moreover, the observed, fractional change of the spectroscopic signal and the total average degree of binding may not, and frequently will not, have a simple and strict proportional relationship (Bujalowski, 2006; Jezewska and Bujalowski, 1996; Szymanski *et al.*, 2010).

For instance, although one can analyze a titration curve by assuming a simple, 1:1 stoichiometry of the binding process, such an analysis is only of a semiquantitative nature, at best, and does not provide any deeper insight into the studied interactions. Figure 2.1A shows a typical titration curve for the ligand binding to six independent sites on the macromolecule, as monitored by the fluorescence emission and performed at a single macromolecule concentration. The selected binding constant, K, and the total macromolecule concentration, $[\mathrm{M}]_\mathrm{T}$, are $1 \times 10^{-6}\,M^{-1}$ and $1 \times 10^{-7}\,M$, respectively. The titration curve is described by Eq. (2.1)

$$\Delta F_{\mathrm{obs}} = \Delta F_{\mathrm{max}} \frac{6K[\mathrm{L}]_\mathrm{F}}{1 + K[\mathrm{L}]_\mathrm{F}} = \Delta F_{\mathrm{max}} \Sigma\Theta_i \tag{2.1}$$

where an obvious strict proportionality exists between the observed signal, ΔF_{obs}, and the total average degree of binding, $\Sigma\Theta_i$. The proportionality constant is the maximum observed signal change, $\Delta F_{\mathrm{max}} = 0.5$, for the ligand binding to each of the six sites. The analogous equation for a single ligand molecule binding to a single site on the macromolecule is

$$\Delta F_{\mathrm{obs}} = \Delta F_{\mathrm{max}} \frac{K[\mathrm{L}]_\mathrm{F}}{1 + K[\mathrm{L}]_\mathrm{F}} = \Delta F_{\mathrm{max}} \Sigma\Theta_i \tag{2.2}$$

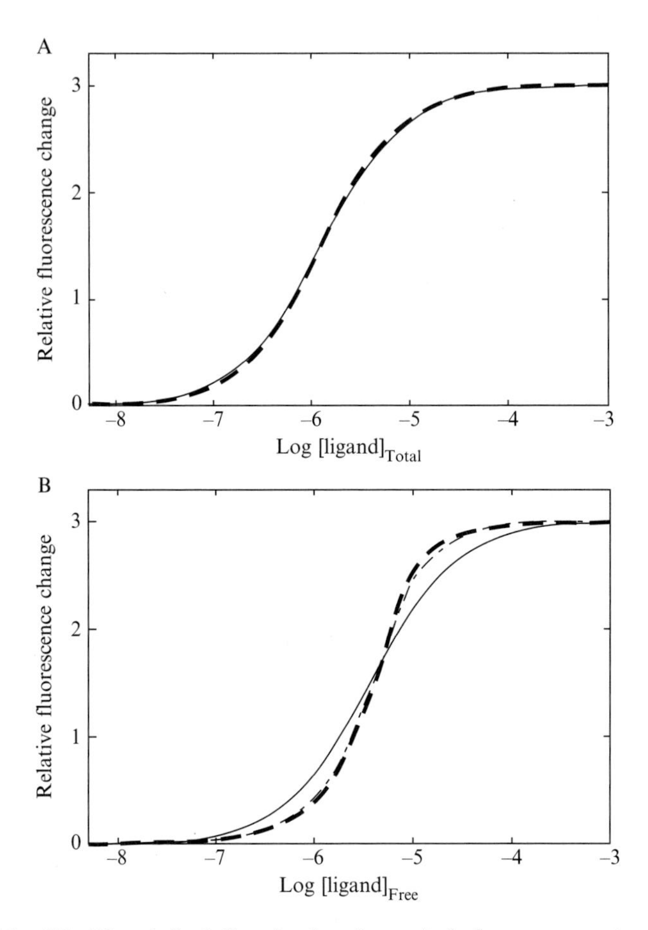

Figure 2.1 (A) The dashed line is the theoretical fluorescence titration of a macromolecule that has six discrete identical binding sites with a ligand at the macromolecule concentration $[M]_T = 1 \times 10^{-7}$ *M*. The intrinsic binding constant, characterizing the binding, $K = 1 \times 10^6$ M^{-1}. Association of the ligand induces at each binding site the relative change of the macromolecule fluorescence, $\Delta F = 0.5$, providing the maximum change of the macromolecule fluorescence at saturation, $\Delta F_{max} = 3.0$. The solid line is the same titration curve but approximated using a single-site binding titration curve, with the binding constant $K = 8 \times 10^5$ M^{-1} and $\Delta F_{max} = 3.0$. (B) The dashed line is the theoretical fluorescence titration of a macromolecule, which has six discrete, identical binding sites, characterized by the intrinsic binding constants, $K = 1 \times 10^6$ M^{-1}, with a ligand. The macromolecule concentration, $[M]_T = 1 \times 10^{-6}$ *M*. The binding of the ligand induces the relative change of the macromolecule fluorescence, at each binding site, $\Delta F = 0.5$, providing the maximum change of the macromolecule fluorescence at saturation, $\Delta F_{max} = 3.0$. The solid line is the same titration curve but approximated using a single-site binding titration curve with the binding constant $K = 2.8 \times 10^5$ M^{-1} and $\Delta F_{max} = 3.0$. The line (– - –) is the approximation of the titration curve using a simple binding model of two cooperative binding sites with the intrinsic binding constant, $K = 1.8 \times 10^5$ M^{-1}, the cooperativity parameter, $\sigma = 4$, and the relative fluorescence change accompanying the binding to each site $\Delta F = 1.5$.

The corresponding plot of the titration curve, defined by Eq. (2.2), is included in Fig. 2.1A with $K = 8 \times 10^5\ M^{-1}$ and $\Delta F_{max} = 3.0$. Within experimental accuracy, the plots are indistinguishable. The error in determining the binding constant (~20%) is significant, but probably not that troublesome in typical biochemical studies. Such a "small" error results from the fact that the concentration of the macromolecule is much lower than the total concentration of the ligand, that is, the concentration of the bound ligand, $[L]_B$, is negligible, as compared to the total ligand concentration, $[L]_T$. However, the character of the binding process, with all conclusions as to the functionality of the macromolecule, is completely missed.

The situation is even worse when the titrations are performed at a single, high macromolecular concentration. In such a case, $[L]_B$ constitutes a significant part of $[L]_T$. In the simplified analysis, which assumes the 1:1 stoichiometry, this would not be *a priori* known to the experimenter and simply ignored. Figure 2.1B shows a simulated titration curve for ligand binding to six independent binding sites on the macromolecule, monitored by the fluorescence emission, with the binding constant, K, and the total macromolecule concentration, $[M]_T$, $1 \times 10^6\ M^{-1}$ and $1 \times 10^{-6}\ M$, respectively. The plot is described by Eq. (2.1). The single-site binding model cannot fit the curve even for a less quantitatively oriented experimenter (Eq. (2.2)). Moreover, the apparent behavior of the plot would suggest a positive cooperativity in the binding process. The experimenter would be then tempted to use the next simplest model, which would be the ligand binding to two sites with cooperative interactions and accompanied by the same fluorescence emission changes, as defined by

$$\Delta F_{obs} = \Delta F_{max} \frac{K[L]_F + 2K^2\sigma[L]_F^2}{1 + K[L]_F + K^2\sigma[L]_F^2} = \Delta F_{max}\Sigma\Theta_i \tag{2.3}$$

This model nicely fits the titration curve, although, once again, completely misses the nature of the occurring binding process, which is an independent binding of six ligand molecules to the macromolecule, defined by Eq. (2.1). The obtained binding constant and cooperativity parameter are fitting constants, not the thermodynamic quantities characterizing the system. The example discussed above indicates some pitfalls, but are not, by any means, the most complex possibilities. It simply reflects the fact that a single titration curve, where a spectroscopic or a physico-chemical signal is used to monitor the binding, is not a binding isotherm (see above).

3. Quantitative Equilibrium Spectroscopic Titrations: Thermodynamic Bases

Quantitative analysis of the binding of a ligand to a macromolecule can be performed using two different types of equilibrium spectroscopic titrations, "normal" or "reverse" titration approach (Boschelli, 1982; Bujalowski, 2006; Bujalowski and Jezewska, 2000, 2009; Lohman and Bujalowski, 1991; Szymanski *et al.*, 2010). In the normal titration, a fluorescing macromolecule is titrated with a nonfluorescing ligand. The total average degree of binding, $\Sigma\Theta_i$, increases as the titration progresses. In the reverse titration approach, the fluorescing ligand is titrated with the nonfluorescing macromolecule and $\Sigma\Theta_i$ decreases throughout the titration (Bujalowski and Jezewska, 2000; Jezewska and Bujalowski, 1997; Lohman and Bujalowski, 1991). Although both types of titration curves can be quantitatively analyzed, we will limit our discussion to the normal titration method, as more commonly occurring in the experimental practice. Moreover, we will discuss specific ligand–macromolecule interactions examined in our laboratory, which include nucleotide binding to DNA helicases and protein–nucleic acid interactions, where the binding processes have been addressed using protein fluorescence, emission of fluorescent analogs of the nucleotides, and fluorescent derivative of homoadenosine oligomers (Bujalowski, 2006; Bujalowski and Jezewska, 2000, 2009; Bujalowski and Klonowska, 1993, 1994a,b; Bujalowski *et al.*, 1994; Jezewska and Bujalowski, 1996, 1997; Jezewska *et al.*, 1996, 1998a,b, 2001, 2003, 2006; Rajendran *et al.*, 1998, 2001; Szymanski *et al.*, 2010). For multiple binding of nucleotides to the protein, it is rather obvious that the nucleotide cofactor is the ligand and the protein is the macromolecule. Analogously, because usually several protein molecules bind to the examined nucleic acids, the protein is treated as a ligand and the nucleic acid is the macromolecule.

The task is to convert the fluorescence titration curve, for example, a change in the monitored fluorescence as a function of the total protein concentration into a model-independent, thermodynamic binding isotherm, which can be analyzed, using an appropriate binding model to extract binding parameters. We start with the signal and mass conservation equations (Bujalowski, 2006; Bujalowski and Jezewska, 2000, 2009; Jezewska and Bujalowski, 1996; Jezewska *et al.*, 1996, 1998a,b, 2001, 2003, 2006; Lohman and Bujalowski, 1991; Rajendran *et al.*, 1998, 2001; Szymanski *et al.*, 2010). For the total macromolecule concentration, $[M]_T$, the equilibrium distribution of the macromolecule among its different states with a different number of bound ligand molecules, M_i, is determined solely by the free protein concentration, $[L]_F$ (Bujalowski, 2006; Bujalowski and Jezewska, 2000, 2009; Jezewska and Bujalowski, 1996; Jezewska and Bujalowski, 1997; Jezewska *et al.*, 1996, 1998a,b, 2001, 2003, 2006;

Lohman and Bujalowski, 1991; Rajendran *et al.*, 1998, 2001; Szymanski *et al.*, 2010). Therefore, at each $[L]_F$, the observed spectroscopic signal, F_{obs}, is the algebraic sum of the macromolecule concentrations in each state, M_i, weighted by the value of the intensive spectroscopic property of that state, F_i. The "signal conservation" equation for the observed fluorescence emission, F_{obs}, is then

$$F_{obs} = F_F[M]_F + \sum F_i[M]_i \tag{2.4}$$

where F_F is the molar fluorescence of the free nucleic acid and F_i is the molar fluorescence of the complex, $[M]_i$, which represents the concentration of the macromolecule with i bound ligand molecules ($i = 1$ to n). Another mass conservation equation relates $[M]_F$ and $[M]_i$ to $[M]_T$ by

$$[M]_T = [M]_F + \sum [M]_i \tag{2.5}$$

Next, we define the partial degree of binding, Θ_i ("i" moles of ligand bound per mole of macromolecule), corresponding to all complexes with a given number "i" of bound ligand molecules as

$$\Theta_i = \frac{i[M]_i}{[M]_T} \tag{2.6}$$

Therefore, the concentration of the macromolecule with "i" protein molecules bound, $[M]_i$, is defined as

$$[M]_i = \left(\frac{\Theta_i}{i}\right)[M]_T \tag{2.7}$$

Introducing Eqs. (2.6) and (2.7) into Eq. (2.4) provides a general relationship for the observed fluorescence, F_{obs}, as

$$F_{obs} = F_F[M]_T + \left[\Sigma(F_i - F_F)\left(\frac{\Theta_i}{i}\right)\right][M]_T \tag{2.8}$$

By rearranging Eq. (2.8), one can define the experimentally accessible quantity, ΔF_{obs}, that is, the fluorescence change normalized with respect to the initial fluorescence intensity of the free macromolecule, as

$$\Delta F_{obs} = \frac{(F_{obs} - F_F[M]_T)}{F_F[M]_T} \tag{2.9}$$

and

$$\Delta F_{\mathrm{obs}} = \sum \left(\frac{\Delta F_i}{i} \right) (\Theta_i) \tag{2.10}$$

The quantity, ΔF_{obs}, is the experimentally determined fractional fluorescence change observed at the selected total ligand and macromolecule concentrations, $[\mathrm{L}]_\mathrm{T}$ and $[\mathrm{M}]_\mathrm{T}$. The quantity, $\Delta F_i/i$, is the average molar fluorescence change per bound ligand in the complex containing "i" ligand molecules. Because $\Delta F_i/i$ is an intensive molecular property of the ligand–macromolecule complex with "i" ligand molecules bound, Eq. (2.10) indicates that ΔF_{obs} is only a function of the free ligand concentration through the total average degree of binding, $\Sigma\Theta_i$. Therefore, for a given and specific value of ΔF_{obs}, the total average degree of binding, $\Sigma\Theta_i$, must be the same for any value of $[\mathrm{L}]_\mathrm{T}$ and $[\mathrm{M}]_\mathrm{T}$ (Bujalowski, 2006; Bujalowski and Jezewska, 2000, 2009; Jezewska and Bujalowski, 1996, 1997; Jezewska *et al.*, 1996, 1998a,b, 2001, 2003, 2006; Lohman and Bujalowski, 1991; Rajendran *et al.*, 1998, 2001; Szymanski *et al.*, 2010). Thus, if one performs a fluorescence titration of the macromolecule with the ligand, at different $[\mathrm{M}]_\mathrm{T}$, the same value of ΔF_{obs} at different $[\mathrm{M}]_\mathrm{T}$s indicates the same physical state of the macromolecule, that is, the same degree of macromolecule saturation with the ligand, that is, the same $\Sigma\Theta_i$. Since $\Sigma\Theta_i$ is a unique function of the free ligand concentration, $[\mathrm{L}]_\mathrm{F}$, the value of $[\mathrm{L}]_\mathrm{F}$, at the same total average degree of binding, must also be the same. An analogous expression can be derived for the case where the spectroscopic signal originates from the ligand and the binding analysis is performed using the reverse titration method (Bujalowski, 2006; Bujalowski and Jezewska, 2000, 2009; Jezewska and Bujalowski, 1996, 1997; Jezewska *et al.*, 1996, 1998a,b, 2001, 2003, 2006; Lohman and Bujalowski, 1991; Rajendran *et al.*, 1998, 2001; Szymanski *et al.*, 2010). Expression (2.10) indicates a clear method of transforming the fluorescence titration curve into a thermodynamic binding isotherm (Bujalowski, 2006; Bujalowski and Jezewska, 2000, 2009; Jezewska and Bujalowski, 1996, 1997; Jezewska *et al.*, 1996, 1998a,b, 2001, 2003, 2006; Lohman and Bujalowski, 1991; Rajendran *et al.*, 1998, 2001; Szymanski *et al.*, 2010). In optimal and minimal case, one can perform only two titrations at two different total concentrations of the macromolecule, that is, $[\mathrm{M}]_{\mathrm{T1}}$ and $[\mathrm{M}]_{\mathrm{T2}}$. At the same value of ΔF_{obs}, the total average degree of binding, $\Sigma\Theta_i$, and the free ligand concentrations, $[\mathrm{L}]_\mathrm{F}$, must be the same for both titration curves (see above).

Two hypothetical fluorescence titration curves are illustrated in Fig. 2.2A for the binding process where six ligand molecules bind to six identical binding sites on the macromolecule, characterized by intrinsic binding constants, K and the relative fluorescence changes, $\Delta F_{\mathrm{max}} = 0.5$, for the ligand binding to each site, respectively, as defined by Eq. (2.1). The values of ΔF_{obs} are plotted as a function of the logarithm of the total ligand

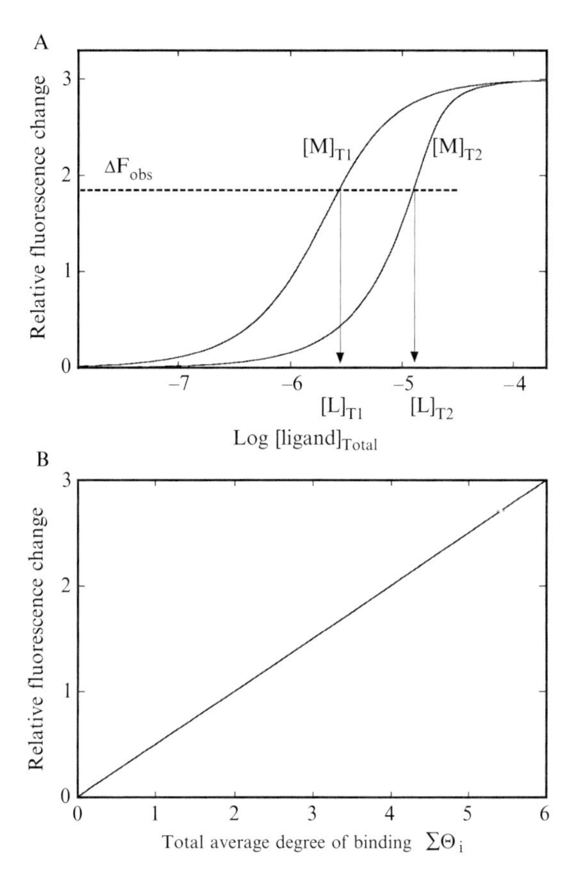

Figure 2.2 (A) Theoretical fluorescence titrations of a macromolecule that has six independent and discrete binding sites with a ligand obtained at two different macromolecule concentrations, $[M]_{T1} = 3 \times 10^{-7}$ M, and $[M]_{T2} = 3 \times 10^{-6}$ M, respectively. The binding sites are characterized by the intrinsic binding constants, $K = 1 \times 10^{5}$ M^{-1}. Association of the ligand induces at each binding site the relative change of the macromolecule fluorescence, $\Delta F = 0.5$, providing the maximum change of the macromolecule fluorescence at saturation, $\Delta F_{max} = 3.0$. The arrows indicates the total ligand concentrations, $[L]_{T1}$ and $[L]_{T2}$, at the same selected value of the observed fluorescence change, marked by the dashed horizontal line, ΔF_{obs}, at which the total average degree of binding of the ligand, $\Sigma\Theta_i$, is the same for both titration curves. (B) Dependence of the relative fluorescence of the macromolecule, ΔF_{obs}, upon the total average degree of binding, $\Sigma\Theta_i$, of the ligand for the binding model in (A). Because the value of $\Delta F = 0.5$ is the same for each binding site, the plot is linear over the entire range of $\Sigma\Theta_i$.

concentration, $[L]_T$. At a higher macromolecule concentration, a given relative fluorescence increase, ΔF_{obs}, is reached at higher ligand concentrations, as more ligand is required to saturate the macromolecule at its higher concentration. A set of values of $(\Sigma\Theta_i)_j$ and $([L]_F)_j$ for the selected "j" value of $(\Delta F_{obs})_j$ is obtained from these data in the following manner. One draws a horizontal line that intersects both titration curves at the same value of $(\Delta F_{obs})_j$; Fig. 2.2A). The point of intersection of the horizontal line with each titration curve defines two values of the total ligand concentration, $([L]_{T1})_j$ and $([L]_{T2})_j$, for which $([L]_F)_j$ and $(\Sigma\Theta_i)_j$ are the same (see above; Bujalowski, 2006; Bujalowski and Jezewska, 2000, 2009; Jezewska and Bujalowski, 1996, 1997; Jezewska *et al.*, 1996, 1998a,b, 2001, 2003, 2006; Lohman and Bujalowski, 1991; Rajendran *et al.*, 1998, 2001; Szymanski *et al.*, 2010). Then, one has two mass conservation equations for the total concentrations of the protein, $([L]_{T1})_j$ and $([L]_{T2})_j$, as

$$\left([L]_{T1}\right)_j = \left([L]_F\right)_j + \left(\sum \Theta_i\right)_j [M]_{T1} \tag{2.11a}$$

and

$$\left([L]_{T2}\right)_j = \left([L]_F\right)_j + \left(\sum \Theta_i\right)_j [M]_{T2} \tag{2.11b}$$

from which one obtains that at a given $(\Delta F_{obs})_j$, the total average degree of binding and the free protein concentration are

$$\left(\sum \Theta_i\right)_j = \frac{\left([L]_{T2}\right)_j - \left([L]_{T2}\right)_j}{[M]_{T1} - [M]_{T2}} \tag{2.12}$$

and

$$\left([L]_F\right)_j = \left([L]_{Tx}\right)_j - \left(\sum \Theta_i\right)_j \left([M]_{Tx}\right) \tag{2.13}$$

where subscript, x, is 1 or 2(Bujalowski, 2006; Bujalowski and Jezewska, 2000, 2009; Jezewska and Bujalowski, 1996, 1997; Jezewska *et al.*, 1996, 1998a,b, 2001, 2003, 2006; Lohman and Bujalowski, 1991; Rajendran *et al.*, 1998, 2001; Szymanski *et al.*, 2010). Performing a similar analysis along the titration curves at a selected interval of the observed signal change, one obtains model-independent values of $([L]_F)_j$ and $(\Sigma\Theta_i)_j$ at any selected "j" value of $(\Delta F_{obs})_j$. The most accurate estimates of $([L]_F)_j$ and $(\Sigma\Theta_i)_j$ are obtained in the region of the titration curves where the concentration of a bound ligand is comparable to its total concentration, $[L]_T$. In our practice, this limits the accurate determination of the total average degree of binding and $[L]_F$ to the region of the titration curves where the concentration of the bound ligand is at least ~10–15% of the $[L]_T$. Selection of suitable concentrations of the macromolecule is of paramount importance for obtaining

$([L]_F)_j$ and $(\Sigma\Theta_i)_j$ over the largest possible region of the titration curves. The selection of the macromolecule concentrations is based on preliminary titrations that provide initial estimates of the expected affinity. Nevertheless, the accuracy of the determination of $(\Sigma\Theta_i)_j$ is mostly affected in the region of the high concentrations of the ligand, that is, where the binding process approaches the maximum saturation.

Usually, the maximum of the recorded fluorescence changes, that is, the saturation of the observed binding process, is attainable with adequate accuracy. What is mostly unknown is the maximum stoichiometry of the complex at saturation (Bujalowski, 2006; Bujalowski and Jezewska, 1995, 2000, 2009; Bujalowski and Klonowska, 1993; Jezewska and Bujalowski, 1996, 1997; Jezewska *et al.*, 1996, 1998a,b, 2001, 2003, 2006; Lohman and Bujalowski, 1991; Rajendran *et al.*, 1998, 2001; Szymanski *et al.*, 2010). Another unknown is the relationship between the fluorescence change, ΔF_{obs}, and the total average degree of binding (Bujalowski, 2006; Bujalowski and Jezewska, 1995, 2000, 2009; Bujalowski and Klonowska, 1993; Jezewska and Bujalowski, 1996, 1997; Jezewska *et al.*, 1996, 1998a,b, 2001, 2003, 2006; Lohman and Bujalowski, 1991; Rajendran *et al.*, 1998, 2001; Szymanski *et al.*, 2010). Both unknowns can be determined by plotting ΔF_{obs} as a function of $\Sigma\Theta_i$, as depicted in Fig. 2.2B, for the hypothetical binding model considered here. The plot is clearly linear because we selected the same value of $\Delta F_{max} = 0.5$ for each of the six binding sites. As mentioned above, in practice, the maximum value of $\Sigma\Theta_i$ cannot be directly determined, due to the inaccuracy at the high-ligand concentration region. However, knowing the maximum increase of the macromolecule fluorescence ΔF_{max}, one can perform a short extrapolation of the plot (ΔF_{obs} vs. $\Sigma\Theta_i$) to this maximum value of the observed fluorescence change, which establishes the maximum stoichiometry of the formed complex at saturation (Bujalowski, 2006; Bujalowski and Jezewska, 1995, 2000, 2009; Bujalowski and Klonowska, 1993; Jezewska and Bujalowski, 1996, 1997; Jezewska *et al.*, 1996, 1998a,b, 2001, 2003, 2006; Lohman and Bujalowski, 1991; Rajendran *et al.*, 1998, 2001; Szymanski *et al.*, 2010).

4. Nucleotide Binding to the RepA Protein of Plasmid RSF1010

As an example of the real analysis, we consider the binding of the nucleotide cofactors to the RepA protein of plasmid RSF1010 (Jezewska *et al.*, 2005a,b). RepA protein is a hexameric helicase specifically involved in the replication of nonconjugative plasmid RSF1010, which confers bacterial resistance to sulfonamides and streptomycin. Binding of the unmodified nucleotide cofactors to the RepA hexamer is not accompanied by a change of the protein fluorescence that is adequate to perform quantitative analysis of the complex binding process. However, binding of nucleotide analogs, TNP-ATP and TNP-ADP, to the RepA protein is accompanied by a strong quenching of the protein fluorescence, providing an excellent signal to monitor the association. The hydrolysis of TNP-ATP is undetectable on the time scale of the binding experiments (Jezewska *et al.*, 2005a,b).

Fluorescence titrations of the RepA hexamer with TNP-ADP, at three different protein concentrations, are shown in Fig. 2.3A. The maximum quenching of the protein fluorescence at saturation is 0.87 ± 0.03. The selected protein concentrations provide the separation of the binding isotherms up to the quenching value of ~0.83 (Jezewska *et al.*, 2005a,b). The spectroscopic titration curves in Fig. 2.3A would indicate an apparent single binding process. However, analysis of the data, using the thermodynamically rigorous approach, as discussed above, reveals a much more complex behavior (Jezewska *et al.*, 2005a,b). Because we have three titration curves, we can select any two of them to determine the total average degree of binding. However, higher accuracy is usually attained with the most separated titration curves, with respect to the total ligand concentrations.

The dependence of the observed fluorescence quenching, ΔF_{obs}, upon the total average degree of binding, $\Sigma\Theta_i$, of the TNP-ADP on the RepA hexamer is shown in Fig. 2.3B. The separation of the binding isotherms allows us to obtain the values of $\Sigma\Theta_i$ up to ~5.1 of the TNP-ADP molecules per RepA hexamer. The plot in Fig. 2.3B is clearly nonlinear. For comparison, the dashed line represents the hypothetical case where a strict proportionality between the total average degree of binding and the quenching of the RepA hexamer fluorescence would exist. Thus, binding of the first three molecules of the nucleotide induces a larger decrease of the protein fluorescence emission than the binding of the subsequent cofactor molecules. Because the maximum quenching is available from the experiment (Fig. 2.3A), short extrapolation to the maximum quenching $\Delta F_{max} = 0.87 \pm 0.03$ shows that at saturation, the RSF1010 RepA hexamer binds 6.0 ± 0.3 molecules of TNP-ADP. In spite of the nonlinear behavior of the plot in Fig. 2.3B, such extrapolation is rather easy because the values of $\Sigma\Theta_i$ could be determined up to ~90% of the observed signal. In our practice, we select the last four to five points of the plot and perform a linear regression. The maximum stoichiometry is determined by the interception of the regression line with the horizontal line, which indicates the maximum value of ΔF. For nonlinear experimental curves, which cover ~70–80% of the observed signal, the error in determination of the maximum stoichiometry is ~10–15% (Bujalowski, 2006; Jezewska and Bujalowski, 1996; Jezewska *et al.*, 2005a,b). The situation is much simpler with the linear plots where the plot of ΔF versus $\Sigma\Theta_i$, covering ~50%, provides the maximum stoichiometry of the formed complex with similar, ~10–15% accuracy.

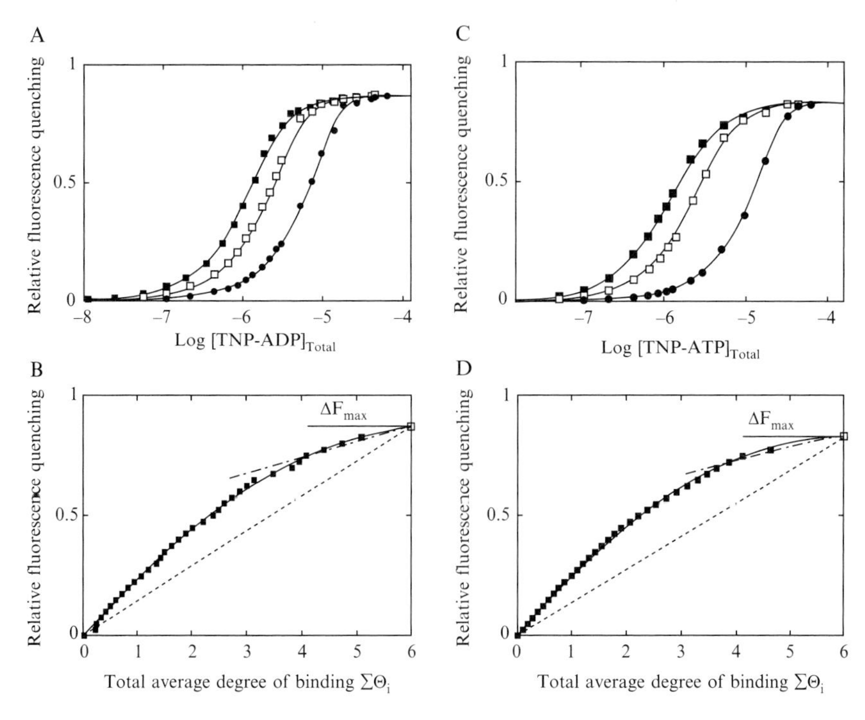

Figure 2.3 (A) Fluorescence titration of the RepA helicase with TNP-ADP in 50 mM Tris/HCl (pH 7.6, 10 °C), containing 10 mM NaCl and 1 mM $MgCl_2$, at different RepA protein concentrations: (■) 5×10^{-7} M; (□) 1×10^{-6} M; (●) 3×10^{-6} M (hexamer). The solid lines are nonlinear least squares fits of the titration curves, according to the hexagon model (Eqs. (2.14)–(2.16)) using a single set of binding parameters with the intrinsic binding constant $K = 8 \times 10^6$ M^{-1} and cooperativity parameter $\sigma = 0.36$(Jezewska *et al.*, 2005a,b). (B) Dependence of the relative fluorescence quenching, ΔF_{obs}, upon the average degree of binding of TNP-ADP on the RepA hexamer, $\Sigma\Theta_i$, (■). The point (□) and the straight horizontal line indicate the location of the maximum value of the observed ΔF. The line (——) is the linear regression using the last five points of the plot. The solid line is the nonlinear least squares fit, using the second-degree polynomial function representing the empirical function (Jezewska *et al.*, 2005a,b). The dashed line is the hypothetical dependence of ΔF_{obs} upon $\Sigma\Theta_i$ that assumes a strict linear relationship between the observed fluorescence quenching and the average degree of binding. The maximum value of ΔF_{max} $= 0.87 \pm 0.03$. (C) Fluorescence titration of the RepA helicase with TNP-ATP in 50 mM Tris/HCl (pH 7.6, 10 °C), containing 10 mM NaCl and 1 mM $MgCl_2$, at different RepA protein concentrations: (■) 5×10^{-7} M; (□) 1×10^{-6} M; (●) 6×10^{-6} M (hexamer). The solid lines are nonlinear least squares fits of the titration curves, according to the hexagon model (Eqs. (2.14)–(2.16)) using a single set of binding parameters with the intrinsic binding constant $K = 2.38 \times 10^6$ M^{-1} and cooperativity parameter $\sigma = 0.37$(Jezewska *et al.*, 2005a,b). (D) Dependence of the relative fluorescence quenching, ΔF_{obs}, upon the average degree of binding of TNP-ADP on the RepA hexamer, $\Sigma\Theta_i$, (■). The point (□) and the straight horizontal line indicate the location of the maximum value of the observed ΔF. The line (– – –) is the linear regression using the last four points of the plot. The solid line is the nonlinear

The determination of the maximum stoichiometry of the nucleoside triphosphate analog, TNP-ATP, binding to the RepA hexamer has been analogously performed. Fluorescence titrations of the RepA protein with TNP-ATP at three different protein concentrations are shown in Fig. 2.3C. At saturation, the maximum quenching of RepA helicase emission is 0.83 ± 0.03. The dependence of the protein fluorescence quenching, ΔF_{obs}, upon the total average degree of binding of TNP-ATP on the RepA hexamer is shown in Fig. 2.3D. The separation of the isotherms allows us to obtain $\Sigma\Theta_i$ up to $\sim$4.6. Similar to the TNP-ADP case (Fig. 2.3B), the plot is nonlinear, indicating that the binding of the first three cofactor molecules is accompanied by a larger quenching of the protein fluorescence than the quenching accompanying the saturation of the remaining nucleotide-binding sites. Extrapolation to the maximum value of $\Delta F_{max} = 0.83 \pm 0.03$ shows that, at saturation, 6.0 ± 0.5 TNP-ATP molecules bind to the RepA hexamer. Because we have quantitatively determined the average degree of binding, $\Sigma\Theta_i$, over $\sim$80% of the spectroscopic titration curves, we can obtain the free nucleotide concentrations, using Eq. (2.13), and construct a true thermodynamic binding isotherm, that is, the average degree of binding, $\Sigma\Theta_i$, as a function of the free nucleotide concentration (Jezewska *et al.*, 2005a,b). The dependence of $\Sigma\Theta_i$ upon the free concentration of the $[\text{TNP-ADP}]_F$ is shown in Fig. 2.4A. An analogous plot of $\Sigma\Theta_i$, as a function of $[\text{TNP-ATP}]_F$, is shown in Fig. 2.4B. In contrast to the fluorescence titrations in Fig. 2.3A and C, the plots in Fig. 2.4A and B are true thermodynamic isotherms.

5. Applying the Statistical Thermodynamic Model for the Nucleotide Binding to the RSF1010 RepA Protein Hexamer

Once the binding isotherms are available, one can address the intrinsic affinities and cooperativities in the system (Bujalowski, 2006; Bujalowski and Jezewska, 2000, 2009; Jezewska and Bujalowski, 1996, 1997; Jezewska *et al.*, 1996, 1998a,b, 2001, 2003, 2005a,b, 2006; Rajendran *et al.*, 1998, 2001; Szymanski *et al.*, 2010). The properties of the binding isotherms (Fig. 2.4A and B) and the known properties of the examined system provide crucial clues as to how to achieve that. A striking feature of the plots in

least squares fit, using the second-degree polynomial function representing the empirical function (Jezewska *et al.*, 2005a,b). The dashed line is the theoretical dependence of ΔF_{obs} upon $\Sigma\Theta_i$ that assumes a strict linear relationship between the observed fluorescence quenching and the average degree of binding. The maximum value of $\Delta F_{max} = 0.83 \pm 0.03$.

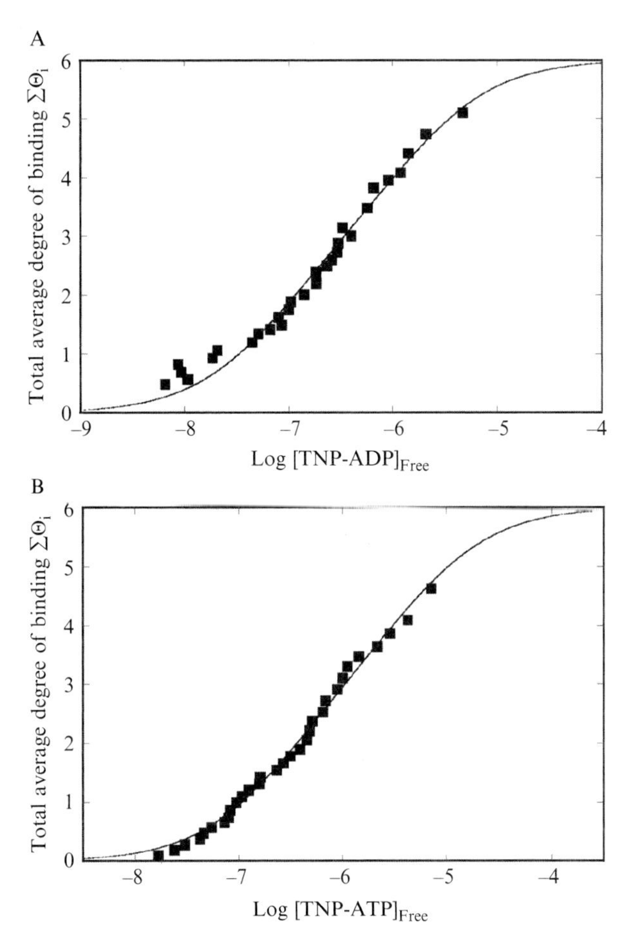

Figure 2.4 (A) The average number of TNP-ADP molecules bound per RepA hexamer as a function of the free concentration of the cofactor. The solid line is a nonlinear least squares fit according to the hexagon model (Eqs. (2.14)–(2.16)) with the intrinsic binding constant $K = 7.7 \times 10^6\ M^{-1}$ and cooperativity parameter $\sigma = 0.38$. (B) The average number of TNP-ATP molecules bound per RepA hexamer as a function of the free concentration of the cofactor. The solid line is a nonlinear least squares fit according to the hexagon model (Eqs. (2.14)–(2.16)) with the intrinsic binding constant $K = 2.38 \times 10^6\ M^{-1}$ and cooperativity parameter $\sigma = 0.4$ (Jezewska *et al.*, 2005a,b).

Fig. 2.4A and B is that they extend over more than two orders of magnitude of the free nucleotide concentration, that is, between ~10% and ~90% of the maximum degree of binding. If the six binding sites had the same affinity for the ligand (nucleotide cofactor) then the free ligand concentration could not span more than two orders of magnitude between 10% and ~90% of the maximum degree of binding (Jezewska and Bujalowski, 1996; Jezewska *et al.*, 2005a,b). In other words, a binding system of multiple discrete binding sites, without heterogeneity in their intrinsic affinities, cannot generate such behavior. Therefore, these data unambiguously show that the macroscopic affinities of the nucleotide cofactors for the RepA hexamer are a decreasing function of $\Sigma\Theta_i$.

The RepA hexamer is built of six identical protomers forming a ring-like structure with the protomer–protomer contacts limited to two adjacent subunits. The thermodynamic data discussed above show that the RSF1010 RepA hexamer binds six nucleotide molecules and the binding is characterized by the decreasing macroscopic affinity with the increase of the degree of binding. The simplest statistical thermodynamic model, that is, the model containing the smallest number of parameters, that can account for such behavior is the RepA hexamer exhibiting nearest-neighbor negative cooperativity among the otherwise identical six nucleotide-binding sites. This model is referred to as the hexagon model (Bujalowski and Klonowska, 1993, 1994a,b; Bujalowski *et al.*, 1994; Jezewska *et al.*, 2005a,b). The other possible simple model of the nucleotide binding to the six sites of the RepA hexamer, also containing only two binding parameters, is that the hexamer possesses two independent classes of independent binding sites, each class encompassing three independent binding sites. However, further analysis of the system, discussed in the original works, showed that this model does not apply to the RepA hexamer (Jezewska *et al.*, 2005a,b).

In the hexagon model, there are only two parameters that characterize the binding, the intrinsic binding constant, K, and the parameter, σ, describing cooperative interactions (Jezewska *et al.*, 2005a,b). The first ligand can bind to any of six, initially equivalent binding sites. The cooperative interactions are limited to only two neighboring sites. The partition functions, Z_H, is then

$$Z_H = 1 + 6x + 3(3 + 2\sigma)x^2 + 2(1 + 6\sigma + 3\sigma^2)x^3 + 3(3\sigma^2 + 2\sigma^3)x^4 + 6\sigma^4x^5 + \sigma^6x^6 \tag{2.14}$$

where $x = K[N]_F$, and $[N]_F$ is the free nucleotide concentration. The total average degree of binding, $\Sigma\Theta_i$, is defined by the standard statistical thermodynamic formula (Bujalowski and Klonowska, 1993, 1994a,b; Bujalowski *et al.*, 1994; Jezewska *et al.*, 2005a,b)

$$\Sigma\Theta_i = \frac{\partial \ln Z_H}{\partial \ln L} \tag{2.15}$$

and

$$\Sigma\Theta_i = \frac{6x + 6(3+2\sigma)x^2 + 6(1+6\sigma+3\sigma^2)x^3 + 12(3\sigma^2+2\sigma^3)x^4 + 30\sigma^4x^5 + 6\sigma^6x^6}{Z_H} \tag{2.16}$$

The solid lines in Fig. 2.4A and B are the nonlinear least square fits of the thermodynamic isotherms for TNP-ADP and TNP-ATP binding the RepA hexamer, according to the hexagon model, with only two binding parameters, using Eqs. (2.14)–(2.16). The obtained values of the intrinsic binding constants and cooperativity parameter σ are $K = (7.7 \pm 1.5) \times 10^6\ M^{-1}$, $\sigma = 0.38 \pm 0.05$ and $K = (2.4 \pm 0.2) \times 10^6\ M^{-1}$, $\sigma = 0.37 \pm 0.05$, for TNP-ADP and TNP-ATP, respectively.

6. Empirical Function Approach

Direct fitting of the spectroscopic titration curves may refine the obtained binding parameters, as it uses a larger span of the experimental curves. For the direct fitting of the spectroscopic titration curves, using Eqs. (2.14)–(2.16), one would have to know all molar fluorescence intensities of all possible RepA–nucleotide complexes. This is rather a hopeless task. Notice, the plots in Fig. 2.3B and D do not provide any information as to what the values of these parameters should be, with the exception of ΔF_{max}. On the other hand, the problem of finding all optical parameters can be avoided by using the empirical function approach (Bujalowski and Klonowska, 1993; Jezewska *et al.*, 2005a,b). This approach can be applied to any ligand–macromolecule systems, where the determination of all optical parameters is practically impossible. It is based on introducing the representation of the observed relative fluorescence change of the macromolecule, ΔF_{obs}, as a function of the total average degree of binding, $\Sigma\Theta_i$, such as in Fig. 2.3B and D, via an empirical function (Bujalowski and Klonowska, 1993; Jezewska *et al.*, 2005a,b). The empirical function is usually a polynomial that accurately relates the experimentally determined ΔF_{obs}, to the experimentally determined, $\Sigma\Theta_i$, as

$$\Delta F_{obs} = \sum_{j=0}^{n} a_j(\Sigma\Theta_i)^j \tag{2.17}$$

where a_j are the fitting constants. The generation of the theoretical titration curve for the selected binding model is then accomplished by first calculating the value of $\Sigma\Theta_i$, for a given free ligand concentration and initial estimates of the binding parameters. Next, the obtained $\Sigma\Theta_i$ is introduced into Eq. (2.17) and the value of ΔF_{obs}, corresponding that value of $\Sigma\Theta_i$, is obtained. These calculations are then performed for the entire titration curve. The solid lines in Fig. 2.3A and B are generated using the empirical function approach based on the plots in Fig. 2.3B and D (Bujalowski and Klonowska, 1993; Jezewska *et al.*, 2005a,b).

7. MCT Method: General Considerations

So far, we discussed the methods of analysis of the titration curves that allow the experimenter to generate a model-independent thermodynamic binding isotherm. However, the analysis requires that the ligand binding is accompanied by significant changes in the spectroscopic signal originating from the macromolecule. On the other hand, many examined systems do not have a convenient signal to monitor the binding. One way to overcome this problem is to use fluorescence derivative of one of the interacting species, for example, the nucleic acid and/or protein, although this may not be feasible in all cases. A modification introduced on a protein may affect its activity. Modifications on the short fragments of a nucleic acid may create additional, undesirable interaction areas. Long polymeric nucleic acids provide an additional problem. For instance, while modification of homoadenosine polymers, to obtain fluorescent etheno-derivative, is relatively an easy reaction, obtaining homogeneously labeled fluorescent derivative of, for example, poly(d*T*) or poly(d*C*) is not (Baker *et al.*, 1978; Bujalowski, 2006; Bujalowski and Jezewska, 1995; Chabbert *et al.*, 1987; Menetski and Kowalczykowski, 1985; Secrist *et al.*, 1972; Tolman *et al.*, 1974). For this purpose, the MCT method provides a way to quantitatively examine binding of multiple protein ligands to unmodified polymer nucleic acid lattices and, in general, to any unmodified macromolecule with multiple binding sites (Bujalowski, 2006; Jezewska and Bujalowski, 1996). Thus, the method is invaluable in addressing base- or sequence-specificity in protein–nucleic acid systems, which are otherwise very difficult to examine. In the first step, the interactions with a modified fluorescent system must be characterized. Next, the examination of interactions with an unmodified component of the reaction can be performed using competition studies (Bujalowski, 2006; Jezewska and Bujalowski, 1996).

The approach is based on the same thermodynamic argument, which leads to Eqs. (2.11a)–(2.13). The titration of a reference fluorescent macromolecule with the ligand is performed in the presence of the second competing, nonfluorescent macromolecule whose interactions with the ligand are to be determined. Although the ligand binds to two different macromolecules, in solution, the observed signal originates only from the fluorescent "reference" macromolecule. The saturation of both macromolecules increases with the increasing concentration of the free ligand.

The entire binding system is composed of two macromolecules, reference and unmodified, competing for the ligand (Bujalowski, 2006; Jezewska and Bujalowski, 1996). In the absence of the competing unmodified macromolecule, the total concentration of the ligand, $[L]_{TR}$, at a particular value of the spectroscopic parameter (emission intensity), ΔF_{obs}, is (Fig. 2.5)

$$[L]_{TR} = [L]_F + [L]_{BR} \tag{2.18}$$

and

$$[L]_{TR} = [L]_F + (\Sigma\Theta_i)[M]_{TR} \tag{2.19}$$

where, $[L]_F$ is the free ligand concentration, $(\Sigma\Theta_i)_R$ is the total average degrees of binding on the fluorescent reference macromolecule and $[M]_{TR}$ is the total concentration of the reference macromolecule, respectively. In the presence of the unmodified macromolecule, the total concentration of the ligand, $[L]_{TS}$, which provides the same value of ΔF_{obs}, is

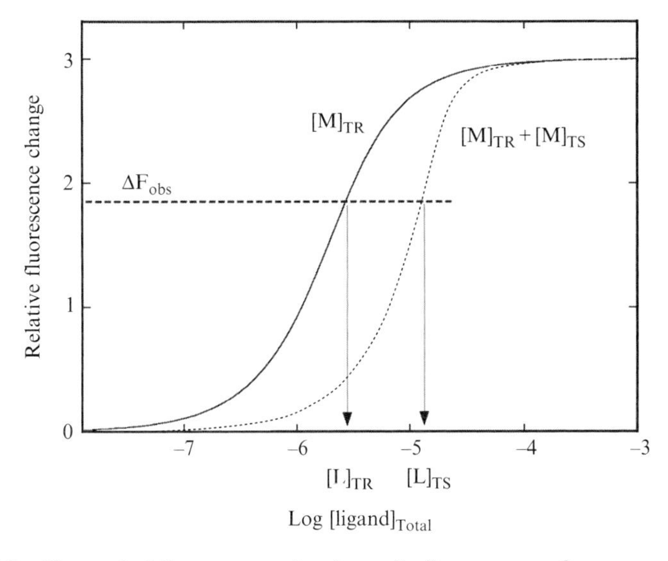

Figure 2.5 Theoretical fluorescence titrations of a fluorescent reference macromolecule with a ligand, at concentration, $[M]_{TR}$, in the absence (solid line) and presence of a competing unmodified macromolecule, at concentration, $[M]_{TS}$ (dashed line). The arrows indicate the total ligand concentrations, $[L]_{TR}$ and $[L]_{TS}$, at the selected value of the observed fluorescence change, marked by the dashed horizontal line, ΔF_{obs}, at which the total average degree of binding of the ligand, $\Sigma\Theta_i$, on the reference fluorescence macromolecule and the free ligand concentration, $[L]_F$, are the same for both titration curves.

$$[L]_{TS} = [L]_F + [L]_{BR} + [L]_{BS} \tag{2.20}$$

and

$$[L]_{TS} = [L]_F + (\Sigma\Theta_i)_R[M]_{TR} + (\Sigma\Theta_i)_S[M]_{TS} \tag{2.21}$$

where,($\Sigma\Theta_i)_S$ and $[M]_{TS}$ are the total average degrees of binding on the unmodified competing macromolecule, and the total concentration of the unmodified macromolecule, respectively. Subtracting Eq. (2.19) from Eq. (2.21) and rearranging provides

$$(\Sigma\Theta_i)_S = \frac{[L]_{TS} - [L]_{TR}}{[M]_{TS}} \tag{2.22}$$

The concentration of the free ligand, $[L]_F$, is then

$$[L]_F = [L]_{TR} - [L]_B = [L]_{TS} - (\Sigma\Theta_i)_R[M]_{TR} \tag{2.23}$$

or

$$[L]_F = [L]_{TS} - [L]_B = [L]_{TS} - (\Sigma\Theta_i)_R[M]_{TR} - (\Sigma\Theta_i)_S[M]_{TS} \tag{2.24}$$

Notice, one may determine $(\Sigma\Theta_i)_S$ without addressing the exact model for the ligand binding to the reference fluorescence macromolecule and treat it only as an indicator of the same free ligand concentration at a given value of the observed relative fluorescence intensity. This can be accomplished by performing titrations of the reference macromolecule at two or more concentrations of the unmodified macromolecule, $[L]_{TS1}$ and $[L]_{TS2}$ (Bujalowski, 2006; Jezewska and Bujalowski, 1996). In such an approach, one has, at a given value of ΔF_{obs},

$$[L]_{TS1} = [L]_F + (\Sigma\Theta_i)_R[M]_{TR} + (\Sigma\Theta_i)_S[M]_{TS1} \tag{2.25}$$

and

$$[L]_{TS2} = [L]_F + (\Sigma\Theta_i)_R[M]_{TR} + (\Sigma\Theta_i)_S[M]_{TS} \tag{2.26}$$

where($\Sigma\Theta_i)_{RS}$ and $[M]_{TR}$ have the same meaning as above, $(\Sigma\Theta_i)_S$, $[M]_{TS1}$, and $[M]_{TS2}$ are the total average degrees of binding on the unmodified competing macromolecule and the two total concentrations of the unmodified macromolecule, respectively. Subtracting Eq. (2.25) from Eq. (2.26) and rearranging provides

$$(\Sigma\Theta_i)_S = \frac{[L]_{TS2} - [L]_{TS1}}{[M]_{TS2} - [M]_{TS2}} \quad (2.27)$$

Nevertheless, to obtain the concentration of the free ligand, $[L]_F$, $(\Sigma\Theta_i)_{RS}$ must be known, providing the free ligand concentration as

$$[L]_F = [L]_{TS1} - [L]_B = [L]_{TS1} - (\Sigma\Theta_i)_R[M]_{TR} - (\Sigma\Theta_i)_S[M]_{TS1} \quad (2.28)$$

or

$$[L]_F = [L]_{TS2} - [L]_B = [L]_{TS2} - (\Sigma\Theta_i)_R[M]_{TR} - (\Sigma\Theta_i)_S[M]_{TS2} \quad (2.29)$$

8. Application of the MCT Method to the Base Specificity Problem in ASFV Pol X–ssDNA System

We describe the application of the MCT method to examine the base specificity in the interactions of the African swine fever virus polymerase X with the ssDNA (Jezewska *et al.*, 2007). ASFV Pol X is a small DNA repair polymerase involved in maintaining the genome stability of the virus (Jezewska *et al.*, 2007; Maciejewski *et al.*, 2001; Showwalter *et al.*, 2003). Binding the enzyme to the etheno-derivative of the homoadenosine nucleic acids is accompanied by a large nucleic acid fluorescence increase. Unfortunately, association of the polymerase with unmodified ssDNA is not accompanied by an adequate fluorescence change of the enzyme. Thus, determination of the ASFV pol X affinity for the unmodified ssDNA oligomers has been performed using the MCT method. In these studies, the fluorescent 16-mer, $d\varepsilon A(p\varepsilon A)_{15}$, has been used as a reference lattice and competition studies are performed with the unmodified ssDNA 20-mers. Both the 16- and 20-mer can only accommodate a single pol X molecule, what greatly simplifies the analysis (Jezewska *et al.*, 2007). Thus, the entire binding system is composed of two short ssDNA lattices, reference and unmodified oligomer, competing for ASFV pol X. The partition function, Z_R, for the reference 16- and unmodified 20-mer are then

$$Z_R = 1 + K_{16}[L]_F \quad (2.30)$$

and

$$Z_S = 1 + K_{20}[L]_F \quad (2.31)$$

where $[L]_F$ is the free ASFV pol X concentration, K_{16} and K_{20} are the macroscopic binding constant for ASFV pol X binding to the 16- and 20-mer, respectively. In the absence and the presence of the competing 20-mer, the total concentrations of the ASFV pol X, which provide the same value of the fluorescence change of the reference fluorescent 16-mer, are defined by Eqs. (2.18) and (2.20). Using the partition functions (Eqs. (2.30) and (2.31)) one has

$$[L]_{TR} = [L]_F + \left(\frac{K_{16}[L]_F}{1 + K_{16}[L]_F}\right)[M]_{TR} \tag{2.32}$$

and

$$[L]_{TS} = [L]_F + \left(\frac{K_{16}[L]_F}{1 + K_{16}[L]_F}\right)[M]_{TR} + \left(\frac{K_{20}[L]_F}{1 + K_{20}[L]_F}\right)[M]_{TS} \tag{2.33}$$

Therefore, the total average degree of binding of ASFV pol X on the unmodified 20-mer is

$$(\Sigma\Theta_i)_S = \frac{K_{20}[L]_F}{1 + K_{20}[P]_F} = \frac{[L]_{TS} - [L]_{TR}}{[M]_{TS}} \tag{2.34}$$

The concentration of the bound ASFV pol X on both nucleic acids is

$$[L]_B = \left(\frac{K_{16}[L]_F}{1 + K_{16}[L]_F}\right)[M]_{TR} + \left(\frac{K_{20}[L]_F}{1 + K_{20}[L]_F}\right)[M]_{TS} \tag{2.35}$$

and the concentration of the free ASFV pol X is in the presence of unmodified 20-mer is

$$[L]_F = [L]_{TS} - [L]_B \tag{2.36}$$

where $[L]_{TS}$ is the total concentration of the ASFV pol X, $[M]_{TR}$ and $[M]_{TS}$ are the total concentrations of the fluorescent reference 16- and the unmodified ssDNA 20-mer, respectively. However, because the system is very simple, we can directly express the observed fluorescence change of the reference 16-mer, ΔF_{obs}, as a function of the ASFV free pol X concentration, as

$$\Delta F_{obs} = \Delta F_{max}\left(\frac{K_{16}[L]_F}{1 + K_{16}[L]_F}\right) \tag{2.37}$$

with $[L]_F$ is defined by Eqs. (2.35) and (2.36).

Fluorescence titrations of d$\varepsilon A(p\varepsilon A)_{15}$, with the ASFV pol X, in the presence of two different concentrations of the 20-mer, d$A(pA)_{19}$, are shown in Fig. 2.6A. For comparison, the titration curve in the absence of the competing unmodified 16-mer is also included. Analogous fluorescence titrations of d$\varepsilon A(p\varepsilon A)_{15}$, with pol X in the presence of two different concentrations of the 20-mer, d$T(pT)_{19}$, are shown in Fig. 2.6B. A large shift of the binding isotherms, in the presence of d$A(pA)_{19}$ or d$T(pT)_{19}$, indicates an efficient competition between dεA(p$\varepsilon A)_{15}$ and the 20-mers for pol X. The solid lines in Fig. 2.6A and B are nonlinear least squares fits of the experimental titration curves, with a single fitting parameter, K_{20}, using Eqs. (2.30)–(2.37) and $K_{16} = 4.1 \times 10^5\ M^{-1}$, independently determined for the reference 16-mer (Jezewska *et al.*, 2007). The obtained binding constants are $K_{20} = (3.0 \pm 0.9) \times 10^7\ M^{-1}$ and $(8.0 \pm 1.8) \times 10^6\ M^{-1}$ for dA $(pA)_{19}$ and d$T(pT)_{19}$, respectively. Analogous studies with d$C(pC)_{19}$ provided $K_{20} = (3.0 \pm 0.6) \times 10^7\ M^{-1}$. Thus, the ASFV pol X does not have a preference for the type of base in the natural ssDNAs (Jezewska *et al.*, 2007).

9. Application of MCT Method to Protein–ssDNA Lattice Binding Systems

The major difference between the analyses discussed above and the ligand binding to long polymer nucleic acids is the replacement of the total average degree of binding, $\Sigma\Theta_i$, by the total average binding density, $\Sigma\nu_i$ (Bujalowski, 2006; Jezewska and Bujalowski, 1996). In other words, one expresses the extent of binding not as the number of moles of the ligand molecules per mole of the macromolecule, but per mole of the monomer of the lattice. Moreover, because more than one ligand molecule bind to the nucleic acid lattice, any statistical thermodynamic binding model must include cooperative interactions. To illustrate the general behavior of such a titration analysis, a series of theoretical spectroscopic titration curves of a reference fluorescent nucleic acid, with the ligand at a constant reference fluorescent nucleic acid concentration, but in the presence of different concentrations of a nonfluorescent, unmodified nucleic acid, is shown in Fig. 2.7. Both the reference and the unmodified nucleic acids are treated as long polymers. The titration curves have been generated, using the combined application of the generalized McGhee–von Hippel approach and the combinatorial theory for large ligand binding to a linear, homogeneous nucleic acid, described below (Bujalowski, 2006; Bujalowski and Lohman, 1986; Crothers, 1968; Epstein, 1978; Jezewska and Bujalowski, 1996; Lohman and Bujalowski, 1991; Lohman and Ferrari, 1994; McGhee and von Hippel, 1974). To simplify the considerations, the protein–nucleic acid complexes for both polymers have the site-size of $n = 20$, cooperativity

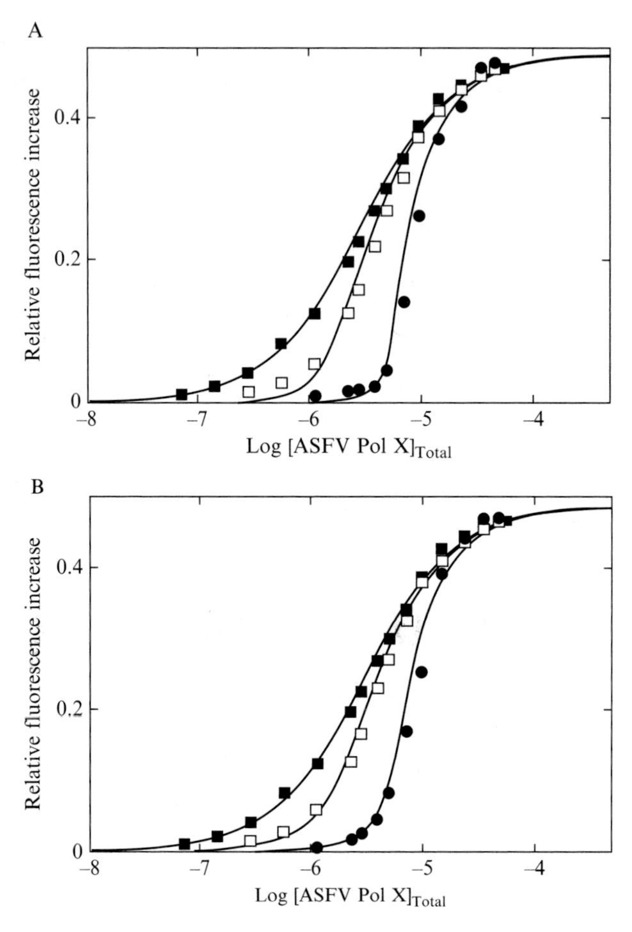

Figure 2.6 (A) Fluorescence titrations of the ssDNA 16-mer, d$\varepsilon A(p\varepsilon A)_{15}$, ($\lambda_{ex} = 325$ nm, $\lambda_{em} = 410$ nm) with the ASFV pol X in 10 mM sodium cacodylate (pH 7.0, 10 °C), containing 100 mM NaCl and 1 mM $MgCl_2$, in the absence (■) and the presence of the unmodified ssDNA 20-mer, d$A(pA)_{19}$. The concentration of d$\varepsilon A(p\varepsilon A)_{15}$ is 1×10^{-6} M. The concentration of unmodified ssDNA oligomer, d$A(pA)_{19}$ are (□) 1×10^{-6} M (●) 5×10^{-6} M (oligomer). The solid lines are nonlinear least squares fits of the experimental fluorescence titration curves, using the MCT method (Jezewska *et al.*, 2007). (B) Fluorescence titrations of the ssDNA 16-mer, d$\varepsilon A(p\varepsilon A)_{15}$, ($\lambda_{ex} = 325$ nm, $\lambda_{em} = 410$ nm) with the ASFV pol X in 10 mM sodium cacodylate (pH 7.0, 10 °C), containing 100 mM NaCl and 1 mM $MgCl_2$, in absence (■) and the presence of the unmodified ssDNA 20-mer, d$T(pT)_{19}$. The concentration of dεA $(p\varepsilon A)_{15}$ is 1×10^{-6} M. The concentrations of unmodified ssDNA oligomer, dT $(pT)_{19}$ are: (□) 1×10^{-6} M (●) 5×10^{-6} M (oligomer). The solid lines are nonlinear least squares fits of the experimental fluorescence titration curves, using the MCT method (Jezewska *et al.*, 2007).

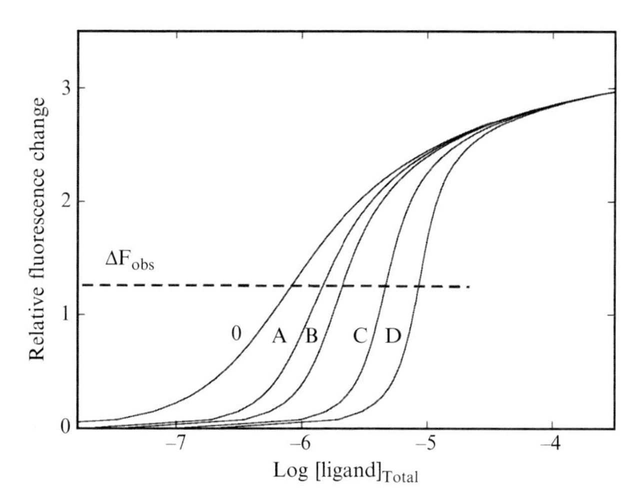

Figure 2.7 Theoretical fluorescence titration curves of the reference fluorescent nucleic acid with the protein ligand in the presence of different concentrations of the competing, nonfluorescent nucleic acid lattice. Binding of the ligand to the reference lattice is described by the McGhee–von Hippel model for large ligand binding to a linear, homogeneous nucleic acid, using intrinsic binding constant $K = 10^5\ M^{-1}$, cooperativity parameter $\omega = 1$, and site-size $n = 20$. The maximum increase of the nucleic acid fluorescence intensity upon saturation with the ligand is $\Delta F_{max} = 3.5$ (Jezewska and Bujalowski, 1996). Binding of the ligand to a competing, nonfluorescent nucleic acid is described by the combinatorial theory using intrinsic binding constant $K = 10^5\ M^{-1}$, cooperativity parameter $\omega = 1$, and site-size $n = 20$. The selected length of the nucleic acid is 1600 nucleotides. The concentration of the competing nucleic acid (nucleotide) is: (*O*) 0; (*A*) 2×10^{-5} *M*; (*B*) 4×10^{-5} *M*; (*C*) 1.2×10^{-4} *M*; (*D*) 2.4×10^{-4} *M*. The concentration of the reference fluorescent nucleic acid is 2×10^{-5} *M* (Nucleotide). The horizontal dashed line connects points on all titration curves characterized by the same value of the relative fluorescence increase, ΔF_i.

parameter $\omega = 10$. The intrinsic binding constants are $K = 10^5\ M^{-1}$ and $K = 10^6\ M^{-1}$ for the reference fluorescent and the nonfluorescent, competing nucleic acid, respectively. Nevertheless, the discussed thermodynamic analysis is independent of any particular binding model for both polymers.

Recall, the measured relative fluorescence increase, ΔF_{obs}, exclusively monitors the saturation of the reference fluorescence nucleic acid with the ligand. Therefore, all curves span the same range of the relative fluorescence change and tend to the same plateau value at selected $\Delta S_{max} = 3.5$. At very high protein concentrations, all curves are superimposed, because the concentration of the bound protein to both lattices becomes negligibly small, when compared to the total ligand concentration, $[L]_T$. The same value of the relative fluorescence change of the reference nucleic acid, in the presence of the protein, means the same protein binding density, $(\Sigma v_i)_R$ on the reference lattice and the same free protein ligand concentration, $[L]_F$, independent of the presence of the competing unmodified nucleic acid. Thus, in the presence of different concentrations of a competing nucleic acid, and the constant concentration of the reference fluorescence lattice, at a given value of ΔF_{obs}, the concentration of the free protein ligand, $[L]_F$, must be the same and independent of the total concentration of the competing nucleic acid, N_{TS} (dashed line in Fig. 2.7). Hence, one can obtain rigorous measurements of the protein total average binding density, $(\Sigma v_i)_S$, on the unmodified competing nucleic acid and the free protein ligand concentration, $[L]_F$, from titrations of samples containing constant concentrations of the reference fluorescent nucleic acid, $[M]_{TR}$, with the protein in the presence of two or more concentrations of the nonfluorescent, competing nucleic acid lattice (Bujalowski, 2006; Jezewska and Bujalowski, 1996).

The experiment provides a set of mass conservation equations for the total protein ligand concentration in solution. In the presence of two different competing nucleic acid concentrations, $[M]_{TS1}$ and $[M]_{TS2}$, the total protein ligand concentrations, $[L]_{T1}$ and $[L]_{T2}$, at which the same relative fluorescence change, ΔF_{obs}, is observed, are defined as

$$[L]_{T1} = (\Sigma v_i)_R[M]_{TR} + (\Sigma v_i)_S[M]_{TS1} + [L]_F \tag{2.38a}$$

and

$$[L]_{T2} = (\Sigma v_i)_R[M]_{TR} + (\Sigma v_i)_S[M]_{TS2} + [L]_F \tag{2.38b}$$

Subtracting Eq. (2.38a) from Eq. (2.38b) and rearranging provides the expression for the total average binding density of the protein on the competing, unmodified lattice, $(\Sigma v_i)_S$, and the free protein concentration, $[L]_F$, in terms of known quantities, $(\Sigma v_i)_R$, total protein, and nucleic acid concentrations as (Bujalowski, 2006; Jezewska and Bujalowski, 1996)

$$(\Sigma v_i)_S = \frac{[L]_{T2} - [L]_{T1}}{[M]_{TS2} - [M]_{TS1}} \tag{2.39a}$$

And

$$[L]_F = [L]_{Tx} - (\Sigma v_i)_S[M]_{TSx} - (\Sigma v_i)_R[M]_{TR} \tag{2.39b}$$

where x is 1 or 2. Notice, the binding density, $(\Sigma v_i)_R$, of the protein on the reference fluorescent nucleic acid, at any value of ΔF_{obs}, can be determined in an independent titration experiment. Calculations of $(\Sigma v_i)_S$ and, subsequently, $[L]_F$ can be performed at any value of the observed fluorescence change, ΔF_{obs}, along the titration curves (Fig. 2.7) generating a thermodynamic binding isotherm for protein binding to the competing unmodified nucleic acid. The approach is demonstrated in Section 10, using the experimental data for the binding of the *Escherichia coli* replicative helicase DnaB protein to nonfluorescent poly(d*A*), in the presence of the reference fluorescent etheno-derivative, poly(dε*A*).

It should be noted that, if the ligand affinities for the reference fluorescent nucleic acid and the competing, nonfluorescent lattice are different, the total average binding densities are different for both nucleic acids, at the same value of the measured relative fluorescence change, ΔF_{obs}. Figure 2.8 shows the theoretical dependence of the observed relative fluorescence change, ΔS_{obs} upon the total average binding density of the protein on the

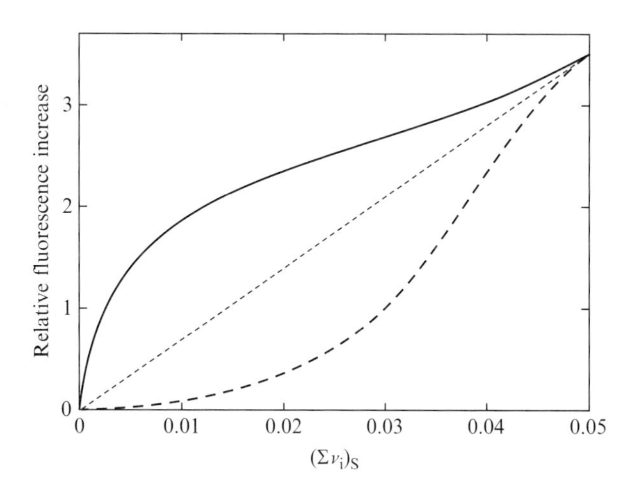

Figure 2.8 Computer simulation of the dependence of the relative fluorescence increase of the reference fluorescent nucleic acid, ΔF, upon the binding density, $(\Sigma v_i)_S$, of the ligand on the competing, nonfluorescent lattice where the competing, nonfluorescent lattice has higher (— — —) and lower (———) affinity toward the ligand. The binding of the ligand to the reference lattice is described by the McGhee–von Hippel model of large ligand binding to a linear, homogeneous nucleic acid, using intrinsic binding constant $K = 10^5\ M^{-1}$, cooperativity parameter $\omega = 1$, and site-size $n = 20$. The maximum increase of the nucleic acid fluorescence intensity upon saturation with the ligand is $\Delta F_{max} = 3.5$. Binding of the ligand to competing, nonfluorescent nucleic acid is described by the combinatorial theory, using $\omega = 1$ and $n = 20$; $K = 10^4\ M^{-1}$ and $10^6\ M^{-1}$ for lower and higher affinity cases, respectively. The selected length of the nucleic acid is 1600 nucleotides (Jezewska and Bujalowski, 1996).

competing unmodified nucleic acid, for two different intrinsic affinities of the protein for the competing unmodified nucleic acid. Protein binding to the reference fluorescent nucleic acid is characterized by $K = 10^5\ M^{-1}$, $\omega = 1$, and $\Delta F_{max} = 3.5$; the ligand binding to the competing lattice is characterized by $K = 10^6\ M^{-1}$, $\omega = 1$, and $K = 10^4\ M^{-1}$, $\omega = 1$, for high and low affinity cases, respectively. For simplicity, the dependence of ΔF_{obs} upon $(\Sigma v_i)_R$ for the reference lattice was selected to be strictly proportional (straight, dashed line). In the case where the macroscopic ligand affinity for the competing lattice is higher than for the reference lattice, the total average binding density, $(\Sigma v_i)_S$, of the ligand on the competing, nonfluorescent lattice is higher, when compared to the reference fluorescent lattice at any value of ΔF_{obs} and the plot is concave down. The opposite is true in the case where the protein ligand affinity is lower for the competing lattice as compared to the reference fluorescent nucleic acid. The plot rises sharply at the initial values of the binding density and levels off at the intermediate range of $(\Sigma v_i)_S$, gradually approaching the maximum possible value of ΔF_{obs}. This behavior results from the fact that ΔF_{obs} solely reflects the binding density on the reference lattice, $(\Sigma v_i)_R$, which, due to the higher affinity of the ligand for the reference nucleic acid, saturates with the ligand in advance of the saturation of the competing unmodified nucleic acid. The plots in Fig. 2.8 indicate that, in order to obtain the most accurate estimate of the stoichiometry of protein–unmodified nucleic acid complex, the affinity of the competing nonfluorescent nucleic acid for the protein should be similar or higher than that of the fluorescent reference lattice.

10. Quantitative Analysis of the Binding of the *E. coli* DnaB Helicase to Unmodified Nucleic Acids Using the MCT Method

The *E. coli* DnaB protein is its primary replicative helicase, that is, the factor that unwinds the duplex DNA in front of the replication fork (Jezewska and Bujalowski, 1996). Binding of the enzyme to unmodified nucleic acids, for example, poly(d*A*), and other polydeoxynucleotides does not cause any significant change in the protein fluorescence (Bujalowski and Jezewska, 1995; Jezewska and Bujalowski, 1996). On the other hand, binding of the DnaB helicase to the fluorescent etheno-derivative of poly (d*A*), poly(dε*A*), induces a strong, ~3.5-fold, relative fluorescence increase of the nucleic acid which allows the precise estimate of the stoichiometry and interaction parameters of the DnaB protein–poly(dε*A*) complex

(Bujalowski and Jezewska, 1995; Jezewska and Bujalowski, 1996). Thus, poly(dεA) can serve as a reference fluorescent nucleic acid in the MCT method.

Figure 2.9 shows the fluorescence titration curves of poly(dεA) with the DnaB protein in the absence and presence of two different concentrations of poly(dA). A significant shift of the titration curves toward higher total $[\text{DnaB}]_T$, in the presence of poly(dA), indicates efficient competition between the two nucleic acids for the enzyme. At the same value of the relative fluorescence increase, the total average binding density, $(\Sigma v_i)_S$, on the competing lattice, poly(dA), and the free DnaB concentration, $[\text{DnaB}]_F$, are the same, independent of the concentration of poly(dA) (Eqs. (2.38a), (2.38b), (2.39a), (2.39b)). Therefore, one can obtain a set of total DnaB concentrations, $[\text{DnaB}]_{T1}$ and $[\text{DnaB}]_{T2}$, which are determined by the intersection of the horizontal line with both titration curves at the same value of the ΔF_{obs} (e.g., horizontal line in Fig. 2.9). Since the total concentrations of poly(dA) are known, one can calculate the true binding density, $(\Sigma v_i)_S$, of the DnaB protein on poly(dA) and the free DnaB concentration, $[\text{DnaB}]_F$, using Eqs. (2.39a) and (2.39b). The same procedure is then repeated over the entire range of ΔF_{obs}, for selected intervals of ΔF_{obs}, providing $(\Sigma v_i)_S$ as a function of $[\text{DnaB}]_F$. In other words, one obtains the thermodynamically rigorous binding isotherm for DnaB helicase association with poly(dA), although the signal used to monitor the binding originates exclusively from the reference fluorescent lattice, poly(dεA).

It should be rather obvious that the site-size of the DnaB protein–unmodified nucleic acid is the same as the site-size of the protein complex with a reference fluorescent nucleic acid, due to the same physical nature of both nucleic acids (ssDNA homopolymers). Nevertheless, for completeness, we include the independent determination of this quantity in our discussion. Figure 2.10 shows the dependence of the observed relative fluorescence change, ΔF_{obs}, upon the total average binding density, $(\Sigma v_i)_S$, of the DnaB helicase on poly(dA). For comparison, the dependence of ΔF_{obs} upon the total average binding density, $(\Sigma v_i)_R$, of the DnaB helicase on the reference lattice, poly(dεA), is also included (dashed straight line). The quantity, $(\Sigma v_i)_S$, could be determined up to the value of ~ 0.044, which corresponds with $\Delta F_{obs} \sim 2.6$. Extrapolation to the maximum possible value of $\Delta F_{max} = 3.6 \pm 0.3$ gives the maximum value of $(\Sigma v_i)_S = 0.05 \pm 0.005$ and the estimation of the site-size poly(dA)–DnaB helicase complex, $\underline{n} = 20 \pm 3$. As expected, this value is the same as the estimated $n = 20 \pm 3$ in identical solution conditions for the poly(dεA)–DnaB complex (Jezewska and Bujalowski, 1996).

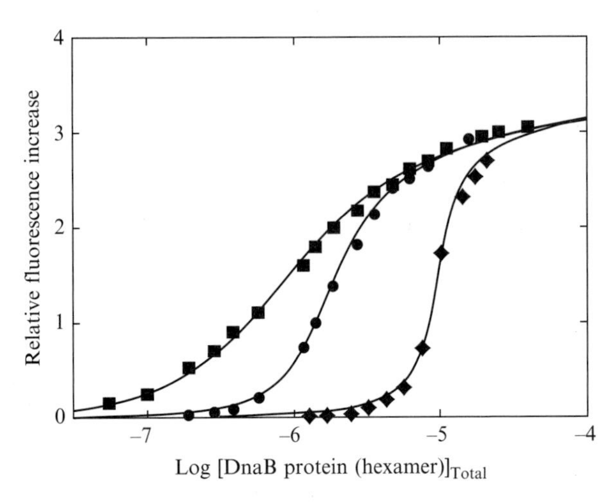

Figure 2.9 Fluorescence titrations (λ_{ex} = 325 nm, λ_{em} = 410 nm) of poly(dε*A*) with the DnaB helicase in 50 m*M* Tris/HCl (pH 8.1, 10 °C) containing 50 m*M* NaCl and 1 m*M* AMP-PNP, in the presence of different concentrations of poly(d*A*) (Nucleotide): (■) 0; (●) 2×10^{-5} *M*; (◆) 1.2×10^{-4} *M*. The concentration of the poly(dε*A*) is 2×10^{-5} *M* (Nucleotide). The solid lines separate the sets of data points and do not have theoretical bases. The horizontal dashed line connects points on all titration curves characterized by the same value of the relative fluorescence increase, ΔF_i. The intersection points of the dashed horizontal line with the titration curves, in the presence of poly(d*A*), define the total DnaB concentrations, P_{T2} and P_{T2}, at which the binding density on poly(dε*A*), $(\Sigma v_i)_R$, the binding density on poly(d*A*), $(\Sigma v_i)_S$, and the free helicase concentration, P_F, are constant. The total concentrations of nucleic acids at each titration are included in the figure (Jezewska and Bujalowski, 1996).

11. Direct Analysis of the Experimental Isotherm of Protein Ligand Binding to Two Competing Nucleic Acid Lattices

For binding of a large protein ligand to a long homogeneous nucleic acid lattice, the McGhee–von Hippel model is the simplest statistical thermodynamic description for the binding process (Bujalowski and Lohman, 1987; Bujalowski *et al.*, 1989; Epstein, 1978; Jezewska and Bujalowski, 1996; McGhee and von Hippel, 1974). The lattice site can only exist in two states, free and bound with the ligand in a single type of the complex. The model takes into account the cooperativity of the protein association and the overlap of the potential binding sites. Although the original work provides two expressions for the noncooperative and cooperative binding, a single, generalized equation for the McGhee–von Hippel model, which can be applied to both cooperative and noncooperative binding systems, has

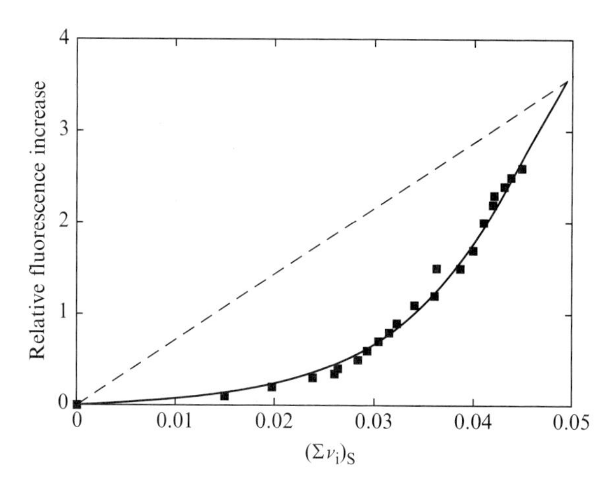

Figure 2.10 The dependence of the relative fluorescence increase of poly(dεA) upon the binding density, $(\Sigma v_i)_S$, of the DnaB helicase on competing poly(dA) in 50 mM Tris/HCl (pH 8.1, 10 °C) containing 50 mM NaCl and 1 mM AMP-PNP. The dependence of the relative fluorescence increase of poly(dεA) upon the binding density, $(\Sigma v_i)_R$, of the DnaB helicase on poly(dεA), in the same buffer conditions, is also included (dashed, straight line). The solid line is the computer fit, using an approach based on the combined McGhee–von Hippel model and the combinatorial theory for binding a large ligand to two different, competing homogeneous lattices (see below), for the simultaneous binding of the DnaB helicase to poly(dεA) and poly(dA) (**30**). The binding of the helicase to the reference poly(dεA) is described by the McGhee–von Hippel model, using the independently determined intrinsic binding constant $K = 1.2 \times 10^5\ M^{-1}$, cooperativity parameter $\omega = 3$, and site-size $n = 20$. Binding the enzyme to the competing poly(dA) is described by the combinatorial theory, using cooperativity parameter $\omega = 4.8$, site-size $n = 20$, and intrinsic binding constant $K = 1.4 \times 10^6\ M^{-1}$.

been derived and is defined by Bujalowski (2006) and Bujalowski *et al.*, (1989).

$$\Sigma v_i = K(1 - n\Sigma v_i)\left\{\frac{[2\omega(1 - n\Sigma v_i)]}{[(2\omega - 1)(1 - n\Sigma v_i) + \Sigma v_i + R]}\right\}^{(n-1)}\left\{\frac{[1 - (n+1)\Sigma v_i + R]}{2(1 - n\Sigma v_i)}\right\}^2 [\mathrm{L}]_\mathrm{F} \qquad (2.40)$$

where K is the intrinsic binding constant, n is the site-size of the protein–nucleic acid complex (number of bases or base pairs occluded by the bound protein in the complex), ω is the cooperative interactions parameter, and $R = \{[1-(n+1)\Sigma v_i]^2 + 4\omega\sum v_i(1-n\Sigma v_i)\}^{0.5}$.

Though it is a simple statistical thermodynamic model, Eq. (2.40) is a formidable expression. In the case of protein binding to two competing and long nucleic acid lattices, the ligand interactions with each nucleic acid are described by an equation in the form of expression (2.40). An attempt to simultaneously use two isotherms, of the type described by Eq. (2.40), is hindered by the fact that they are complex, polynomial implicit functions of the total average binding density, Σv_i, and free ligand concentration. Thus, in order to simulate, or fit, the competition titration curves of a large ligand binding to two competing nucleic acid lattices, complex and cumbersome numerical calculations are required.

It should be stressed that the McGhee–von Hippel model is equivalent to the combinatorial model as the length of the homogenous lattice approaches infinity (Bujalowski *et al.*, 1989; Epstein, 1978; McGhee and von Hippel, 1974). To overcome the problem of resorting to complex numerical calculations and to obtain a general method to analyze simultaneous binding of a protein ligand to two or more polymer nucleic acids, one can apply a combined application of the generalized McGhee–von Hippel model, as defined by Eq. (2.40), and the combinatorial theory for large ligand binding to a linear, homogeneous lattice (Bujalowski, 2006; Jezewska and Bujalowski, 1996). In this approach, the ligand binding to the reference fluorescent nucleic acid is described by the generalized McGhee–von Hippel equation. Binding of the protein to the competing nonfluorescent lattice is described by the combinatorial theory for cooperative binding of a large ligand to a linear finite homogeneous nucleic acid. For a long enough lattice, the isotherm generated using the combinatorial theory, will be, within experimental accuracy, indistinguishable from the isotherm generated, using the generalized Eq. (2.40). For instance, in the case of a protein like the DnaB helicase ($n = 20$), a lattice, which can accommodate ≥ 40 protein molecules (800 nucleotides), represents an "infinite" lattice for any practical purpose.

In the combinatorial theory, the partition function of the protein ligand–nucleic acid system, Z_S, is defined by

$$Z_S = \sum_{k=0}^{g} \sum_{j=0}^{k-1} S_N(k,j)\left(K_S[L]_F\right)^k \omega_S^j \tag{2.41}$$

where g is the maximum number of ligand molecules which may bind to the finite nucleic acid lattice (for the nucleic acid lattice N residues long $g = M/n$), K_S is the intrinsic binding constant characterizing interactions with the unmodified lattice, ω is the corresponding cooperative interaction parameter, k is the number of ligand molecules bound, and j is the number of cooperative contacts between k bound ligand molecules in a particular configuration on the lattice. The combinatorial factor $S_N(k, j)$ is the number of distinct ways that k ligands bind to a lattice, with j cooperative contacts, and is defined by

$$S_N(k,j) = \frac{[(M - nk + 1)!(k - 1)!]}{[(M - nk + j + 1)!(k - j)!j!(k - j - 1)!]} \quad (2.42)$$

The total average binding density, $(\Sigma v_i)_S$ is then obtained by using the standard statistical thermodynamic expression, $(\Sigma v_i)_S = \partial \ln Z_S / \partial \ln L_F$, as Fried and Crothers (1981), Jezewska and Bujalowski (1996), and Bujalowski (2006).

$$(\Sigma v_i)_S = \frac{\sum_{k=1}^{g} \sum_{j=0}^{k-1} k S_N(k,j) \left(K_S [L]_F\right)^k \omega_S{}^j}{\sum_{k=0}^{g} \sum_{j=0}^{k-1} S_N(k,j) \left(K_S [L]_F\right)^k \omega_S{}^j} \quad (2.43)$$

Expressions (2.41)–(2.43) describe the binding of a large ligand to a finite, linear homogeneous nucleic acid. Contrary to the generalized McGhee–von Hippel model (Eq. (2.40)), expression (2.43) is an explicit function of the free ligand concentration, which allows us to calculate the binding density directly for the known K, ω, n, and $[L]_F$. Because free protein ligand concentrations can be explicitly calculated, using an infinite lattice model through Eq. (2.40), combining both equations offers a simple way of fitting the simultaneous binding of a large ligand to two (or more) competing, different linear lattices, for example, a reference fluorescent nucleic acid in the presence of a competing, unmodified nucleic acid. This is accomplished by first applying Eq. (2.40) to the reference fluorescent nucleic acid, calculating the free ligand concentration, L_F, for given values of the parameters K, ω, and n, with $(\Sigma v_i)_R$ as the independent variable. Subsequently, the obtained value of $[L]_F$ is introduced into Eq. (2.43) and the binding density, $(\Sigma v_i)_S$, is calculated for the given K_S, ω_S, and n_S, which characterize the binding of the protein to the competing nonfluorescent lattice. The calculations are repeated for the entire range of the $(\Sigma v_i)_R$, generating the required $(\Sigma v_i)_S$ for the competing nucleic acid lattice, as a function of $[L]_F$. The experimental binding isotherm, which is the observed relative fluorescence change ΔF_{obs} as a function of the total protein concentration $[L]_T$, is then obtained by calculating for each value of ΔF_{obs} the total protein concentration $[L]_T$, by introducing $(\Sigma v_i)_S$, $(\Sigma v_i)_R$, and $[L]_F$ into expression

$$[L]_T = [L]_F - (\Sigma v_i)_R [M]_{TR} - (\Sigma v_i)_S [M]_{TS} \quad (2.44)$$

The solid lines in Fig. 2.9 are nonlinear least squares fits of the simultaneous binding of the DnaB helicase to two competing nucleic acids, using the procedure outlined above (Jezewska and Bujalowski, 1996). The binding of the DnaB helicase to the fluorescent poly(dεA) has been independently determined and described by using the generalized McGhee–von Hippel (Eq. (2.40)). The binding of the protein to the competing poly(dA) has been described, using the combinatorial theory (Eqs. (2.31)–(2.43)). Because the site-size of the DnaB protein–poly(dεA) complex, $n = 20 \pm 3$, has been independently estimated, there are only two parameters, the intrinsic binding constant, K_S, and the cooperativity parameter, ω_S, to be determined (Jezewska and Bujalowski, 1996). It should be pointed out that the analysis of simultaneous binding of a large ligand to competing nucleic acid lattices could also be performed by applying Eqs. (2.31)–(2.43) of the combinatorial theory, to both the reference and the competing nucleic acids. This is a completely equivalent approach to the combined application of the McGhee–von Hippel and combinatorial models. However, the major advantage of using the combined McGhee–von Hippel and the combinatorial theories lies in the tremendously decreased computational time, particularly, if a very long lattice is used.

12. Using a Single Concentration of a Nonfluorescent Unmodified Nucleic Acid

Using several concentrations of the unmodified nucleic acid increases the accuracy of the binding experiment. However, it may be very time consuming and often impractical, particularly if the availability of the specific nucleic acid substrate is limited. However, the thermodynamic binding density, $(\Sigma \nu_i)_S$, and the free ligand concentration, L_F, can also be obtained by using a single titration of the fluorescent nucleic acid, at a single concentration of a competing nonfluorescent nucleic acid, in direct reference to the titration curve of the fluorescent reference nucleic acid alone (Jezewska and Bujalowski, 1996). In Fig. 2.9, one can obtain $(\Sigma \nu_i)_S$ and $[\mathrm{DnaB}]_F$ by simultaneously analyzing titration curves 1 and 2 or 1 and 3, instead of 2 and 3. For the analysis using a single concentration of the nonfluorescent nucleic acid, the total protein concentrations, at which the same value of the relative fluorescence increase, ΔF_{obs}, of the reference fluorescent nucleic acid is observed, in the absence and presence of competing nonfluorescent lattice, $[L]_{TR}$, and $[L]_{TSx}$, respectively, are defined as

$$[L]_{TR} = (\Sigma \nu_i)_R [M]_{TR} + [L]_F \tag{2.45a}$$

and

$$[L]_{TS} = (\Sigma \nu_i)_R [M]_{TR} + (\Sigma \nu_i)_S [M]_{TS} + [L]_F \tag{2.45b}$$

Solving the set of Eqs. (2.45a) and (2.45b), for $(\Sigma \nu_i)_S$ and $[L]_F$, provides

$$(\Sigma \nu_i)_S = [L]_{TS} - \left[\frac{[L]_{TR}}{[M]_{TS}} \right] \quad (2.46a)$$

and

$$[L]_F = [L]_{TS} - (\Sigma \nu_i)_R [M]_{TR} - (\Sigma \nu_i)_S [M]_{TS} \quad (2.46b)$$

13. Using Short Fluorescent Oligonucleotides in Competition with the Polymer Nucleic Acid

As long as two chemical species compete for the same binding site, they can be used in the competition titration experiments. In the case of the protein nucleic acid interactions, the MCT analysis can be simplified by using short fragments of a nucleic acid, as a reference fluorescent lattice, in competition titrations with various DNA substrates (forks, template-primers, gapped substrates, long nucleic acid polymers, etc.), which share the same binding site on the ligand (protein). The optimal reference choice is a short nucleic acid fragment, which forms a simple 1:1 complex with the protein and is long enough to exactly span the total site-size of the protein–nucleic acid complex. A selection of the length of the fluorescent reference fragment can be based on the initial estimate of the stoichiometry of the protein–nucleic acid complex obtained using a modified fluorescent polymer nucleic acid or a series of fluorescent oligomers (Bujalowski, 2006; Jezewska and Bujalowski, 1996). In such studies, the degree of saturation of the nucleic acid oligomer with the protein, $\Sigma\Theta_R$, and the fluorescence change, ΔF_{obs}, accompanying the formation of the complex, are described by

$$(\Sigma \Theta_i)_R = \frac{K_o [L]_F}{1 + K_o [L]_F} \quad (2.47a)$$

and

$$\Delta F_{obs} = \Delta F_{max} \left(\frac{K_o [L]_F}{1 + K_o [L]_F} \right) \quad (2.47b)$$

where ΔF_{max} is the maximum observed fluorescence change of the short nucleic acid lattice upon saturation with the protein and K_o is the macroscopic binding constant of the protein for the reference fluorescent oligomer. Therefore, in the presence of for example, a competing, unmodified polymer nucleic acid, the observed fluorescence change of the short nucleic acid fragment is thermodynamically linked with the total average binding density, $(\Sigma\nu_i)_S$, on the competing polymer through the free protein ligand concentration, $[L]_F$, as defined by

$$[L]_F = [L]_{TS} - (\Sigma\Theta_i)_R[O]_{TR} - (\Sigma\nu_i)_S[M]_{TS} \tag{2.48}$$

In practice, two experimental situations usually arise. The site-size of the protein on the polymer nucleic acid, or the maximum stoichiometry on a specific DNA substrate is known and the experimenter wants to examine the intrinsic affinities and cooperativities in some specific solution conditions. The competition titration curve, which is ΔF_{obs} as a function of the total protein concentration, $[L]_{TS}$, in the presence of the competing nucleic acid, is then fitted to Eq.2.47b), with $[L]_F$ treated as independent variable. The values of $[L]_{TS}$ are then calculated using Eq. (2.48) and the selected statistical thermodynamic binding model for $(\Sigma\nu_i)_S$. However, if the binding to the competing unmodified nucleic acid is unknown, then a complete analysis is in place. As in the case of the reference polymer nucleic acid, the same value of ΔF_{obs} at different competing nucleic acid concentrations, corresponds with the same degree of saturation of the short nucleic acid fragment and the same free concentration of the protein, $[L]_F$, that is, also the same $(\Sigma\nu_i)_S$. Therefore, performing two fluorescence titrations of a short reference nucleic acid with the protein, at the same total concentration of the reference oligomer, $[O]_{TR}$, but in the presence of two different total concentrations of the competing nucleic acid, $[N]_{TS1}$ and $[N]_{TS2}$, allows the experimenter to obtain the total average binding density, $(\Sigma\nu_i)_S$, and free protein concentration, $[L]_F$, analogously, as described above for a polymer reference lattice. Expressions (2.38a) and (2.38b) are

$$[L]_{T1} = (\Sigma\Theta_i)_R[O]_{TR} + (\Sigma\nu_i)_S[M]_{TS1} + [L]_F \tag{2.49a}$$

and

$$[L]_{T2} = (\Sigma\Theta_i)_R[O]_{TR} + (\Sigma\nu_i)_S[M]_{TS2} + [L]_F \tag{2.49b}$$

The total average binding density, $(\Sigma\nu_i)_S$, and the free protein concentration are described by Eqs. (2.39a) and (2.39b). Analogously, if the reference fluorescent oligomer is titrated with the protein in the presence of a single concentration of the competing, long polymer nucleic acid at the concentration $[N]_{TS}$, the total concentrations of the protein ligand in the absence and presence of the competing nucleic acid and at the same value of the fluorescence intensity, ΔF_{obs}, are (Jezewska and Bujalowski, 1996)

$$[L]_{TR} = (\Sigma\Theta_i)_R[O]_{TR} + [L]_F \quad (2.50a)$$

and

$$[L]_{TS} = (\Sigma\Theta_i)_R[O]_{TR} + (\Sigma\nu_i)_S[M]_{TS} + [L]_F \quad (2.50b)$$

The total average binding density, $(\Sigma\nu_i)_S$, obtained by subtracting Eq. (2.50a) from Eq. (2.50b), and the free protein concentration are described by (Eq. (2.46a) and (2.46b))

$$(\Sigma\nu_i)_S = \frac{[L]_{TS} - [L]_{TR}}{[M]_{TS}} \quad (2.51a)$$

and

$$[L]_F = [L]_{TS} - (\Sigma\Theta_i)_R[O]_{TR} - (\Sigma\nu_i)_S[M]_{TS} \quad (2.51b)$$

14. Conclusions

The MCT method enables one to obtain, rigorously and quantitatively, interaction parameters of protein–nucleic acid interactions, which may not be available by other methods, particularly for the unmodified long polymer lattices and specific DNA substrates, if the binding is not accompanied by adequate spectroscopic signal changes. Nevertheless, the method is not limited by the size of the reference nucleic acid. A simple analysis of the competition titration experiments can be performed in which the fluorescent, short fragment of nucleic acid, spanning the exact site-size of the protein–nucleic acid complex, and binding with only a 1:1 stoichiometry to the protein, is used as a reference macromolecule. We have discussed the MCT method, predominantly, as applied to studying protein–nucleic acid interactions. However, the method can generally be applied to any ligand–macromolecule system, by monitoring the binding using the spectroscopic signal originating from a reference macromolecule, in the presence of the competing macromolecule whose interaction parameters with the ligand are to be determined, and whose spectroscopic properties do not change upon complex formation with the ligand.

Acknowledgments

We thank Gloria Drennan Bellard for reading the chapter. This work was supported by NIH Grants GM46679 and GM58565 (to W. B.).

REFERENCES

Baker, B. M., Vanderkooi, J., and Kallenbach, N. R. (1978). Base stacking in A fluorescent dinucleoside monophosphate: εApεA. *Biopolymers* **17,** 1361.

Boschelli, F. J. (1982). Lambda phage cro repressor. Non-specific DNA binding. *Mol. Biol.* **162,** 267.

Brenowitz, M., Senear, D. F., Shea, M. A., and Ackers, G. K. (1986). *Methods Enzymol.* **130,** 132.

Bujalowski, W. (2006). *Chem. Rev.* **106,** 556.

Bujalowski, W., and Jezewska, M. J. (1995). *Biochemistry* **34,** 8513.

Bujalowski, W., and Jezewska, M. J. (2000). *In* "Spectrophotometry and Spectrofluorimetry: A Practical Approach," (M. G. Gore, ed.), pp. 141–165. IRL Press (Oxford University Press), Oxford, UK, Chapter 5.

Bujalowski, W., and Jezewska, M. J. (2009). *Methods Enzymol.* **466,** 294.

Bujalowski, W., and Klonowska, M. M. (1993). *Biochemistry* **32,** 5888.

Bujalowski, W., and Klonowska, M. M. (1994a). *Biochemistry* **33,** 4682.

Bujalowski, W., and Klonowska, M. M. (1994b). *J. Biol. Chem.* **269,** 31359.

Bujalowski, W., and Lohman, T. M. (1986). *Biochemistry* **25,** 7779.

Bujalowski, W., and Lohman, T. M. (1987). *Biochemistry* **26,** 3099.

Bujalowski, W., Lohman, T. M., and Anderson, C. F. (1989). *Biopolymers* **28,** 1637.

Bujalowski, W., Klonowska, M. M., and Jezewska, M. J. (1994). *J. Biol. Chem.* **269,** 31350.

Chabbert, M., Cazenave, C., and Hélène, C. (1987). *Biochemistry* **26,** 2218.

Crothers, D. M. (1968). *Biopolymers* **6,** 575.

deHaseth, P. L., Gross, C. A., Burgess, R. R., and Record, M. T., Jr. (1977). *Biochemistry* **16,** 4777.

Draper, D. E., and von Hippel, P. H. (1978). *J. Mol. Biol.* **122,** 321.

Epstein, I. R. (1978). *Biophys. Chem.* **8,** 327.

Fernando, T., and Royer, C. (1992). *Biochemistry* **31,** 3429.

Fried, M., and Crothers, D. M. (1981). *Nucleic Acids Res.* **9,** 6505.

Garner, M. M., and Revzin, A. (1981). *Nucleic Acids Res.* **9,** 3047.

Heyduk, T., and Lee, J. C. (1990). *Proc. Natl. Acad. Sci. USA* **87,** 1744.

Jezewska, M. J., and Bujalowski, W. (1996). *Biochemistry* **35,** 2117.

Jezewska, M. J., and Bujalowski, W. (1997). *Biophys. Chem.* **64,** 253.

Jezewska, M. J., Kim, U.-S., and Bujalowski, W. (1996). *Biochemistry* **35,** 2129.

Jezewska, M. J., Rajendran, S., and Bujalowski, W. (1998a). *Biochemistry* **37,** 3116.

Jezewska, M. J., Rajendran, S., and Bujalowski, W. (1998b). *J. Mol. Biol.* **284,** 1113.

Jezewska, M. J., Rajendran, S., and Bujalowski, W. (2001). *Biochemistry* **40,** 3295.

Jezewska, M. J., Galletto, R., and Bujalowski, W. (2003). *Biochemistry* **42,** 5955.

Jezewska, M. J., Lucius, A. L., and Bujalowski, W. (2005a). *Biochemistry* **44,** 3865.

Jezewska, M. J., Lucius, A. L., and Bujalowski, W. (2005b). *Biochemistry* **44,** 3877.

Jezewska, M. J., Marcinowicz, A., Lucius, A. L., and Bujalowski, W. (2006). *J. Mol. Biol.* **356,** 121.

Jezewska, M. J., Bujalowski, P. J., and Bujalowski, W. (2007). *Biochemistry* **46,** 12909.

Kowalczykowski, S. C., Lonberg, N., Newport, J. W., and von Hippel, P. H. (1981). *J. Mol. Biol.* **145,** 75.

Kowalczykowski, S. C., Paul, L. S., Lonberg, N., Newport, J. W., McSwiggen, J. A., and von Hippel, P. H. (1986). *Biochemistry* **25,** 1226.

Lohman, T. M., and Bujalowski, W. (1991). *Methods Enzymol.* **208,** 258.

Lohman, T. M., and Ferrari, M. E. (1994). *Annu. Rev. Biochem.* **63,** 527.

Maciejewski, M., Shin, R., Pan, B., Marintchev, A., Denninger, A., Mullen, M. A., Chen, K., Gryk, M. R., and Mullen, G. P. (2001). *Nat. Struct. Biol.* **8,** 936.

McGhee, J. D., and von Hippel, P. H. (1974). *J. Mol. Biol.* **86,** 469.

McSwiggen, J. A., Bear, D. G., and von Hippel, P. H. (1988). *J. Mol. Biol.* **199,** 609.
Menetski, J. P., and Kowalczykowski, S. C. (1985). *J. Mol. Biol.* **181,** 281.
Porschke, D., and Rauh, H. (1983). *Biochemistry* **22,** 4737.
Rajendran, S., Jezewska, M. J., and Bujalowski, W. (1998). *J. Biol. Chem.* **273,** 31021.
Rajendran, S., Jezewska, M. J., and Bujalowski, W. (2001). *J. Mol. Biol.* **308,** 477.
Revzin, A., and von Hippel, P. H. (1977). *Biochemistry* **16,** 4769.
Scatchard, G. (1949). *Ann. N.Y. Acad. Sci.* **51,** 660.
Secrist, J. A., Bario, J. R., Leonard, N. J., and Weber, G. (1972). *Biochemistry* **11,** 3499.
Showwalter, A. K., Byeon, I.-J., Su, M.-I., and Tsai, M.-D. (2003). *Nat. Struct. Biol.* **8,** 942.
Szymanski, M. R., Jezewska, M. J., and Bujalowski, W. (2010). *J. Mol. Biol.* **398,** 8.
Tolman, G. L., Barrio, J. R., and Leonard, N. J. (1974). *Biochemistry* **13,** 4869.

CHAPTER THREE

Analysis of PKR–RNA Interactions by Sedimentation Velocity

C. Jason Wong,* Katherine Launer-Felty,* *and* James L. Cole*,†

Contents

Abstract

PKR is an interferon-induced kinase that plays a pivotal role in the innate immunity pathway for defense against viral infection. PKR is activated to undergo autophosphorylation upon binding to RNAs that contain duplex regions. Some highly structured viral RNAs do not activate and function as PKR inhibitors. In order to define the mechanisms of activation and inhibition of PKR by RNA, it is necessary to characterize the stoichiometries, affinities, and free energy couplings governing the assembly of the relevant complexes. We have found sedimentation velocity analytical ultracentrifugation to be particularly useful in the study of PKR–RNA interactions. Here, we describe protocols for designing and analyzing sedimentation velocity experiments that are generally applicable to studies of protein–nucleic acid interactions. Initially, velocity data obtained at multiple protein:RNA ratios are analyzed using the dc/dt method's to define the association model and to test whether the system is kinetically limited. The sedimentation velocity data obtained at multiple loading concentrations are then globally fitted to this model to determine the relevant

* Department of Molecular and Cell Biology, University of Connecticut, Storrs, Connecticut, USA
† Department of Chemistry, University of Connecticut, Storrs, Connecticut, USA

Methods in Enzymology, Volume 488
ISSN 0076-6879, DOI: 10.1016/B978-0-12-381268-1.00003-3

© 2011 Elsevier Inc.
All rights reserved.

association constants. The frictional ratios of the complexes are calculated using the fitted sedimentation coefficients to determine whether the hydrodynamic properties are physically reasonable. We demonstrate the utility of this approach using examples from our studies of PKR interactions with simple dsRNAs, the HIV TAR RNA, and the VAI RNA from adenovirus.

1. Introduction

Protein kinase R (PKR) is an interferon-induced kinase that plays a key role in the innate immunity response to viral infection (Toth *et al.*, 2006). PKR is induced in a latent form that is activated by binding dsRNA to undergo autophosphorylation and subsequently phosphorylate cellular substrates. PKR contains an N-terminal dsRNA binding domain (dsRBD), consisting of two tandem copies of the dsRNA binding motif (dsRBM) (Tian *et al.*, 2004), and a C-terminal kinase domain, with a ~90 amino acid unstructured linker lying between these domains (Fig. 3.1). The structures of the isolated dsRBD (Nanduri *et al.*, 1998) and the kinase domain (Dar *et al.*, 2005) have been solved. The linker is flexible and PKR adopts multiple compact and extended conformations in solution (VanOudenhove *et al.*, 2009). Crystallographic and NMR studies indicate that the dsRBM binds to one face of the dsRNA helix, spanning ~16 bp (Tian *et al.*, 2004). The interaction is not sequence specific; however, there are some reports of selective binding of the dsRBM to specific RNA secondary structural features (Ben-Asouli *et al.*, 2002; Leulliot *et al.*, 2004; Liu *et al.*, 2000; Nagel and Ares, 2000; Spanggord and Beal, 2001).

Although PKR is known to bind to dsRNAs as short as 15 bp (Bevilacqua and Cech, 1996; Schmedt *et al.*, 1995; Ucci *et al.*, 2007), at least 30 bp are required for activation (Lemaire *et al.*, 2008; Manche *et al.*, 1992). These data support an activation model where the role of the dsRNA is to bring two or more PKR monomers in close proximity to enhance dimerization via the kinase domain (Cole, 2007). Consistent with this model, HIV TAR RNA, a 23-bp hairpin with three bulges, binds a single PKR and does not activate. In contrast, a dimer of TAR binds two PKRs and activates (Heinicke *et al.*, 2009). However, for more complex RNAs the structural features that distinguish activators of PKR from those that fail to activate are not yet well understood. Incorporation of G–I mismatches blocks PKR activation (Minks *et al.*, 1979) but PKR is activated by RNAs containing tandem A–G mismatches and noncontiguous helices, provided that the RNA adopts an overall A-form geometry (Bevilacqua *et al.*, 1998). Some highly structured viral RNAs do not activate and function as PKR inhibitors *in vivo* (Langland *et al.*, 2006).

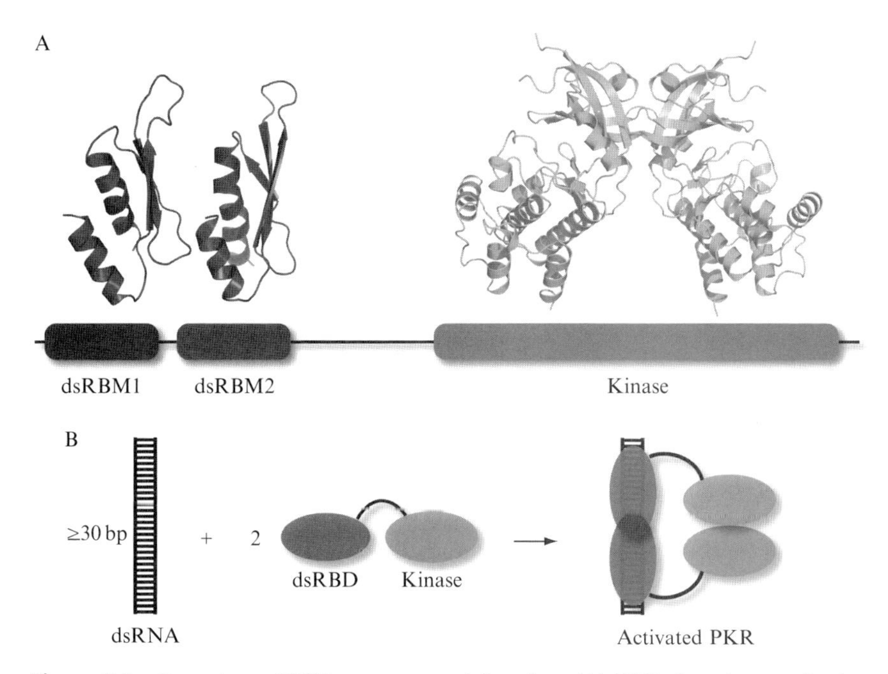

Figure 3.1 Overview of PKR structure and function. (A) PKR domain organization and structure. The N-terminal regulatory domain comprises two dsRBMs, dsRBM1 and dsRBM2, connected by an unstructured linker. Each of these motifs adopts the canonical $\alpha\beta\beta\beta\alpha$ fold in the NMR structure of dsRBD (PDB:1QU6). In the crystal structure of a complex of the PKR kinase domain and eIF2α (PDB:2A1A), the kinase domain has the typical bilobal structure observed in other eukaryotic protein kinases and dimerizes via the N-terminal lobe. (B) Dimerization model for PKR activation by dsRNA. Binding to dsRNA induces PKR dimerization via the kinase domain, resulting in activation.

In order to define the mechanisms of activation and inhibition of PKR by RNAs, it is necessary to characterize the stoichiometries, affinities, and free energy couplings governing the assembly of the relevant macromolecular complexes. Protein–nucleic acid interactions are typically measured using electrophoretic mobility shift (Carey, 1991; Hellman and Fried, 2007) and nitrocellulose filter binding assays (Wong and Lohman, 1993) that are performed under nonequilibrium conditions. Although these assays can be used to accurately define binding parameters under carefully controlled conditions, nonspecific interactions generally have lower affinity and higher dissociation rates than specific interactions and are thus particularly susceptible to artifacts associated with dissociation of complexes during the measurement. Thus, analyses of nonspecific and weaker specific interactions should be performed using free solution biophysical methods. One such method, analytical ultracentrifugation (AUC), has historically played an important role in quantitative analysis of protein–nucleic acid binding reactions. Early studies utilized sedimentation velocity methods to obtain the free and bound concentrations of the reactants (Draper and von Hippel, 1979; Goodman *et al.*, 1984; Jensen and von Hippel, 1977; Lohman *et al.*, 1980; Revzin and Woychik, 1981) and one study suggested potential applications of equilibrium methods (Lanks and Eng, 1976). With the advent of the XL-A analytical ultracentrifuge (Giebeler, 1992) and increasing computational power, it became feasible to globally analyze sedimentation equilibrium data for protein–nucleic acid systems obtained at multiple wavelengths and concentrations (Bailey *et al.*, 1996; Daugherty and Fried, 2005; Laue *et al.*, 1993; Lewis *et al.*, 1994; Wojtuszewski *et al.*, 2001). We have found these methods useful in the analysis of PKR–RNA interactions (Ucci and Cole, 2004; Ucci *et al.*, 2007) and we have developed a software package, HeteroAnalysis, that facilitates global analysis of multiwavelength sedimentation equilibrium data (Cole, 2004). However, these analyses can be challenging. In some cases, equilibrium is not achievable due to slow precipitation of protein–RNA complexes. Global analysis of multiwavelength data requires extremely accurate extinction coefficients that are difficult to achieve due to the poor wavelength reproducibility of the XL-A. Finally, the data reduction and analysis process can be fairly tedious due to the large number of files and parameters and the necessity to carefully specify the linkages among them. Recently, we have taken advantage of "whole boundary" methods for the direct analysis of sedimentation velocity profiles of interacting systems (Correia and Stafford, 2009; Dam *et al.*, 2005; Demeler *et al.*, 2010; Stafford and Sherwood, 2004). These approaches work very well for characterization of PKR–RNA interactions. The measurements are fast and less affected by precipitation and the analysis is generally more straightforward than corresponding sedimentation equilibrium experiments. Below, we provide a protocol for the design and analysis of experiments to characterize protein–RNA (or protein–DNA) interactions using sedimentation velocity with absorption detection. The protocol is summarized in the flowchart in Fig. 3.2.

2. Reagents and Cells

Buffer used for sedimentation velocity measurements should not contain detergents and should have minimal absorbance at the wavelength of interest. Typically, we monitor RNA absorbance at 260 nm. The inclusion of a reducing agent is often required and tris(2-carboxyethyl)phosphine (TCEP) is preferred over dithiothreitol due to its lower absorbance at 260 nm. Particularly when working with highly charged macromolecules such as RNA and DNA, it is critical to include an electrolyte in the buffer to suppress the primary charge effect that slows sedimentation. Usually,

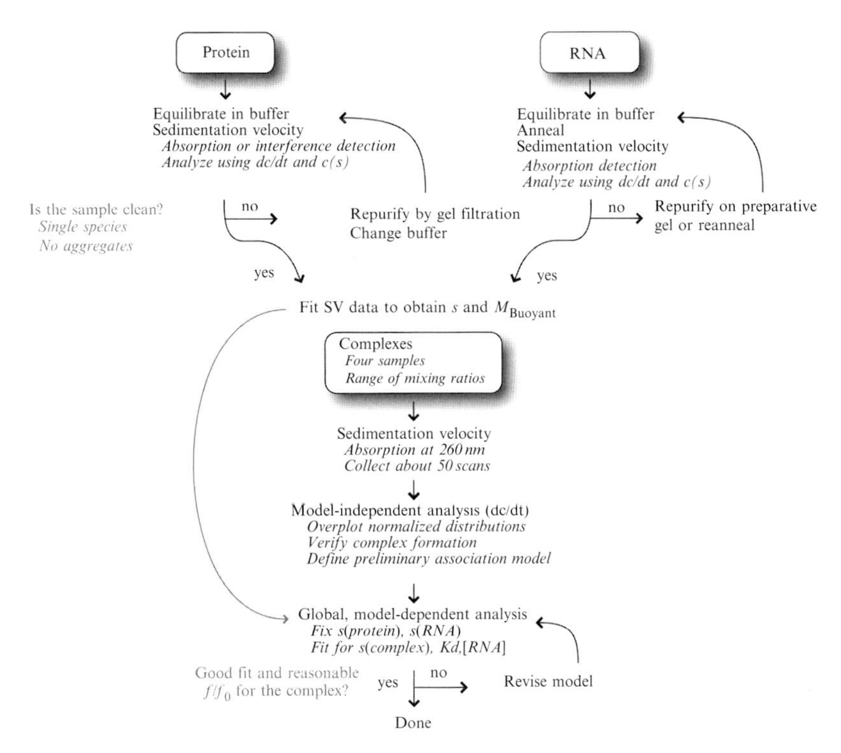

Figure 3.2 Experimental flowchart. For details see the text.

10–50 m*M* of monovalent salt is sufficient. The ionic strength is also a critical parameter that modulates binding affinity and may also affect solubility of the protein and the protein–nucleic acid complexes. Note that high solute concentrations can result in dynamic density and viscosity gradients that will affect sedimentation (Schuck, 2004). RNA is very susceptible to hydrolysis catalyzed by ribonucleases that are ubiquitous in the lab. For RNA work, we use sterile tips and tubes and autoclaved as well as sterile-filtered buffers. We do not find it necessary to resort to baking glassware or using diethylpyrocarbonate-treated water. If ribonuclease contamination is a problem, it may be helpful to add a commercially available ribonuclease inhibitor to the RNA samples. The inhibitor should be kept at low concentrations so that it does not contribute to the sample absorbance. RNAs have the propensity to adopt alternative secondary structures so that it may be necessary to use an annealing or a snap-cooling procedure to prepare a sample with a homogeneous secondary structure. In particular, self-complementary hairpin RNAs have the propensity to form stable dimers and it may be necessary to anneal samples at low concentrations to obtain homogeneous monomers or to purify the species of interest on native gels (Heinicke *et al.*, 2009).

Both the RNA and protein should be equilibrated in the sample buffer via dialysis, gel filtration, or spin column chromatography. The contribution of buffer mismatch to the sedimentation velocity profiles is less of a concern with absorbance optics than with interference optics, but equilibration ensures that contaminating low-molecular weight solutes are not carried over from the RNA or protein stocks. Standard double sector centerpieces or meniscus-matching centerpieces are used with quartz windows. The windows and centerpieces are treated with RNaseZap (Ambion, CA) and then rinsed extensively with sterile deionized water to prevent RNA degradation. It may be useful to reserve some cells exclusively for RNA experiments to ensure that the centerpieces and windows do not become contaminated with ribonucleases.

3. Experimental Design

It is always useful to fix as many parameters as possible in global data analysis methods. Thus, the sedimentation coefficients of the free RNA and protein should be determined independently by analyzing each component separately. Multiple concentrations that span at least a threefold range and encompass the concentration range to be used in the binding experiments should be run to detect potential self-association. The molecular masses of the protein and RNA are obtained from sequence. The partial specific volume ($\bar{\upsilon}$) and extinction coefficient (ε) of the protein and the buffer density (ρ) are calculated using SEDNTERP (Laue *et al.*, 1992). There is no reliable method to calculate $\bar{\upsilon}$ for RNA oligonucleotides. Instead, this parameter can be determined indirectly from the buoyant molecular weight (M^*) measured in the sedimentation velocity experiment:

$$M^* = M(1 - \bar{\upsilon}\rho) \tag{3.1}$$

Typically, we collect data at a single RNA concentration chosen to provide an absorbance at 260 nm between 0.3 and 0.8, where the best sensitivity is achieved, using a broad range of protein:RNA ratios (typically from 0.5:1 to 6:1) to populate all the species participating in the equilibrium. Of course, if the relevant binding reactions are too strong, it will not be possible to resolve the K_ds with reasonable precision at the accessible reagent concentrations using the absorption optics. It can be helpful to simulate the experiment using the estimated dissociation constants with realistic noise levels to provide guidance on the optimal reagent concentrations and to determine whether it is possible to resolve the K_ds with sufficient accuracy and precision. Alternatively, fluorescence optics can be used with labeled RNAs to greatly enhance sensitivity.

3.1. Data collection

General information on how to perform sedimentation velocity experiment methods can be found in several recent reviews (Balbo *et al.*, 2009; Brown *et al.*, 2008; Cole *et al.*, 2008; Stafford, 2009). As mentioned above, we typically collect absorbance data at 260 nm for analysis of protein–RNA interactions. Although data collected at additional wavelengths would be extremely useful in global analysis, particularly at 230 and 280 nm where the relative contribution of the protein is greater, the slow scan rate of the XL-A absorbance system and the poor wavelength reproducibility preclude this option. Rapid scan absorbance systems that overcome these limitations are under development (Colfen *et al.*, 2009). Although simultaneous collection of interference and absorbance data is feasible, we have not implemented this approach because the concentrations of protein and RNA are too low in our experiments for interference. The total number of scans that one can collect per run is limited by the data collection rate of the absorbance optics on the XL-A. For our instruments, each absorbance scan requires ~1.5 min (0.003 cm/point, continuous mode). About 50 scans per cell is appropriate for reliable global data analysis. Given the slow data acquisition, a maximum of four cells is recommended, which gives a total run time of about 5 h. For larger complexes, with sedimentation coefficients greater than about 5 S, one can reduce the rotor speed from the maximum (50,000 RPM for the 8 hole rotor) to slow the sedimentation rate. The rotor speed (in RPM) can be estimated using the following formula:

$$\text{Speed} = 9.7 \times 10^4 \sqrt{s} \tag{3.2}$$

where s is the sedimentation coefficient of the species of interest. Alternatively, the throughput can be increased by loading sample into both sectors of the cell, recording raw intensities and analyzing pseudo-absorbance data (Kar *et al.*, 2000).

3.2. Data analysis

Initially, velocity data are analyzed using a model-independent approach to help define the correct binding model and to determine whether the system is kinetically limited. We typically employ the time derivative method (Stafford, 1992) using the program DCDT+ (Philo, 2006) to obtain a $g(s^*)$ distribution for each sample. By inspecting an overlay of the normalized $g(s^*)$ distributions for all of the samples, one can immediately verify complex formation by the appearance of features at higher s than the free protein or RNA. The shape of the boundary will depend on the kinetics of association and dissociation. For systems that equilibrate slowly on the timescale of the sedimentation run, features will be present in the boundary corresponding to each of the species: free protein (provided the concentration is high enough to detect at 260 nm), free RNA, and the protein–RNA complex. Their amplitudes but not positions will be protein-concentration dependent. In contrast, for rapidly reversible systems, complex boundary shapes are observed due to reequilibration during sedimentation (Cann, 1970; Gilbert and Jenkins, 1959). One feature occurs at the position of one of the free components and the other feature is a reaction boundary with an apparent sedimentation coefficient lying between the faster sedimenting component and the protein–RNA complex. The sedimentation coefficient of the reaction boundary increases with increasing protein concentration. If binding is sufficiently tight, one can use the limiting sedimentation coefficient of the reaction boundary at high-protein concentrations to estimate the stoichiometry of the largest complex. We use SEDNTERP (Laue *et al.*, 1992) to calculate frictional ratios (f/f_0) based on the sedimentation coefficient, predicted mass, and weight average $\bar{\upsilon}$, assuming alternative stoichiometries for the largest protein–RNA complex. For typical protein–RNA complexes, f/f_0 lies between 1.2 and 1.6; values outside this range likely indicate an incorrect stoichiometry. Later, one can test the model more rigorously by repeating this calculation using the sedimentation coefficient of the complex obtained from global analysis (see below).

Having defined the binding model, we then globally fit the sedimentation velocity data obtained at multiple loading concentrations to determine the relevant association constants as well as the hydrodynamic properties of each complex. Several powerful software packages are available for sedimentation velocity analysis of interacting systems, including SEDPHAT (Dam *et al.*, 2005), SEDANAL (Correia and Stafford, 2009; Stafford and Sherwood, 2004), and ULTRASCAN (Demeler *et al.*, 2010). We find SEDANAL to be particularly convenient for analysis of PKR–RNA interactions. Association models for particular systems can be defined by the user with the model editor. It is also simple to apply constraints to the fitted parameters, such as preset limits on their allowable ranges or imposing particular relationships between parameters. As mentioned above, parameters for the free RNA and protein (sedimentation coefficient, molecular weight, density increment $(1-\bar{\upsilon}\rho)$, and mass extinction coefficient) are fixed based on prior calculation and previous experiments. We usually find it necessary to treat the RNA loading concentrations as adjustable parameters because they often decrease upon addition of protein due to precipitation of complexes. However, the protein loading concentration, or alternatively, the protein:RNA ratio, should be held fixed, as the protein contribution of absorbance at 260 nm is typically too low for these parameters to be well determined from the data. Of course, the association constants and sedimentation coefficients are allowed to float during the fit. The programs for sedimentation velocity analysis of interacting systems mentioned above can be configured to fit the data assuming either rapid equilibration or kinetically limited reactions. In all of the PKR–RNA interactions that we have examined so far, the rate of complex equilibration is fast on the timescale of sedimentation such that reaction boundaries are detected in the $g(s^*)$ distributions and good fits are obtained in global analyses assuming rapid equilibration.

As is true for all nonlinear least squares problems, a good fit is defined by the absence of systematic trends in the residuals and an overall RMS deviation consistent with the stochastic noise level. It is useful to verify that the best fit corresponds to a global minimum in the error surface by repeating the process using different initial values for the parameters being floated or using the "perturb fit and redo" feature in SEDANAL. Once the best fit has been defined, confidence intervals of the fitted parameters should be obtained based on the *F*-statistic, Monte-Carlo, or bootstrap with replacement methods. One can also recalculate f/f_0 for the complexes, using the fitted sedimentation coefficients to determine whether the hydrodynamic properties are physically reasonable.

The parameters associated with higher-order binding events are often poorly defined due to low population of the relevant species as well as strong cross-correlation between association constants and sedimentation coefficients. Inclusion of data from samples containing higher protein:RNA ratios often solves this problem by enhancing the population of the higher-order complexes. One can also constrain the ratio of successive association constants or fix the sedimentation coefficient of the complexes based on a prediction of f/f_0 to reduce the number of adjustable parameters. Another problem that we have encountered is the formation of higher-order, nonspecific complexes that are not accounted for in the association model, leading to poor quality fits. SEDANAL only displays the first and last difference curve from each channel but by default it saves all of the curves in text files. We have written macros for IGOR Pro that automatically load and process these files that can be downloaded at http://www.biotech.uconn.edu/auf/?i=aufftp. Examples of the output are shown below in Figs. 3.3C and 3.4C.

4. Examples

4.1. 20 bp dsRNA

We have employed sedimentation velocity to define the interactions of PKR with a series of nonactivating and activating dsRNAs (Lemaire *et al.*, 2008). Figure 3.3A shows a small (20 bp) nonactivating dsRNA used in these studies and Fig. 3.3B shows the normalized $g(s^*)$ distributions obtained from a titration of this RNA with PKR. In the absence of PKR, a single feature is observed near $s = 2.5$ S corresponding to the free RNA. Upon

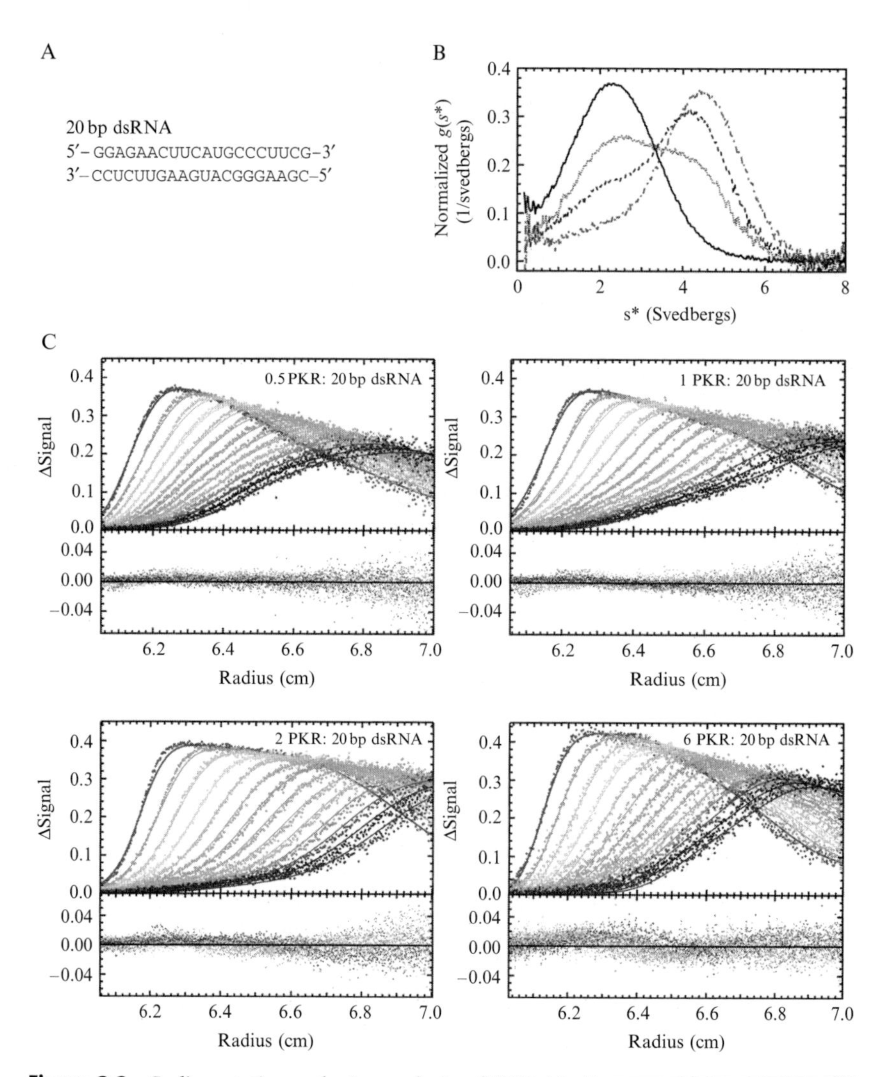

Figure 3.3 Sedimentation velocity analysis of PKR binding to a 20 bp dsRNA. (A) RNA structure. (B) Normalized $g(s^*)$ distributions of 1 μM dsRNA (black, solid), dsRNA + 0.5 equiv. PKR (green, dot), dsRNA + 1 equiv. PKR (blue, dash), dsRNA + 2 equiv. PKR (red, dot–dash). The distributions are normalized by area. (C) Global analysis of sedimentation velocity difference curves. The data were subtracted in pairs and four data sets at the indicated ratios of PKR:dsRNA were fit to 1:1 binding stoichiometry model. The top panels show the data (points) and fit (solid lines) and the bottom panels show the residuals (points). For clarity, only every 2nd difference curve is shown. Conditions: rotor speed, 50,000 RPM; temperature, 20 °C; wavelength, 260 nm. (See Color Insert.)

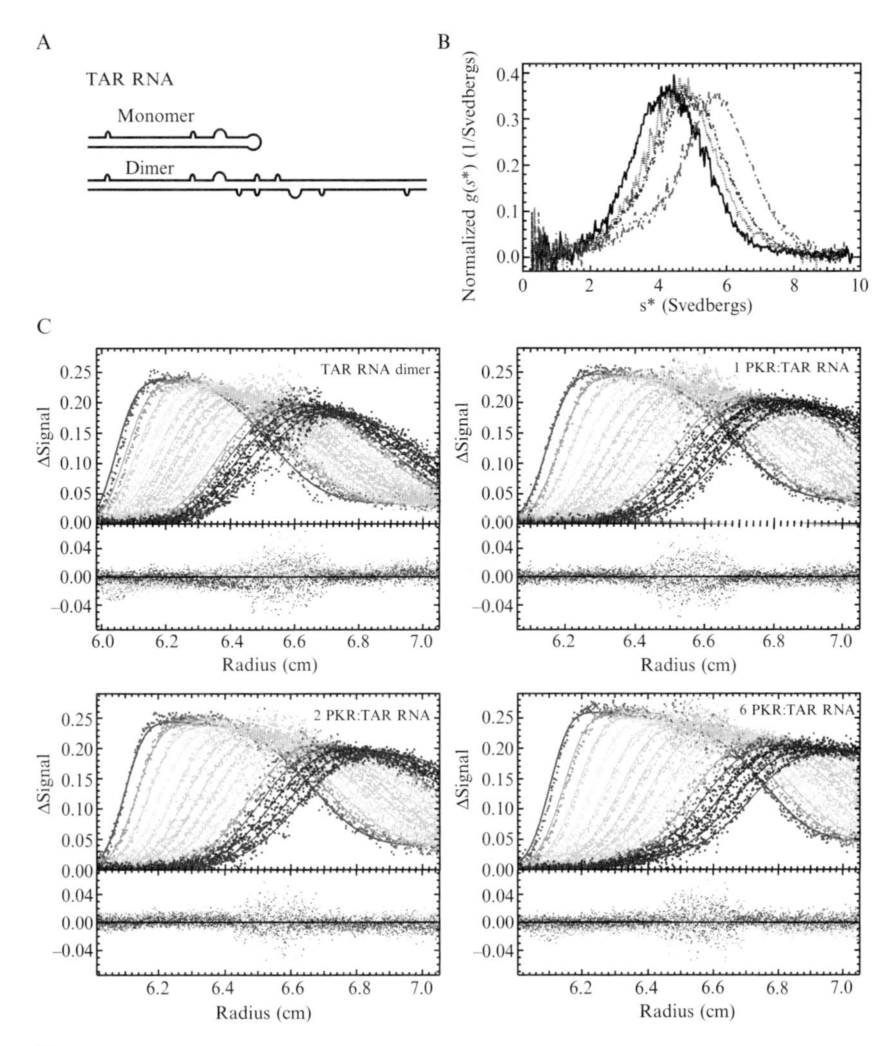

Figure 3.4 Sedimentation velocity analysis of PKR binding to HIV TAR RNA dimer. Measurements were performed in AU200 buffer at 20 °C and 40,000 RPM. (A) Structure of TAR monomer and dimer. (B) Plot of normalized $g\,(s^{*})$ distributions for 0.5 μ*M* TAR dimer (black, solid) and 1 μ*M* TAR dimer plus 1 equiv. PKR (green, dot), 2 equiv. PKR (blue, dash), or 6 equiv. PKR (red, dot–dash). The distributions are normalized by area under the curve. (C) Global analysis of the sedimentation difference curves. Scans within each dataset were subtracted in pairs to remove time-invariant background and fit to a 1:2 binding model using SEDANAL. Top panels show data points and the solid lines represent fitting results using the parameters presented in Table 3.1. Residuals for each fit are shown in the bottom panels. Only every 2nd difference curve is shown for clarity. Measurements were performed in AU200 buffer at 20 °C and 40,000 RPM using absorbance detection at 260 nm. (See Color Insert.)

addition of PKR, the dsRNA peak decreases in amplitude and a new feature develops which shifts to the right with increasing protein concentration. This behavior is consistent with rapidly reversible formation of an RNA–PKR complex. Assuming 1:1 binding, the limiting sedimentation coefficient of ~5 S corresponds to $f/f_0 = 1.3$, which is quite reasonable. Model-dependent analysis was performed in SEDANAL using the model

$$R + P \overset{K_{d1}}{\rightleftarrows} RP \tag{3.3}$$

In this program, the sedimentation velocity scans are subtracted in pairs to remove systematic noise and the resulting difference scans are fit to Lamm equation solutions incorporating reversible association. A disadvantage of fitting difference scans is that the noise level is increased slightly. However, this method for removing systematic noise has the advantage of being model-independent, unlike the algebraic noise decomposition method used in SEDFIT and SEDPHAT (however, see Schuck, 2010 for a contrary view). Figure 3.3C shows a global fit of the same data sets depicted in Fig. 3.3B to this model and Table 3.1 summarizes the results. The data fit well to this model with mostly random residuals and an RMS deviation close to the intrinsic noise level of the optical system. The best-fit K_d of 859 nM is fairly weak and provides an important baseline that we have used to compare dissociation constants for PKR binding to longer dsRNAs and structured RNAs (Heinicke *et al.*, 2009; Launer-Felty *et al.*, 2010). PKR binds more strongly to longer dsRNAs. This decrease in K_d is consistent with the expected statistical effects for nonspecific protein–nucleic acid interactions where the number of binding configurations increases with the length of the dsRNA lattice (Cole, 2004).

4.2. TAR RNA dimer

The TAR RNA stem-loop has been used as a model system for studying PKR regulation by viral RNAs. However, earlier analyses have been complicated by the propensity of this RNA to form dimers (Fig. 3.4A). We have examined PKR binding and activation by several TAR monomer and dimer constructs (Heinicke *et al.*, 2009). For example, Fig. 3.4B shows normalized $g(s^*)$ distributions obtained from a titration of wild-type TAR dimer with PKR. The pattern is more complex than observed for PKR binding to the 20 bp dsRNA. The maximum shifts from ~4.3 S for the free RNA up to ~5 S upon addition of 1–2 equiv. PKR then shifts more dramatically to ~6 S upon addition of 6 equiv. of PKR. This behavior is consistent with sequential binding of two PKRs according to the model

Table 3.1 PKR–RNA interaction parameters derived from sedimentation velocity analysis

RNA	Model	K_{d1} (n*M*)	K_{d2} (n*M*)	s(RNA)[a]	s(RP)[a]	$s(RP_2)$[a]	RMS[b]
20 bp	$R + P \leftrightarrow RP$	859 (746, 987)	–	2.52	5.14 (5.03, 5.25)	–	0.00961
TAR dimer	$R + P \leftrightarrow RP$ $RP + P \leftrightarrow RP_2$	404 (257, 648)	2760 (1170,4540)	4.25	5.51 (5.02, 5.95)	7.70 (6.97, 8.60)	0.00794
A34U:U37A TAR dimer	$R + P \leftrightarrow RP$ $RP + P \leftrightarrow RP_2$	331 (226, 490)	1470 (910, 2430)	4.31	5.29 (5.13, 5.47)	7.18 (6.86, 7.66)	0.00581
A34U:U37A TAR dimer	$R + P \leftrightarrow RP$ $2RP \leftrightarrow (RP)_2$	980	2.58×10^6	4.31	6.75	11.85	0.00785
VAI	$R + P \leftrightarrow RP$ $RP + P \leftrightarrow RP_2$	14 (3,41)	601 (359,1200)	5.34	6.78 (6.63, 7.04)	8.54 (8.30, 9.00)	0.00841
VAI+ 5 m*M* Mg^{2+}	$R + P \leftrightarrow RP$	334 (278, 401)	–	5.29	7.18 (7.07, 7.29)	–	0.00691

Measurements were performed in AU 200 buffer or AU 200 buffer + Mg^{2+} at 20 °C. Parameters were obtained by global nonlinear least squares using SEDANAL. The values in parentheses correspond to the 95% joint confidence intervals obtained using the *F*-statistic to define a statistically significant increase in the variance upon adjusting each parameter from its best-fit value.

[a] Uncorrected sedimentation coefficient (Svedbergs).

[b] RMS deviation of the fit in absorbance units.

$$\begin{aligned} R + P &\overset{K_{d1}}{\rightleftarrows} RP \\ RP + P &\overset{K_{d2}}{\rightleftarrows} RP_2 \end{aligned} \tag{3.4}$$

The data fit well to this model (Fig. 3.4C) and the dissociation constants and sedimentation coefficients are shown in Table 3.1. This model was confirmed using a double mutant of TAR (A34U:U37A) that has an enhanced propensity to dimerize. It is interesting to note that the second PKR binds weaker than the first. This reduced affinity is not due to negative cooperativity but is a statistical effect and arises due to the reduction of the number of binding configurations upon binding of the first PKR. In both cases, the ratio K_{d2}:K_{d1} is close to the value four as predicted for a simple model of a pair of noninteracting, identical sites in the TAR dimer.

An alternative model has been proposed in which binding of PKR to TAR enhances protein dimerization, resulting in formation of an $(RP)_2$ complex (McKenna *et al.*, 2007a,b)

$$\begin{aligned} R + P &\overset{K_{d1}}{\rightleftarrows} RP \\ 2RP &\overset{K_{d2}}{\rightleftarrows} (RP)_2 \end{aligned} \tag{3.5}$$

This model does not fit the sedimentation velocity data for PKR binding to dimeric TAR. For example, the fit of the A34U:U37A data gives a higher RMSD relative to the sequential binding model (Eq. 3.4) and an exceptionally weak K_{d2} that is incompatible with activation data (Heinicke *et al.*, 2009). These results highlight the capability of sedimentation velocity measurements to distinguish among closely related binding models.

Analysis of the distribution of free RNA, protein, and complexes as a function of PKR concentrations provides some insight into how we are to resolve the two binding events with good precision for the fitted parameters. Figure 3.5B shows the species distribution using the best-fit parameters in Table 3.1 with the actual PKR concentrations used for the analysis in Fig. 3.4 (shown as gray lines). Figure 3.5B and C shows the expected contributions of each species to the sedimentation velocity absorbance profile at the approximate midpoint of the run for each of the three RNA–protein mixtures. Reliable characterization of binding energetics requires that each of the species that participates in the equilibrium is present at measurable concentrations. At the lowest protein concentration (1 equiv.), the concentrations of R, P, and the RP complex are approximately equal (Fig. 3.5A), facilitating the measurement of K_{d1}. However, owing to the low extinction coefficient of PKR at 260 nm, this species does not contribute much to the sedimentation velocity profile. This problem is mitigated to some extent in the global analysis in Fig. 3.4 by fixing the protein:RNA ratios. At the two higher protein concentrations (2 and 6

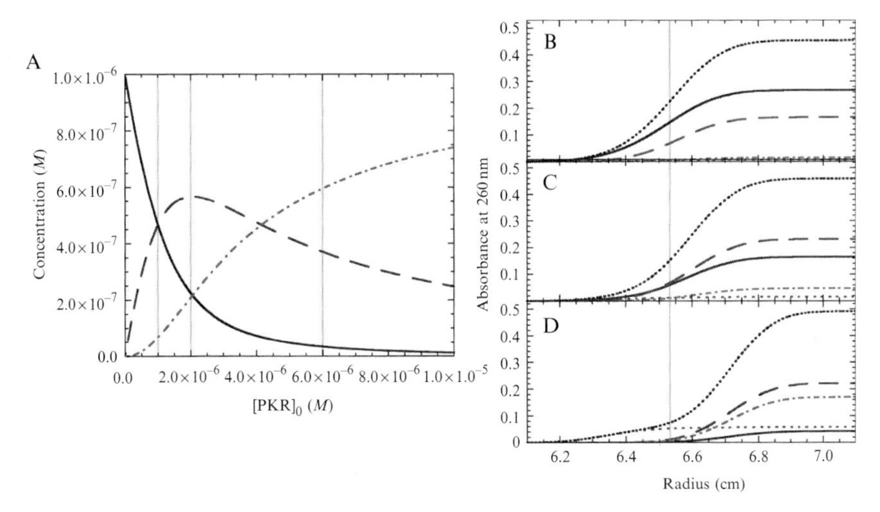

Figure 3.5 Species distributions for binding of PKR to TAR RNA dimer. Concentrations were calculated based on the best-fit parameters in Table 3.1 and the extinction coefficients of TAR and PKR at 260 nm. (A) Molar concentration distributions for TAR RNA dimer (blue, solid), RP complex (green, dash), and RP_2 complex (tan, dot–dash). The vertical gray lines indicate the three PKR concentrations used in the experiment depicted in Fig. 3.4. (B, C) Sedimentation velocity absorption profiles for TAR RNA dimer (blue, solid), PKR (red, dot), RP complex (green, dash), RP_2 complex (tan, dot–dash), and the total absorbance (black, dot) calculated for 1 equiv. PKR (B), 2 equiv. PKR (C), and 6 equiv. PKR (D). The profile was simulated at a time corresponding to the middle of the sedimentation run. The gray line indicates the midpoint of the total absorbance curve in (B). (See Color Insert.)

equiv.), the RP_2 species becomes substantially populated, thereby facilitating measurement of K_{d2}.

4.3. VAI

Adenovirus encodes VAI, a highly structured RNA inhibitor that binds PKR but fails to activate. VAI contains three major domains: the terminal stem, a complex central domain, and an apical stem-loop (Fig. 3.6A). We have characterized the stoichiometry and affinity of PKR binding to define the mechanism of PKR inhibition by VAI. Early enzymatic probing measurements suggested that Mg^{2+} alters VAI conformation (Clarke and Mathews, 1995), so we have characterized PKR binding in the absence and presence of divalent ion (Launer-Felty *et al.*, 2010). Figure 3.6B shows that VAI has a sedimentation coefficient near 5 S and the peak shifts to the right upon binding PKR. At the highest PKR concentration (6 equiv.), the main peak is located near 8.5 S with a shoulder near 3.5 S, corresponding to free PKR.

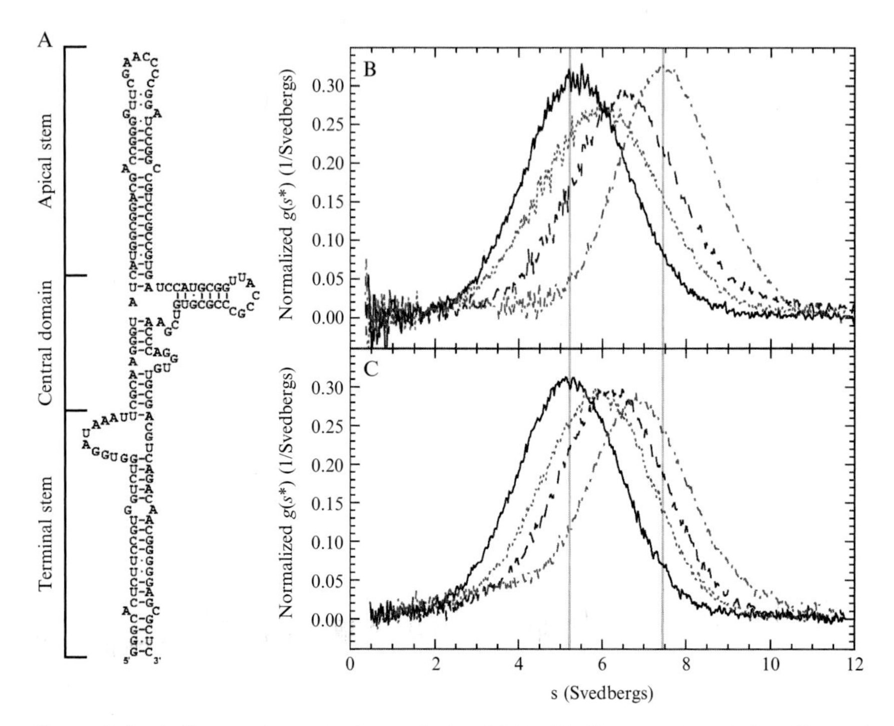

Figure 3.6 Sedimentation velocity analysis of PKR binding to VAI RNA: effects of divalent ion. (A) Structure of VAI. (B) Normalized $g(s^*)$ distributions obtained in the absence of Mg^{2+}: 0.4 μM VAI (black, solid), VAI + 1 equiv. PKR (green, dot), VAI + 2 equiv. PKR (blue, dash), and VAI + 6 equiv. PKR (red, dot–dash). The vertical gray line on the left corresponds to the peak of the distribution for the free VAI RNA and the line on the right corresponds to the peak in the presence of 6 equiv. of PKR. (C) Normalized $g(s^*)$ distributions obtained in the presence of 5 mM Mg^{2+}. The labeling is the same as in part B. (See Color Insert.)

Assuming a model of a single PKR binding leads to an estimate of a frictional ratio (f/f_0) of 1.28, which is much lower than the RNA alone or other PKR–RNA complexes. Thus, we fit the data to a model of sequential binding of two PKRs (Eq. (3.4)) and the results are summarized in Table 3.1. The first PKR binds with quite high affinity ($K_d = 14$ nM) and the second binds with lower affinity ($K_d = 601$ nM). The uncorrected sedimentation coefficient of VAI decreases very slightly in the presence of Mg^{2+} (Table 3.1) but the effect is entirely due to the increase in buffer density and viscosity. Thus, Mg^{2+} does not induce a large-scale change in VAI conformation. The sedimentation velocity results are supported by small angle X-ray scattering studies, where the radius of gyration (R_g) of VAI increases very slightly upon addition of Mg^{2+} (Launer-Felty *et al.*, 2010). Although divalent ion does not induce a dramatic structural change in VAI, it does affect PKR binding.

The magnitude of the shift in the $g(s^*)$ is significantly reduced (Fig. 3.6B) and the data fitted to a model of a single PKR binding with an affinity about 20-fold less than in absence of Mg^{2+}. The reduction in binding affinity is much more than the two- to threefold expected based on nonspecific effects (Launer-Felty *et al.*, 2010). We propose that VAI acts as an *in vivo* inhibitor of PKR because it binds only a single PKR under physiological conditions where divalent cation is present.

5. Conclusions

We have described protocols for designing and analyzing sedimentation velocity experiments for characterizing the stoichiometries and affinities of protein–nucleic acid interactions using several examples from our studies of the binding of PKR to RNA activators and inhibitors. Although conventional gel shift and filter binding measurements can be used to accurately define binding parameters, they are susceptible to artifacts associated with dissociation of complexes during the measurements. This can be a particular problem in the analysis of nonspecific interactions and we have emphasized free solution methods such as AUC in our PKR work. Recently, we have taken advantage of efficient algorithms for fitting sedimentation velocity data to Lamm equation solutions incorporating reversible association. Although multiwavelength sedimentation equilibrium measurements work well for characterizing protein–nucleic acid interactions, the velocity experiments have several advantages: they are faster, less sensitive to formation of aggregates that often accompanies formation of protein–nucleic acid complexes, and generally easier to analyze. We have found that data collected at a single wavelength (typically, 260 nm) provides sufficient information to characterize multistep association reactions. This is fortunate, given the limitations in the current absorbance system in the XL-A. The ability to collect absorbance data more rapidly at multiple wavelengths in new instruments (Colfen *et al.*, 2009) will certainly enhance the power of sedimentation velocity to define more complex interactions. A disadvantage of the velocity approach is the necessity to fit for the sedimentation coefficient of the complex. This complicates data analysis by requiring additional adjustable parameters that can be strongly correlated with the dissociation constant if it is not possible to achieve high population of the complex. On the other hand, the resulting hydrodynamic information can be valuable in developing structural models of the complex.

Acknowledgments

This work was supported by grant number AI-53615 from the NIH to J. L. C.

REFERENCES

Bailey, M. F., Davidson, B. E., Minton, A. P., Sawyer, W. H., and Howlett, G. J. (1996). The effect of self-association on the interaction of the *Escherichia coli* regulatory protein TyrR with DNA. *J. Mol. Biol.* **263,** 671–684.

Balbo, A., Zhao, H., Brown, P. H., and Schuck, P. (2009). Assembly, loading, and alignment of an analytical ultracentrifuge sample cell. *J. Vis. Exp.* **33,** doi: 10.3791/1530.

Ben-Asouli, Y., Banai, Y., Pel-Or, Y., Shir, A., and Kaempfer, R. (2002). Human interferon-gamma mRNA autoregulates its translation through a pseudoknot that activates the interferon-inducible protein kinase PKR. *Cell* **108,** 221–232.

Bevilacqua, P. C., and Cech, T. R. (1996). Minor-groove recognition of double-stranded RNA by the double-stranded RNA-binding domain of the RNA-activated protein kinase PKR. *Biochemistry* **35,** 9983–9994.

Bevilacqua, P. C., George, C. S., Samuel, C. E., and Cech, T. R. (1998). Binding of the protein kinase PKR to RNAs with secondary structure defects: Role of the tandem A-G mismatch and noncontinguous helices. *Biochemistry* **37,** 6303–6316.

Brown, P. H., Balbo, A., and Schuck, P. (2008). Characterizing protein–protein interactions by sedimentation velocity analytical ultracentrifugation. *Curr. Protoc. Immunol.* 15, Chapter 18, Unit 18.

Cann, J. R. (1970). Interacting Macromolecules. Academic Press, New York.

Carey, J. (1991). Gel retardation. *Methods Enzymol.* **208,** 103–117.

Clarke, P. A., and Mathews, M. B. (1995). Interactions between the double-stranded RNA binding motif and RNA: Definition of the binding site for the interferon-induced protein kinase DAI (PKR) on adenovirus VA RNA. *RNA* **1,** 7–20.

Cole, J. L. (2004). Analysis of heterogeneous interactions. *Methods Enzymol.* **384,** 212–232.

Cole, J. L. (2007). Activation of PKR: An open and shut case? *Trends Biochem. Sci.* **32,** 57–62.

Cole, J. L., Lary, J. W., Moody, T. P., and Laue, T. M. (2008). Analytical ultracentrifugation: Sedimentation velocity and sedimentation equilibrium. *Methods Cell Biol.* **84,** 143–179.

Colfen, H., Laue, T. M., Wohlleben, W., Schilling, K., Karabudak, E., Langhorst, B. W., Brookes, E., Dubbs, B., Zollars, D., Rocco, M., and Demeler, B. (2009). The open AUC project. *Eur. Biophys. J.* **39,** 347–359.

Correia, J. J., and Stafford, W. F. (2009). Extracting equilibrium constants from kinetically limited reacting systems. *Methods Enzymol.* **455,** 419–446.

Dam, J., Velikovsky, C. A., Mariuzza, R. A., Urbanke, C., and Schuck, P. (2005). Sedimentation velocity analysis of heterogeneous protein–protein interactions: Lamm equation modeling and sedimentation coefficient distributions c(s). *Biophys. J.* **89,** 619–634.

Dar, A. C., Dever, T. E., and Sicheri, F. (2005). Higher-order substrate recognition of eIF2alpha by the RNA-dependent protein kinase PKR. *Cell* **122,** 887–900.

Daugherty, M. A., and Fried, M. G. (2005). *In* "Modern Analytical Ultracentrifugation: Techniques and Methods," (D. J. Scott, ed.)pp. 195–209. Royal Society of Chemistry, Oxford.

Demeler, B., Brookes, E., Wang, R., Schirf, V., and Kim, C. A. (2010). Characterization of reversible associations by sedimentation velocity with UltraScan. *Macromol. Biosci.* **10,** 775–782.

Draper, D. E., and von Hippel, P. H. (1979). Measurement of macromolecular equilibrium binding constants by a sucrose gradient band sedimentation method. Application to protein–nucleic acid interactions. *Biochemistry* **18,** 753–760.

Giebeler, R. (1992). *In* "Analytical Ultracentrifugation in Biochemistry and Polymer Science," (S. E. Harding, A. J. Rowe, and J. C. Horton, eds.), pp. 16–25. Royal Society of Chemistry, Cambridge.

Gilbert, G. A., and Jenkins, R. C. (1959). Sedimentation and electrophoresis of interacting substances. II. Asymptotic boundary shape for two substances interacting reversibly. *Proc. R. Soc. A* **253,** 420–437.

Goodman, T. C., Nagel, L., Rappold, W., Klotz, G., and Riesner, D. (1984). Viroid replication: Equilibrium association constant and comparative activity measurements for the viroid-polymerase interaction. *Nucleic Acids Res.* **12,** 6231–6246.

Heinicke, L. A., Wong, C. J., Lary, J., Nallagatla, S. R., Diegelman-Parente, A., Zheng, X., Cole, J. L., and Bevilacqua, P. C. (2009). RNA dimerization promotes PKR dimerization and activation. *J. Mol. Biol.* **390,** 319–338.

Hellman, L. M., and Fried, M. G. (2007). Electrophoretic mobility shift assay (EMSA) for detecting protein–nucleic acid interactions. *Nat. Protoc.* **2,** 1849–1861.

Jensen, D. E., and von Hippel, P. H. (1977). A boundary sedimentation velocity method for determining nonspecific nucleic acid–protein interaction binding parameters. *Anal. Biochem.* **80,** 267–281.

Kar, S. R., Kingsbury, J. S., Lewis, M. S., Laue, T. M., and Schuck, P. (2000). Analysis of transport experiments using pseudo-absorbance data. *Anal. Biochem.* **285,** 135–142.

Langland, J. O., Cameron, J. M., Heck, M. C., Jancovich, J. K., and Jacobs, B. L. (2006). Inhibition of PKR by RNA and DNA viruses. *Virus Res.* **119,** 100–110.

Lanks, K. W., and Eng, R. K. (1976). Detection of nucleic acid–protein complexes by equilibrium ultracentrifugation. *Res. Commun. Chem. Pathol. Pharmacol.* **15,** 377–380.

Laue, T. M., Shah, B. D., Ridgeway, T. M., and Pelletier, S. L. (1992). *In* "Analytical Ultracentrifugation in Biochemistry and Polymer Science," (S. Harding, A. Rowe, and J. Horton, eds.), pp. 90–125. Royal Society of Chemistry, Cambridge.

Laue, T. M., Senear, D. F., Eaton, S., and Ross, J. B. (1993). 5-hydroxytryptophan as a new intrinsic probe for investigating protein–DNA interactions by analytical ultracentrifugation. Study of the effect of DNA on self-assembly of the bacteriophage lambda cI repressor. *Biochemistry* **32,** 2469–2472.

Launer-Felty, K., Wong, C. J., Wahid, A. M., Conn, G. L., and Cole, J. L. (2010). Magnesium-dependent interaction of PKR with adenovirus VAI RNA. *J. Mol. Biol.* **402,** 638–644.

Lemaire, P. A., Anderson, E., Lary, J., and Cole, J. L. (2008). Mechanism of PKR activation by dsRNA. *J. Mol. Biol.* **381,** 351–360.

Leulliot, N., Quevillon-Cheruel, S., Graille, M., van Tilbeurgh, H., Leeper, T. C., Godin, K. S., Edwards, T. E., Sigurdsson, S. T., Rozenkrants, N., Nagel, R. J., Ares, M., and Varani, G. (2004). A new alpha-helical extension promotes RNA binding by the dsRBD of Rnt1p RNAse III. *EMBO J.* **23,** 2468–2477.

Lewis, M. S., Shrager, R. I., and Kim, S.-J. (1994). *In* "Modern Analytical Ultracentrifugation," (T. M. Shuster and T. M. Laue, eds.), pp. 94–115. Birkhauser, Boston.

Liu, Y., Lei, M., and Samuel, C. E. (2000). Chimeric double-stranded RNA-specific adenosine deaminase ADAR1 proteins reveal functional selectivity of double-stranded RNA binding domains from ADAR1 and protein kinase PKR. *Proc. Natl. Acad. Sci. USA* **97,** 12541–12546.

Lohman, T. M., Wensley, C. G., Cina, J., Burgess, R. R., and Record, M. T., Jr. (1980). Use of difference boundary sedimentation velocity to investigate nonspecific protein–nucleic acid interactions. *Biochemistry* **19,** 3516–3522.

Manche, L., Green, S. R., Schmedt, C., and Mathews, M. B. (1992). Interactions between double-stranded RNA regulators and the protein kinase DAI. *Mol. Cell. Biol.* **12,** 5238–5248.

McKenna, S. A., Lindhout, D. A., Kim, I., Liu, C. W., Gelev, V. M., Wagner, G., and Puglisi, J. D. (2007a). Molecular framework for the activation of RNA-dependent protein kinase. *J. Biol. Chem.* **282,** 11474–11486.

McKenna, S. A., Lindhout, D. A., Shimoike, T., Aitken, C. E., and Puglisi, J. D. (2007b). Viral dsRNA inhibitors prevent self-association and autophosphorylation of PKR. *J. Mol. Biol.* **372,** 103–113.

Minks, M. A., West, D. K., Benvin, S., and Baglioni, C. (1979). Structural requirements of double-stranded RNA for the activation of 2'–5'-Oligo(A) polymerase and protein kinase of interferon-treated HeLa cells. *J. Biol. Chem.* **254,** 10180–10183.

Nagel, R., and Ares, M., Jr. (2000). Substrate recognition by a eukaryotic RNase III: The double-stranded RNA-binding domain of Rnt1p selectively binds RNA containing a 5'-AGNN-3' tetraloop. *RNA* **6,** 1142–1156.

Nanduri, S., Carpick, B. W., Yang, Y., Williams, B. R., and Qin, J. (1998). Structure of the double-stranded RNA binding domain of the protein kinase PKR reveals the molecular basis of its dsRNA-mediated activation. *EMBO J.* **17,** 5458–5465.

Philo, J. S. (2006). Improved methods for fitting sedimentation coefficient distributions derived by time-derivative techniques. *Anal. Biochem.* **354,** 238–246.

Revzin, A., and Woychik, R. P. (1981). Quantitation of the interaction of *Escherichia coli* RNA polymerase holoenzyme with double-helical DNA using a thermodynamically rigorous centrifugation method. *Biochemistry* **20,** 250–256.

Schmedt, C., Green, S. R., Manche, L., Taylor, D. R., Ma, Y., and Mathews, M. B. (1995). Functional characterization of the RNA-binding domain and motif of the double-stranded RNA-dependent protein kinase DAI (PKR). *J. Mol. Biol.* **249,** 29–44.

Schuck, P. (2004). A model for sedimentation in inhomogeneous media. I. Dynamic density gradients from sedimenting co-solutes. *Biophys. Chem.* **108,** 187–200.

Schuck, P. (2010). Some statistical properties of differencing schemes for baseline correction of sedimentation velocity data. *Anal. Biochem.* **401,** 280–287.

Spanggord, R. J., and Beal, P. A. (2001). Selective binding by the RNA binding domain of PKR revealed by affinity cleavage. *Biochemistry* **40,** 4272–4280.

Stafford, W. F. (1992). Boundary analysis in sedimentation transport experiments: A procedure for obtaining sedimentation coefficient distributions using the time derivative of the concentration profile. *Anal. Biochem.* **203,** 295–301.

Stafford, W. F., III (2009). Protein–protein and ligand–protein interactions studied by analytical ultracentrifugation. *Methods Mol. Biol.* **490,** 83–113.

Stafford, W. F., and Sherwood, P. J. (2004). Analysis of heterologous interacting systems by sedimentation velocity: Curve fitting algorithms for estimation of sedimentation coefficients, equilibrium and kinetic constants. *Biophys. Chem.* **108,** 231–243.

Tian, B., Bevilacqua, P. C., Diegelman-Parente, A., and Mathews, M. B. (2004). The double-stranded RNA binding motif: Interference and much more. *Nat. Rev. Mol. Cell Biol.* **5,** 1013–1023.

Toth, A. M., Zhang, P., Das, S., George, C. X., and Samuel, C. E. (2006). Interferon action and the double-stranded RNA-dependent enzymes ADAR1 adenosine deaminase and PKR protein kinase. *Prog. Nucleic Acid Res. Mol. Biol.* **81,** 369–434.

Ucci, J. W., and Cole, J. L. (2004). Global analysis of non-specific protein–nucleic interactions by sedimentation equilibrium. *Biophys. Chem.* **108,** 127–140.

Ucci, J. W., Kobayashi, Y., Choi, G., Alexandrescu, A. T., and Cole, J. L. (2007). Mechanism of interaction of the double-stranded RNA (dsRNA) binding domain of protein kinase R with short dsRNA sequences. *Biochemistry* **46,** 55–65.

VanOudenhove, J., Anderson, E., Krueger, S., and Cole, J. L. (2009). Analysis of PKR structure by small-angle scattering. *J. Mol. Biol.* **387,** 910–920.

Wojtuszewski, K., Hawkins, M. E., Cole, J. L., and Mukerji, I. J. (2001). HU Binding to DNA: Evidence for multiple complex formation and DNA bending. *Biochemistry* **40,** 2588–2598.

Wong, I., and Lohman, T. M. (1993). A double-filter method for nitrocellulose-filter binding: Application to protein–nucleic acid interactions. *Proc. Natl. Acad. Sci. USA* **90,** 5428–5432.

CHAPTER FOUR

Structural and Thermodynamic Analysis of PDZ–Ligand Interactions

Tyson R. Shepherd* *and* Ernesto J. Fuentes*,†

Contents

Abstract

Tiam-family guanine exchange proteins are activators of the Rho GTPase Rac1 and critical for cell morphology, adhesion, migration, and polarity. These modular proteins contain a variety of signaling domains, including a single *p*ostsynaptic density-95/*d*iscs large/*z*onula occludens-1 (PDZ) domain. Here, we show how structural and thermodynamic approaches applied to the Tiam1 PDZ domain can be used to gain unique insights into the affinity and specificity of PDZ–ligand interactions with peptides derived from Syndecan1 and Caspr4 proteins. First, we describe a fluorescence anisotropy-based assay that can be used to determine PDZ–ligand interactions, and describe important considerations in designing binding experiments. Second, we used site-specific mutagenesis in combination with double-mutant cycle analysis to probe the binding energetics and cooperativity of residues in two ligand binding pockets (S_0 and

* Department of Biochemistry, Roy J. and Lucille A. Carver College of Medicine, University of Iowa, Iowa City, Iowa, USA
† Holden Comprehensive Cancer Center, Iowa City, Iowa, USA

Methods in Enzymology, Volume 488
ISSN 0076-6879, DOI: 10.1016/B978-0-12-381268-1.00004-5
© 2011 Elsevier Inc.
All rights reserved.

S_{-2}) that are involved in Tiam1 PDZ–ligand interactions. Peptide ligand binding results and double-mutant cycle analysis revealed that the S_0 pocket was important for Syndecan1 and Caspr4 peptide interactions and that the S_{-2} pocket provided selectivity for the Syndecan1 ligand. Finally, we devised a "peptide evolution" strategy whereby a Model consensus peptide was "evolved" into either the Syndecan1 or Caspr4 peptide by site-directed mutagenesis. These results corroborated the PDZ mutational analysis of the S_0 pocket and identified the P_{-4} position in the ligand as critical for Syndecan1 affinity and selectivity. Together, these studies show that a combined structural and thermodynamic approach is powerful for obtaining insights into the origin of Tiam1 PDZ–ligand domain affinity and specificity.

Abbreviations

CADM1	cell adhesion molecule-1
Caspr4	Contactin-associated protein-like 4
NMR	nuclear magnetic resonance
PDZ	postsynaptic density-95/discs large/zonula occludens-1

1. Introduction

A chief goal in the field of biochemistry is to quantitatively describe protein–protein and protein–ligand interactions, and to determine how the physical parameters that describe these interactions relate to biological function. A common method for obtaining this information is by using a reductionist approach, whereby a large protein is divided into its individual domains, and the function of each is investigated separately. Subsequent studies in the context of the full-length protein can then be used to infer regulatory mechanisms and domain–domain interactions. Eukaryotic signaling proteins are often composed of multiple domains and hence are amenable to this approach. Here, we describe the application of this approach to the guanine nucleotide exchange factor (GEF) protein, T-cell lymphoma invasion and metastasis-1 (Tiam1), to gain insight into the affinity and specificity of its *p*ostsynaptic density-95/*d*iscs large/*z*onula occludens-1 (PDZ) domain (Iden and Collard, 2008; Mertens *et al.*, 2006).

Tiam1 and its homolog Tiam2 are GEF proteins that specifically activate the Rho-family GTPase Rac1 (Mertens *et al.*, 2003). Because Rac1 activation is tightly regulated, GEF proteins themselves must be spatially and temporally controlled. This control occurs primarily through chemical signals such as phosphorylation and protein–protein interactions. Tiam1 and Tiam2 possess similar domain architecture and cellular function (Matsuo *et al.*, 2002, 2003). Both contain a Pleckstrin homology-coiled coil-extra (PH_n-CC-Ex) region, followed by a Ras-binding domain (RBD), a PDZ domain, and a Dbl homology–Pleckstrin homology ($DH–PH_c$) catalytic domain cassette. While most interaction domains in Tiam1 and Tiam2 have high sequence identity, the PDZ domains share only ~28% identity, suggesting that there might be functionally important differences in their specificities (Hoshino *et al.*, 1999).

PDZ domains are small protein–protein interaction domains of ~90 amino acids in length. They fold into a compact β-barrel structure formed by six β-strands and two α-helices and typically bind the 5–10 extreme C-terminal residues of their interaction partner by forming a β-sheet contiguous with the β2 strand of the PDZ domain (Fig. 4.1; Shepherd *et al.*, 2010). Individual pockets within the PDZ domain are used to accommodate the side chains of ligand residues and are the source of the exquisite specificity of PDZ–ligand interactions. This specificity lends itself to the scaffolding and signaling function of PDZ domains, which are abundant in eukaryotes and prokaryotes (Ponting *et al.*, 1997). For example, the human genome encodes an estimated ~214 PDZ-containing proteins, which collectively harbor ~440 PDZ domains (Schultz *et al.*, 2000; SMART database, June 2010). To date, the structures of nearly 200 PDZ domains have been solved, either alone or in complex with a ligand (reviewed by Lee and Zheng, 2010). In addition, several studies investigating the specificity of PDZ–ligand interactions have uncovered the general rules that define PDZ–ligand specificity (Chen *et al.*, 2008; Songyang *et al.*, 1997; Stiffler *et al.*, 2007; Tonikian *et al.*, 2008). Despite this knowledge, gaps in our understanding of the thermodynamics of PDZ–ligand interactions remain. Here, we highlight structural and thermodynamic methods for studying PDZ–ligand affinity and specificity using the Tiam1 PDZ domain as a model system.

2. Structural Studies of the Tiam1 PDZ Domain

Several studies have identified peptides that bind the Tiam1 PDZ domain. Early studies by Songyang *et al.* (1997) identified a synthetic peptide capable of binding the Tiam1 PDZ domain. This peptide ligand ($SSRKEYYA_{COOH}$, henceforth referred to as the Model peptide) was shown to bind the Tiam1 PDZ domain with low affinity (K_d ~ 112 μM). More recently, the cell–cell and cell–matrix adhesion receptors Syndecan1 and CADM1 were identified as physiological Tiam1 PDZ domain binding proteins (Masuda *et al.*, 2010; Shepherd *et al.*, 2010). In addition, bioinformatic

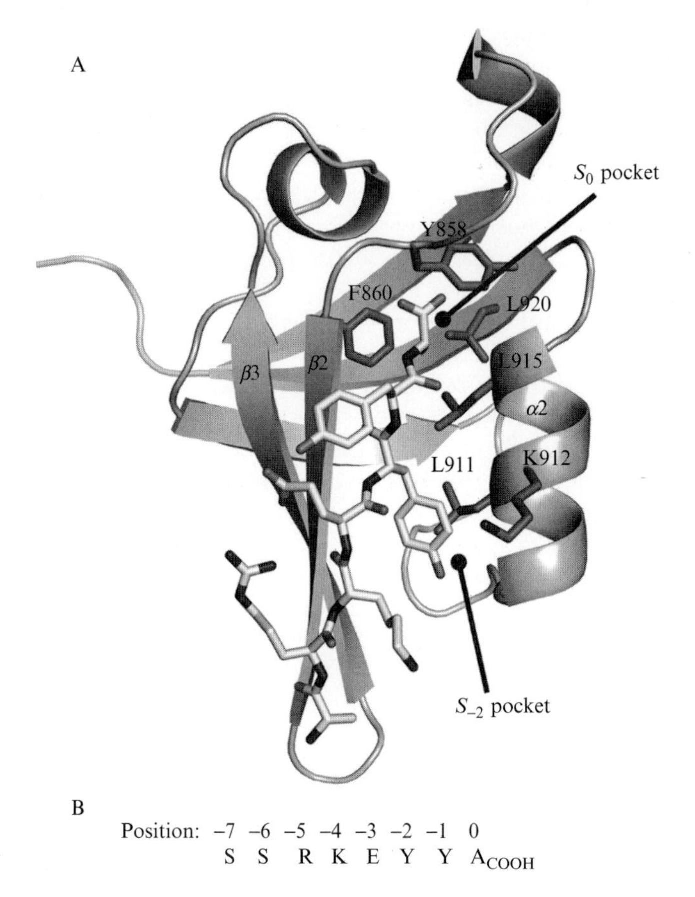

B

Position: −7 −6 −5 −4 −3 −2 −1 0
S S R K E Y Y A_{COOH}

Figure 4.1 The structure of the Tiam1 PDZ domain bound to a model consensus peptide. (A) A cartoon representation of the PDZ/Model structure (chain B, PDB code 3KZE) is shown. The secondary structure of the PDZ domain is shown in gray and the Model peptide is colored in yellow. The specificity pockets of the PDZ domain (S_0 and S_{-2}) are labeled. Residues discussed in the text are colored red. (B) The sequence of the Model peptide used in the crystal structure determination is shown with the ligand position (P_x) shown above each residue (Shepherd *et al.*, 2010). (See Color Insert.)

analysis identified several putative Tiam-family PDZ-binding proteins, including the neuronal cell–cell receptor proteins Contactin-associated protein-like 4 (Caspr4) and Neurexin1 (Shepherd, Hard, Murray, Pei and Fuentes, unpublished data). The C-terminal sequences of these proteins are similar but contain differences at key positions that result in distinct PDZ-binding properties. For instance, the Syndecan1 ($TKQEEFYA_{COOH}$) and Caspr4 ($ENQKEYFF_{COOH}$) peptides bound the Tiam1 PDZ domain, but the Neurexin1 ($NKDKEYYV_{COOH}$) peptide did not. Subsequent studies with the Tiam2 PDZ domain indicated a different pattern of specificity, where the Caspr4 and Neurexin1 peptides bound the PDZ domain tightly, but the Syndecan1 peptide did not (Shepherd, Hard, Murray, Pei and Fuentes, unpublished data).

We determined the crystal structure of the Tiam1 PDZ domain bound to the Model peptide ligand to gain insight into the molecular details of Tiam-family PDZ/ligand specificity. In particular, two specificity pockets, S_0 and S_{-2}, were identified (Fig. 4.1A). The side chains of residues Y858, F860, L915, and L920 form the S_0 pocket that accommodates the C-terminal residue of the ligand. This pocket is relatively shallow, partly explaining the preference for an Ala residue at the P_0 position of the ligand (where P_0 is the most C-terminal residue and P_{-n} denotes the residue position n amino acids from the C-terminus). The structure also showed how the S_{-2} pocket can accommodate a large hydrophobic side chain in the ligand at P_{-2} and implicated residues L911 and K912 as determinants of ligand specificity (Fig. 4.1).

To understand how the Tiam1 PDZ domain achieves specificity for several distinct peptide ligands, we performed NMR-based titration experiments (Shepherd *et al.*, 2010). NMR spectroscopy is a powerful technique because it can provide site-specific information regarding the structure, dynamics, and energetics of protein–ligand interactions. In the titration experiments, we added either the Caspr4 or Syndecan1 peptide to a sample containing the U–^{15}N-labeled PDZ domain of Tiam1 and then acquired a ^{1}H–^{15}N HSQC spectrum at each ligand concentration. Because the position, or chemical shift, of each peak within the HSQC spectrum reflects the chemical environment surrounding that amide group, it is possible to determine which residues participate in the binding reaction by monitoring the changes in the chemical shift of individual cross-peaks as a function of peptide ligand. Figure 4.2 summarizes the titration data for Caspr4 and Syndecan1 peptides, highlighting differences in chemical shifts between these two PDZ–ligand complexes (Shepherd *et al.*, 2010). Figure 4.2A shows the chemical shift changes of each complex mapped onto the surface of the Tiam1 PDZ/Model structure. It is clear that both ligands recognize the same binding cleft. However, the data also identify significant differences in α2 and β6 that likely reflect interactions responsible for binding specificity. In particular, L911, K912, L915, and L920 in the S_0 and S_{-2} pockets were identified as candidate residues that might determine the specificity of the Tiam1 PDZ–ligand interaction (Figs. 4.1 and 4.2). Interestingly, none of these four residues are conserved between the Tiam1 and Tiam2 PDZ domains (Fig. 4.2C), consistent with the notion that these two homologous PDZ domains may have distinct specificities. Here, we focus on the Tiam1 PDZ domain and show how thermodynamic analyses can be used to assess the role of PDZ and ligand residues hypothesized to be important in PDZ–ligand interactions.

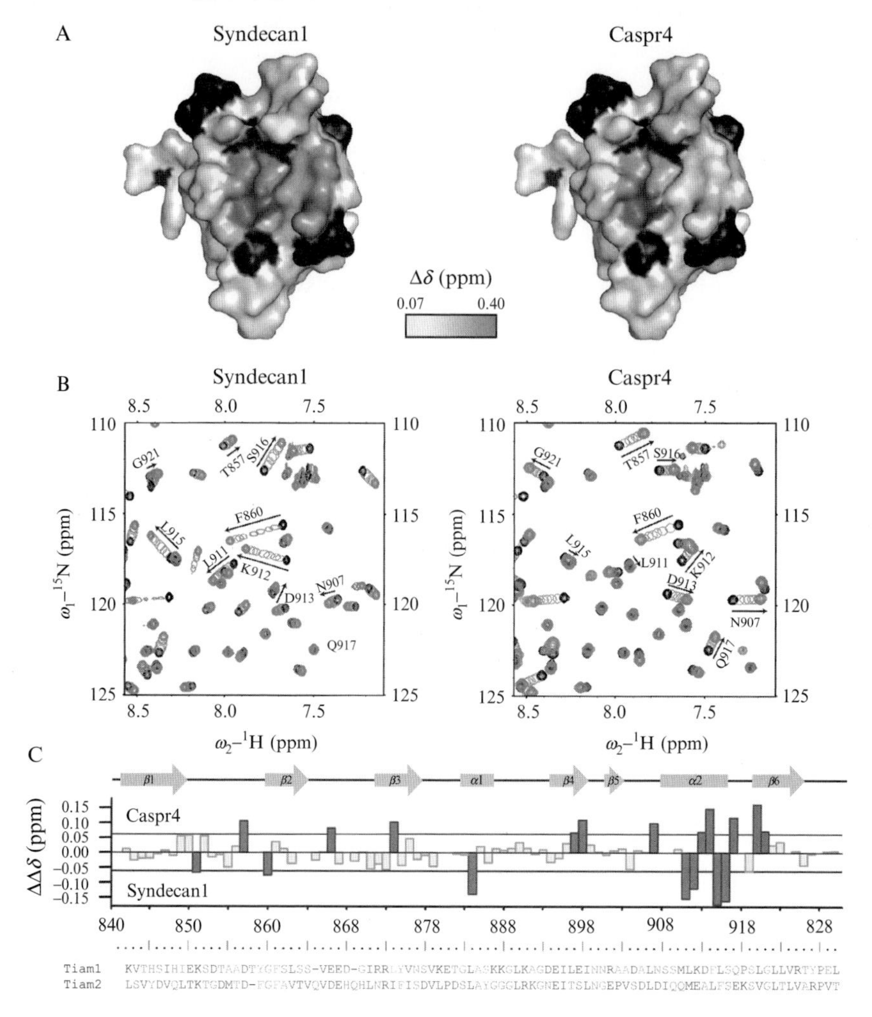

Figure 4.2 Structural analysis of the Tiam1 PDZ domain identifies residues important for specificity. (A) The surface representation of the Tiam1 PDZ domain structure is color coded to indicate the extent of chemical shift changes [$\Delta\delta(^1\text{H},^{15}\text{N}) = (\Delta\delta(^1\text{H})^2 + (0.4\ \Delta\delta(^{15}\text{N}))^2)^{1/2}$] upon titration with Syndecan1 or Caspr4 peptides. Residues shown in black could not be assigned and those in gray had a change in chemical shift that was less than the global average. Residues that had a significant change in chemical shift (greater than the global average) are colored continuously from yellow to red, where red indicates maximal changes. (B) An expanded view of ^{15}N–HSQC spectra obtained during a titration series for the Tiam1 PDZ domain with Syndecan1 and Caspr4 peptides. Labeled residues are implicated in Tiam1 PDZ specificity. (C) A histogram plot summarizing the chemical shift changes per residue in the Tiam1 PDZ domain upon titration with Syndecan1 (lower bars) and Caspr4 (upper bars). The value of each bar represents the absolute difference in chemical shift between the two complexes [$\Delta\Delta\delta = |\Delta\delta(\text{Caspr4})| - |\Delta\delta(\text{Syndecan1})|$]. A value of zero indicates that changes in chemical shift in the two complexes were identical in magnitude. Differences indicate unique chemical shifts changes in either complex. Residues in red underwent changes greater than 1σ from the average, while those in yellow underwent changes of less than 1σ from the average. Data taken from (Shepherd *et al.*, 2010). (See Color Insert.)

3. Fluorescence Anisotropy Methods for Measuring the Energetics of PDZ–Ligand Interactions

Several techniques are available for determining the energetics of protein–protein and protein–ligand interactions, including isothermal titration calorimetry, surface plasmon resonance, NMR, and fluorescence-based methods. Each technique has its relative strengths and weaknesses and can potentially contribute unique information. Fluorescence-based methods rely on measuring the change in fluorescence intensity or anisotropy that occurs upon ligand binding. This method is sensitive, can be carried out relatively quickly, requires only modest amounts of protein, and is highly reproducible. In this section, we describe a fluorescence anisotropy binding assay for studying PDZ–ligand interactions.

3.1. Theoretical background

Fluorescence anisotropy has been extensively reviewed elsewhere (Eccleston *et al.*, 2005; Eftink, 1997; Lakowicz, 1999; Royer and Scarlata, 2008); here we provide a basic overview to introduce key aspects that are important for experimental design. Fluorescence anisotropy relies on selectively exciting a subpopulation of fluorophores with polarized light and monitoring polarized emission. Because a fluorophore has a defined excited-state lifetime that is on a timescale (ns) similar to that which molecules tumble (i.e., experience rotational diffusion), the polarity of the emitted light is sensitive to molecular size. This property allows one to monitor a protein/ligand binding reaction based on accompanying changes in polarization (or anisotropy).

The fundamental equation describing fluorescence anisotropy (r) is

$$r = \frac{\left(I_{||} - I_{\perp}\right)}{\left(I_{||} + 2I_{\perp}\right)} = \frac{I_{||}/I_{\perp} - 1}{I_{||}/I_{\perp} + 2} \tag{4.1}$$

where $I_{||}$ is the intensity of the detected light when the excitation and emission polarization is parallel and $I_{\perp}$ is the intensity of detected light when the excitation and emission polarization is perpendicular. Thus, fluorescence anisotropy is the ratio of the difference in intensity between the emitted parallel and perpendicular polarized light ($I_{||} - I_{\perp}$) to the total intensity of polarized light emitted by the sample ($I_{||} + 2I_{\perp}$) (Lakowicz, 1999), and is consequently dimensionless and independent of the concentration of the fluorophore. In the experimental setting, we use a laboratory reference frame, where I_{VV} is defined as the intensity of light when both the excitation and emission polarizers are mounted vertically, and I_{VH} is the intensity of light when the excitation polarizer is mounted vertically and the emission polarizer is mounted horizontally. Because the experimental sensitivity in these channels may not be identical, a correction factor G (or G-factor) is introduced

$$\frac{I_{VV}}{I_{VH}} = G\frac{I_{\|}}{I_{\perp}} \tag{4.2}$$

Equation (4.2) alone does not uniquely determine the G-factor because of its dependence on the intensity of $I_{\|}$ and $I_{\perp}$; therefore, we must rely on an alternate experimental configuration. With horizontally polarized light, both the excitation and emission components of the excited-state distribution are equal and proportional to $I_{\perp}$ (Lakowicz, 1999). Thus, measured changes between horizontally (I_{HH}) and vertically (I_{HV}) polarized light in the emission path can be used to experimentally determine the G-factor as defined by

$$\frac{I_{HV}}{I_{HH}} = G\frac{I_{\perp}}{I_{\perp}} = G \tag{4.3}$$

Combining Eqs. (4.1) and (4.2) yields the fluorescence anisotropy in measurable quantities

$$r = \frac{(I_{VV}/I_{VH})G - 1}{(I_{VV}/I_{VH})G + 2} = \frac{I_{VV} - GI_{VH}}{I_{VV} + 2GI_{VH}} \tag{4.4}$$

Furthermore, the Perrin equation (Eq. 4.5) (Perrin, 1926) relates fluorescence anisotropy to the rotational correlation time (τ_c) of the labeled molecule and the fluorescence lifetime of the fluorophore (τ)

$$r = \frac{r_0}{\left(1 + \frac{\tau}{\tau_c}\right)} \tag{4.5}$$

where r_0 is the upper limit of anisotropy when the flourophore is "frozen in" (i.e., no motion during the excited-state lifetime). In the approximation of a spherical molecule, the rotational correlation time (τ_c) is related to the molecular weight (M) of the molecule of interest according to

$$\tau_c = \frac{\eta M \bar{v}_h}{RT} \tag{4.6}$$

where η is the solution viscosity, $\bar{v}_h$ the specific volume of a hydrated protein, R the gas constant, and T is the absolute temperature. By combining and rearranging Eqs. (4.5) and (4.6), we have

$$r = \frac{r_0 \eta M \bar{v}_h}{\eta M \bar{v}_h + \tau RT} \tag{4.7}$$

This equation shows that anisotropy increases hyperbolically with molecular mass of the fluorophore-bound molecule (Fig. 4.3; Lakowicz, 1999). Hence, a binding reaction that has a reasonable change in mass (or τ_c) between the free and bound states can be detected by changes in anisotropy. Equation (4.7) is dependent on viscosity and temperature; therefore, these parameters should be controlled during the titration experiment. Figure 4.3 also shows simulated curves for fluorescence lifetimes of 4 and 13 ns that indicate that relatively small changes in the fluorophore lifetime can significantly affect the absolute anisotropy up to a value of ~0.1 (Fig. 4.3, inset). Moreover, these simulations indicate that analyzing PDZ–ligand interactions (changes in molecular weight from ~1 to 12 kDa upon peptide binding) requires a fluorophore with a fluorescence lifetime (τ) on the order of ~1–4 ns to obtain a significant change in anisotropy between the free and bound state. A few assumptions are implicit in Eq. (4.7) as applied to protein–ligand

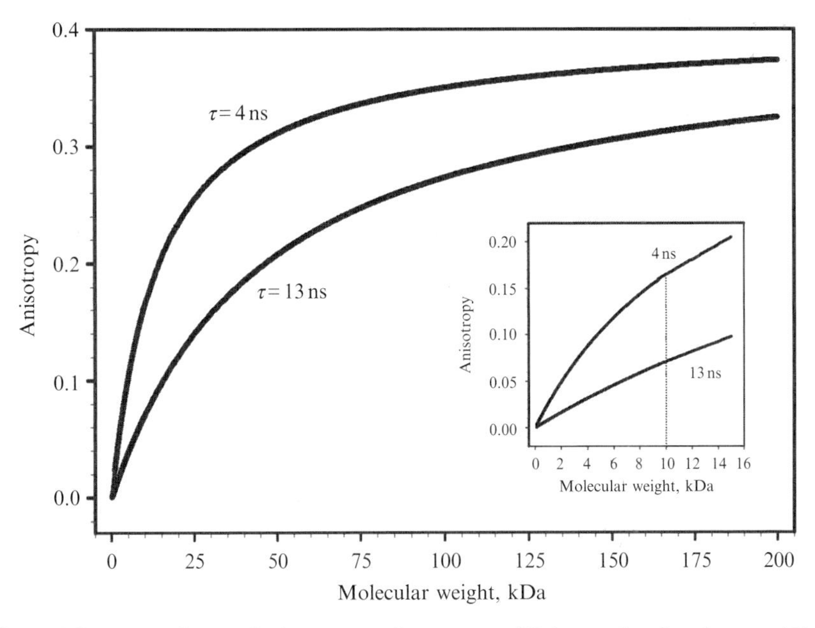

Figure 4.3 Dependence of anisotropy on fluorescence lifetime and molecular mass. The fluorescence anisotropy was simulated using Eq. (4.7) for two fluorescence lifetimes (4 and 13 ns). The parameters used in the simulation where $r_0 = 0.4$, $\nu_h = 0.96$ cm^3/g, $R = 8.314$ cm^3 MPa K^{-1} mol^{-1}, $T = 298.15$ K, and $\eta = 0.94 \times 10^{-6}$ MPa s (Lakowicz, 1999). Data were simulated using molecular weights ranging from 0.1 to 200 kDa, and the inset expands the region from 0.1 to 16 kDa. For reference, a dotted line is placed at 10 kDa, the approximate molecular weight of an individual PDZ domain.

interactions. First, it is assumed that the fluorescence lifetime or quantum yield does not appreciably change upon binding ligand. Second, it is assumed that the fluorophore does not tumble independently of the ligand to which it is bound. If these conditions are met, it is possible to use changes in anisotropy to monitor the fraction bound of a protein–ligand complex.

In a typical experiment, the solution of fluorophore-labeled molecule is contained within a quartz cuvette. The *G*-factor is determined using Eq. (4.3), by exciting the sample with horizontally polarized light and detecting the intensity of the polarized light that is emitted horizontally (I_{HH}) and vertically (I_{HV}). The nonfluorescing titrant (the PDZ domain in the present case) is then added to the analyte (peptide) solution, and the entire sample is exposed to vertically polarized excitation light, while horizontally and vertically emitted light is detected at each titration step. The anisotropy at each titration step is then calculated using Eq. (4.4). Plotting the data as the concentration of the titrant versus the change in fluorescence anisotropy provides a binding curve that can be fit to a hyperbolic binding model (Eq. 4.8)

$$r - r_i = \frac{(r_{max} - r_i)[PDZ]}{K_d + [PDZ]} \tag{4.8}$$

where r is the anisotropy at each titration step, r_i the initial anisotropy of analyte alone, r_{max} the maximum anisotropy when the binding is saturated, K_d the dissociation constant, and [PDZ] is the total concentration of the PDZ domain protein in the analyte solution.

In the studies described here, nonlinear regression analysis was used to fit the fluorescence anisotropy data. Using the known quantities ($r - r_i$) and [PDZ], ($r_{max} - r_i$) and K_d were fit to Eq. (4.8) (Sigma Plot; SPPS Inc.). Equation (4.8) is valid only if several conditions are met. First, a 1:1 stoichiometry is assumed, and in this case consistent with the crystal structure of the Tiam1 PDZ/Model complex (Shepherd *et al.*, 2010). Second, we assume that the concentration of free PDZ domain is on the order of the concentration of the total PDZ domain. Finally, significant changes in the lifetime or intensity of the fluorophore in the bound state require that a correction factor be applied (Eftink, 2000; Mocz *et al.*, 1998). In our studies, we did not see a significant change in fluorescence intensity ($<10\%$) upon PDZ binding and no corrections were applied.

3.2. Experimental design considerations

To maximize the change in mass and hence the anisotropy upon binding, we chose to fluorescently label the peptide and titrate this species with the larger PDZ domain. Additional considerations for the design of fluorescence anisotropy-based binding assays are also required. First, one must select the fluorophore to be used. While measuring intrinsic fluorescence using tryptophan or tyrosine may be an option, their relatively low quantum yield and scarcity within typical PDZ ligands may be limiting. Several extrinsic fluorophores—such as dansyl, rhodamine, and fluorescein—are useful because of their short lifetimes (1–4 ns). Other fluorophores can be considered, but their quantum yield and fluorescence lifetime should be used to guide the choice for particular applications (Owicki, 2000). The characteristic excitation and emission wavelengths of each fluorophore may also influence this decision, because the individual capability of fluorimeters is variable. Other parameters of the fluorophore, such as solubility, as well as sensitivity to changes in pH, salt concentration, and photo bleaching, should also be considered. Another consideration is the position of the fluorophore linkage to the peptide. In PDZ–ligand studies, peptide labeling is restricted to the N-terminus because of the necessity for an unmodified carboxyl terminus in PDZ–ligand interactions.

The second major experimental consideration is the source of the peptide ligand. One possibility is to use recombinant methods to express the peptide using any of the myriad of available biological overexpression systems. However, this can be laborious and inefficient. The convenience and availability of custom peptide synthesis makes it a highly cost-effective alternative for obtaining small peptides, and provides the opportunity to chemically link an extrinsic fluorescent chromophore.

A third experimental design consideration is the length of the peptide. With recent reports of PDZ–ligand interactions involving up to P_{-10} in the ligand (Feng *et al.*, 2008; Tyler *et al.*, 2010), it may be necessary to test several peptides of differing length to optimize ligand design. Too short of a peptide may result in strong fluorophore–PDZ interactions and too long of a peptide can result in a fluorophore that tumbles independently of the peptide, artificially lowering the experimental anisotropy.

For our studies, we chose to have all peptides commercially synthesized as 8-residue peptides modified at their N-termini with the dansyl fluorophore. The choice of an 8-residue peptide was guided by the crystal structure of the Tiam1 PDZ/Model peptide, which showed no interactions beyond position P_{-5} (Shepherd *et al.*, 2010). The choice of the dansyl fluorophore was made in part because this moiety could be readily attached to the amino-termini of the synthetic peptides. Moreover, this fluorophore is pH-independent from pH 5–8 and has a reasonably short excited-state fluorescence lifetime ($\sim$4 ns) in aqueous solutions (Hoenes *et al.*, 1986). Finally, the relatively small size of the dansyl fluorophore minimizes the potential that it will interact with the PDZ domain.

3.3. Experimental procedures

3.3.1. Synthetic peptides

To minimize any contaminating source of background fluorescence, we used reagents with the highest optical purity available. Peptides were commercially synthesized, labeled with an N-terminal dansyl moiety, and judged $\geq$95%

pure by analytical HPLC (GenScript USA, Inc.). The identity of each peptide was confirmed by mass spectrometry (GenScript USA, Inc.). Stock peptide solutions were made by resuspending lyophilized peptides in binding buffer (20 m*M* PO_4, 50 m*M* NaCl, and 0.5 m*M* EDTA at pH 6.8) at a concentration of 1 m*M*. Special attention was paid to adjusting the pH of the peptide solution because trace acid may remain from the original HPLC purification. In our studies, the buffer was chosen to match the buffer system used in the NMR studies. However, in general, care should be taken in the choice of buffer to ensure that precipitation of both the protein and peptide is minimized and that the fluorophore is not quenched by buffer components. Finally, all buffers should be filtered and thoroughly degassed to minimize light scattering from aggregates and/or gas (bubbles), which interfere with anisotropy measurements. The concentration of each dansylated peptide was calculated based on A_{280} measurements and the molar extinction coefficient of the peptide, which was estimated by summing the contributions of the dansyl (1596 M^{-1} cm^{-1}) moiety and peptide at this wavelength. Concentrated stock peptides and working dilutions were stored in light-resistant tubes to avoid photo bleaching and were frozen at −20 or −80 °C for long-term storage.

3.3.2. Protein expression and purification

The Tiam1 PDZ domain expression plasmid was a modified pET21a (Novagen) vector that contained an N-terminal 6× histidine tag and tobacco etch virus (rTEV) protease cleavage site (Shepherd *et al.*, 2010). All protein expression was conducted in BL21(DE3) (Invitrogen) *Escherichia coli* cells. *E. coli* cells were grown at 37 °C in Luria–Bertani medium supplemented with ampicillin (100 μg/mL) under vigorous agitation until an A_{600} of ∼0.6–1.0 was reached. Cultures were cooled to 25 °C and protein expression was induced by the addition of isopropyl 1-thio-D-galactopyranoside to a final concentration of 1 m*M*. Induced cells were incubated for an additional 6–8 h at 25 °C and harvested by centrifugation.

The histidine-tagged Tiam1 PDZ domain was purified by nickel-chelate (GE-Healthcare) and size-exclusion chromatography (G-50 or S-75). The rTEV protease was used to remove the N-terminal His_6 affinity tag. The digested PDZ domain was isolated from undigested fusion protein, cleaved His_6 tag, and histidine-tagged rTEV by nickel-chelate chromatography by collecting the flow through fractions. The concentration of all PDZ proteins was calculated based on the predicted extinction coefficient determined using the program SEDNTERP (v1.09). The final yield of pure PDZ protein was ∼20 mg/L of culture. The protein purity was ∼95% as judged by SDS-PAGE. Samples were used immediately or stored at −20 °C.

3.3.3. Equilibrium fluorescence binding assays

All binding experiments were conducted in 1.3 mL of binding buffer containing peptide at a concentration of either 2 or 5 μ*M*. Measurements were made in a 2-mL quartz cuvette that was stirred and maintained at constant temperature

(25 °C). The anisotropy measurements were recorded on a Fluorolog-3 (Horiba Jobin Yvon, NJ) spectrofluorimeter with polarizers in excitation and emission channels. The maximum excitation and emission wavelength for each dansylated peptide was determined by obtaining a preliminary fluorescence spectrum. The individual peptides had nearly identical excitation and emission wavelengths ($\lambda_{ex} = 340$ and $\lambda_{em} = 555$ nm) and these were used for all fluorescence measurements. The slit widths for the control of excitation and emission intensity were adjusted to optimize the signal-to-noise ratio and maximum intensity and set to 3 and 8 nm, respectively. To allow convenient and reliable delivery of at least 3 μL of solution per titration step, 1:100 and 1:10 dilutions of the stock PDZ protein solution (~1 m*M*) were prepared in binding buffer. For each experiment, 20–30 individual titration steps were performed until the sample had little or no change in anisotropy. The change in fluorescence anisotropy was plotted against protein concentration and fit to a standard hyperbolic ligand-binding curve (Eq. 4.8). Each titration was carried out in triplicate, and the reported values in Table 4.1 are the average K_d and standard deviation of the mean. The change in free energy (ΔG_b) of the PDZ–ligand binding was calculated from the dissociation constant and the error in free energy was obtained by error propagation.

4. Double-Mutant Cycle Analysis of PDZ-Binding Pockets

The use of double-mutant cycles was pioneered by Fersht and colleagues (Carter *et al.*, 1984; Horovitz and Fersht, 1990) and has been applied to many proteins, including PDZ domains (Saro *et al.*, 2007). This technique measures the coupling between two distinct perturbations in a system by comparing the thermodynamics of each perturbation individually and then together. In proteins, the most common perturbation is a site-specific mutation and the monitored thermodynamic process is generally protein folding or ligand binding (Fersht, 1998). In studying PDZ–ligand binding events, we begin by measuring the free energy of binding for the wild-type protein (ΔG_{WT}), two single mutants (ΔG_{M1}, ΔG_{M2}), and a double mutant (ΔG_{DM}) that combines the mutations present in the two single mutants. Cooperativity or coupling is assessed by comparing the sum of the binding free energies of the single mutants (ΔG_{M1} and ΔG_{M2}) to those of the wild-type and double mutant (ΔG_{WT} and ΔG_{DM}), or equivalently by evaluating two legs of a thermodynamic box (Eq. (4.9)).

$$\begin{aligned}\Delta\Delta\Delta G_{int} &= \Delta\Delta G_{WT-M1} - \Delta\Delta G_{M2-DM} \\ &= (\Delta G_{WT} + \Delta G_{DM}) - (\Delta G_{M1} - \Delta G_{M2})\end{aligned} \tag{4.9}$$

Table 4.1 Wild-type Tiam1 PDZ domain binding and double-mutant cycles with evolved peptides

Peptide	Sequence	K_d (μM)	ΔG_b[a]	$\Delta\Delta G_b$[b]	$\Delta\Delta\Delta G_{int}$	$\sum$ Singles
Model peptide	SSRKEYYA	112 ± 5[c]	-5.38 ± 0.05			
Core peptide	SSQKEYYA	119 ± 6	-5.35 ± 0.05			
$K_{P-4}E$–Sdc1	SSQ**E**EYYA	40 ± 2	-5.99 ± 0.06	-0.65 ± 0.07		
$Y_{P-2}F$–Sdc1	SSQKE**F**YA	106 ± 7	-5.42 ± 0.07	-0.07 ± 0.08		
Syndecan1	TKQ**EEF**YA	26.9 ± 0.9^2	-6.23 ± 0.03	-0.88 ± 0.06	-0.17 ± 0.10	-0.71 ± 0.11
$Y_{P-1}F$–Caspr4	SSQKEY**F**A	60.1 ± 0.6	-5.75 ± 0.01	-0.40 ± 0.05		
$A_{P0}F$–Caspr4	SSQKEYY**F**	89 ± 5	-5.52 ± 0.06	-0.17 ± 0.08		
Caspr4	ENQKEY**FF**	19.0 ± 0.4	-6.43 ± 0.02	-1.09 ± 0.05	-0.51 ± 0.08	-0.57 ± 0.09

b, binding; int, interaction.

[a] units of kcal/mol.

[b] $\Delta\Delta G_b = \Delta G_b(\text{Core}) - \Delta G_b(\text{peptide})$.

[c] Data taken from Shepherd *et al.* (2010).

Thus, if the difference in the binding free energies between the first mutant and the wild-type ($\Delta\Delta G_{WT-M1} = \Delta G_{WT} - \Delta G_{M1}$) is equivalent to the difference between the second mutant and the double mutant ($\Delta\Delta G_{M2-DM} = \Delta G_{M2} - \Delta G_{DM}$), then the energy of interaction ($\Delta\Delta\Delta G_{int}$) is additive and no coupling is observed. However, if $\Delta\Delta\Delta G_{int}$ is nonzero, then the two mutations are energetically coupled (or cooperative), and together they influence ligand binding to a greater extent than either one alone. The sign of the coupling is important and indicates whether the cooperativity enhances ($\Delta\Delta\Delta G_{int} < 0$) or reduces ($\Delta\Delta\Delta G_{int} > 0$) binding compared to that of the single mutants.

Double-mutant cycle analysis was applied to the Tiam1 PDZ domain to gain insight into the energetics of ligand binding and PDZ specificity. Specifically, we were interested in determining whether residues within the S_0 (L915 and L920) and S_{-2} (L911 and K912) binding pockets act cooperatively in recognizing ligands. We probed the importance of these residues by using site-directed mutagenesis to change each residue to the corresponding amino acid in the Tiam2 PDZ domain (see Fig. 4.1C) and subjecting the purified protein to a fluorescence anisotropy ligand-binding assay, as described in Section 3.3. Double-mutant cycle analysis indicates that residues L915 and L920 in the S_0 pocket are coupled in the context of binding to both Syndecan1 and Caspr4 (Shepherd, Hard, Murray, Pei and Fuentes, unpublished data). In contrast, residues L911 and K912 were variably coupled depending on the sequence of the ligand: they were not coupled in binding to Syndecan1, as the total change in energy was determined by the single mutant K912E. However, residues L911 and K912 were clearly coupled when binding to the Caspr4 peptide (Shepherd, Hard, Murray, Pei and Fuentes, unpublished data). In this case, the L911M mutation counteracted the negative effect on affinity of the K912E mutation. These data indicate that residues within each pocket work coordinately to modulate and fine-tune the affinity and exquisite selectivity of the PDZ domains.

5. Peptide Evolution as a Tool for Probing PDZ Specificity

5.1. The peptide evolution strategy

The Model peptide ($SSRKEYYA_{COOH}$) used in our structural analysis of the Tiam1 PDZ served as a valuable tool for understanding PDZ–ligand interactions. Interestingly, this peptide has high sequence identity to both the Syndecan1 and Caspr4 peptides. We took advantage of this fact and devised a "peptide evolution" strategy to convert the Model peptide into the Syndecan1 or Caspr4 ligand while assessing the energetic consequences for PDZ binding at each step in the evolution. As depicted in Fig. 4.4, all

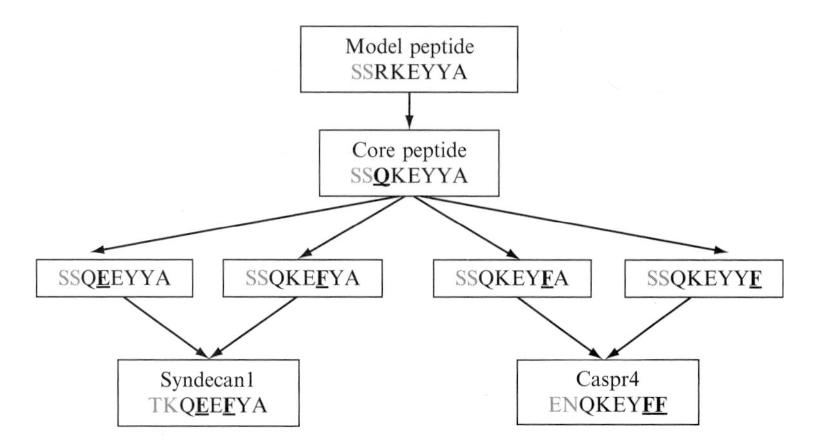

Figure 4.4 A schematic of the peptide evolution strategy. The path for each evolved peptide is shown with the mutated residue(s) underlined and highlighted in bold.

three peptides have the identical core sequence from P_0 to P_{-4}, while the Model ligand contains an Arg at P_{-5}. By mutating the −5 position from Arg to Gln ($R_{P-5}Q$), we created a peptide with the sequence **Q**KEYYA-$_{COOH}$, which corresponds to the "Core" sequence of the evolved peptides. This change had only a small effect on the dissociation constant measured for the Tiam1 PDZ domain compared to that previously reported for the Model peptide (Table 4.1). Conveniently, the Core peptide can be changed to either the Caspr4 or Syndecan1 peptide (Fig. 4.4) by two distinct mutations. Mutating the Core peptide at P_{-2} from Tyr to Phe ($Y_{P-2}F$–Sdc1) and at P_{-4} from Lys to Glu ($K_{P-4}E$–Sdc1) yields a peptide whose final six C-terminal residues are identical to those in the Syndecan1 peptide (Q**E**E**F**YA$_{COOH}$), whereas mutating both P_0 and P_{-1} to Phe ($A_{P0}F$–Caspr4 and $Y_{P-1}F$–Caspr4) yields a peptide with a sequence whose last six amino acids are identical to those of the Caspr4 peptide (QKEY**FF**$_{COOH}$) (Fig. 4.4). The residues at positions P_{-6} and P_{-7} were not considered further because no electron density was evident for these residues in the Tiam1 PDZ/Model peptide crystal structure, suggesting that they do not interact with the PDZ domain.

5.2. Double-mutant cycle analysis of evolved peptides

The Core peptide and each peptide in the evolution path were synthesized with an N-terminal dansyl adduct to facilitate fluorescence anisotropy measurements. The free energy of binding for each peptide was determined, as outlined in Section 3.3. Figure 4.5A and B shows representative binding curves for each peptide. Because the peptide evolution strategy incorporates

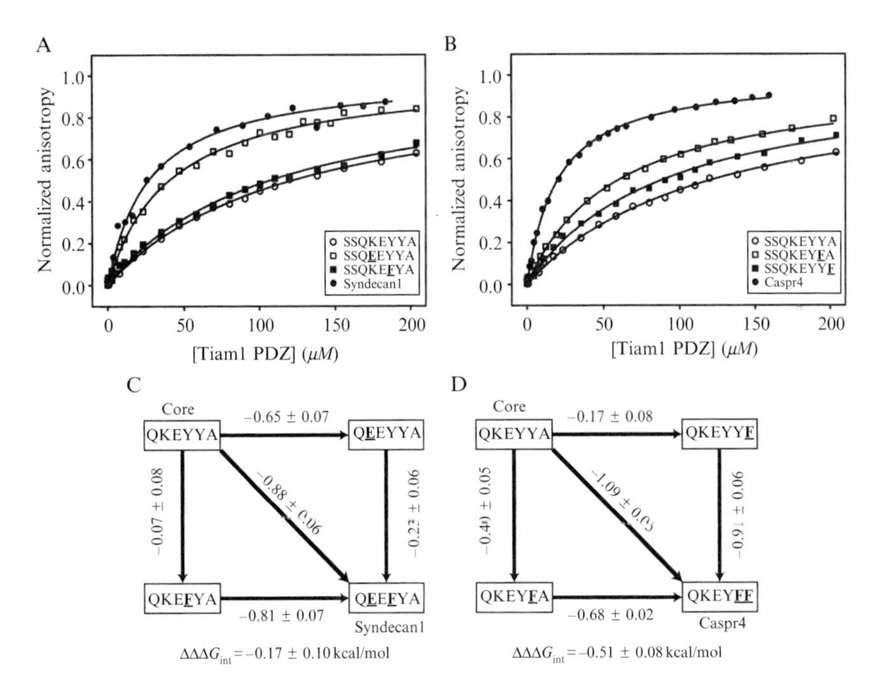

Figure 4.5 Fluorescence titration curves and thermodynamic cycles for binding of the Tiam1 PDZ domain to evolved peptides. Binding curves for peptides that had been converted to the (A) Syndecan1 or (B) Caspr4 sequences via the evolution approach. (C) A summary of the binding energetics for peptides evolved into Syndecan1 (P_{-2} and P_{-4} ligand positions). (D) A summary of the binding energetics for peptides evolved into the Caspr4 peptide (P_{-0} and P_{-1} ligand positions).

single and double mutants, we constructed double-mutant thermodynamic cycles to assess whether the individual residue changes lead to cooperativity in binding the Tiam1 PDZ domain.

The fitted binding parameters for each titration series are summarized in Table 4.1. The $Y_{P-2}F$–Sdc1 peptide was found to have very little effect on binding to the PDZ domain (Fig. 4.5A and Table 4.1). However, the binding affinity of the $K_{P-4}E$–Sdc1 peptide was two-fold higher compared to that of the Core peptide (Fig. 4.5A), indicating that a significant amount of the binding energy comes from this residue. We constructed a double-mutant cycle from the binding results of these two mutant peptides and the Syndecan1 peptide (Shepherd *et al.*, 2010; Table 4.1 and Fig. 4.5C). Analysis of these data indicates that P_{-2} and P_{-4} in the Syndecan1 peptide are not cooperative ($\Delta\Delta\Delta G_{\text{int}} = 0.17 \pm 0.13$ kcal/mol).

Our examination of the energetics of binding (Table 4.1) indicates that the affinity of the Syndecan1 peptide is mainly determined by position −4 in the ligand. Consistent with this, examination of the Tiam1 PDZ/Model structure (Fig. 4.1) suggests that the side chain of residue K912 of the S_{-2} pocket may interact with the Glu side chain at P_{-4} of the Syndecan1 ligand. Indeed, the affinity of the K912E (S_{-2} pocket) PDZ mutant for the Syndecan1 peptide was reduced and near the levels of the affinities of the Model and Core peptides for the wild-type PDZ domain (Shepherd, Hard, Murray, Pei and Fuentes, unpublished). Moreover, the difference in binding free energy ($\Delta\Delta G_b$) for the K912E PDZ/Syndecan1 compared to wild-type PDZ/Syndecan1 was very similar to that for the wild-type PDZ/K_{P-4}E–Sdc1 versus the Core peptide. Thus, ~1 kcal/mol of binding energy results from this Lys/Glu ion pair, and changes in the Lys/Glu ion pair due to changes in residues in either the ligand or the PDZ domain result in similar effects on the binding energetics. Collectively, these results suggest that K912 interacts with the Glu at P_{-4} of Syndecan1, and this interaction is critical for the PDZ–Syndecan1 interaction.

Use of the peptide evolution strategy also allowed for the investigation of the origin of the Tiam1 PDZ domain specificity for the Caspr4 peptide. The Y_{P-1}F–Caspr4 peptide had an approximately three-fold negative effect on ligand affinity (Table 4.1 and Fig. 4.5B). Interestingly, the K_d determined for this peptide was nearly identical to that found for the Caspr4(F → A) C-terminal mutation (ENQKEYF**A**$_{COOH}$; $K_d = 64.8 \pm 5.9$ μM) (Shepherd *et al.*, 2010), verifying that N-terminal residues of the ligand are not critical for binding. The A_{P0}F–Caspr4 peptide had an approximately two-fold increase in ligand affinity, but did not restore binding to the level of that of the Caspr4 peptide (Table 4.1 and Fig. 4.5B). Together, the Y_{P-1}F–Caspr4 and A_{P0}F–Caspr4 mutations recreate the Caspr4 peptide, and it was possible to analyze this system by double-mutant cycle analysis (Table 4.1 and Fig. 4.5D). From this analysis, it is clear that the two C-terminal residues (P_0 and P_{-1}) are important for affinity and that they act cooperatively ($\Delta\Delta\Delta G_{int} = -0.51$ kcal/mol) in binding the Tiam1 PDZ domain (Fig. 4.5D). The molecular origin of this cooperativity is not understood and awaits further structural analysis.

6. Conclusions

In this chapter, we show how a combined approach using structural and thermodynamic methods can provide unique insights into PDZ–ligand specificity. We describe a fluorescence anisotropy-based binding assay and highlight important considerations in designing binding experiments. We apply this fluorescence anisotropy assay to determine the binding energetics for PDZ–ligand interactions. In particular, we probed residues in the S_0 and S_{-2} pockets of the Tiam1 PDZ domain. The binding results and double-mutant cycle analysis revealed that the S_0 pocket was important for PDZ interactions with Syndecan1 and Caspr4 and that the S_{-2} pocket provided selectivity for the Syndecan1 ligand. In addition, we introduced a "peptide evolution" strategy and showed that in combination with double-mutant cycles, it can be used to probe the origin of specificity from the perspective of the ligand. These results corroborated the PDZ mutational analysis of the S_0 pocket and identified the P_{-4} position in the ligand as critical for Syndecan1 affinity and selectivity. Future application of this combined structural and thermodynamic approach to other PDZ domains should provide additional insights into the origin of affinity and specificity of PDZ domains.

ACKNOWLEDGMENTS

The authors thank members of the Fuentes laboratory, Dr Todd Washington, and Dr Madeline Shea for helpful discussions and comments on the chapter. E. J. F. was supported by funds from the National Science Foundation (MCB-0624451) and the American Heart Association (0835261N). T. R. S. was supported in part by a National Institutes of Health graduate training grant in Pharmacology (GM067795) and by a University of Iowa Graduate Student Fellowship sponsored by the Center for Biocatalysis and Bioprocessing.

REFERENCES

Carter, P. J., Winter, G., Wilkinson, A. J., and Fersht, A. R. (1984). The use of double mutants to detect structural-changes in the active-site of the tyrosyl-transfer RNA-synthetase (*Bacillus-stearothermophilus*). *Cell* **38,** 835–840.

Chen, J. R., Chang, B. H., Allen, J. E., Stiffler, M. A., and MacBeath, G. (2008). Predicting PDZ domain-peptide interactions from primary sequences. *Nat. Biotechnol.* **26,** 1041–1045.

Eccleston, J. F., Hutchinson, J. P., and Jameson, D. M. (2005). Fluorescence-based assays. *Prog. Med. Chem.* **43,** 19–48.

Eftink, M. R. (1997). Fluorescence methods for studying equilibrium macromolecule-ligand interactions. *Methods Enzymol.* **278,** 221–257.

Eftink, M. R. (2000). Use of fluorescence spectroscopy as thermodynamics tool. *Methods Enzymol.* **323,** 459–473.

Feng, W., Wu, H., Chan, L. N., and Zhang, M. (2008). Par-3-mediated junctional localization of the lipid phosphatase PTEN is required for cell polarity establishment. *J. Biol. Chem.* **283,** 23440–23449.

Fersht, A. R. (1998). *Structure and Mechanism in Protein Science* W.H. Freeman and Company, New York.

Hoenes, G., Hauser, M., and Pfleiderer, G. (1986). Dynamic total fluorescence and anisotropy decay study of the dansyl fluorophor in model compounds and enzymes. *Photochem. Photobiol.* **43,** 133–137.

Horovitz, A., and Fersht, A. R. (1990). Strategy for analysing the co-operativity of intramolecular interactions in peptides and proteins. *J. Mol. Biol.* **214,** 597–629.

Hoshino, M., Sone, M., Fukata, M., Kuroda, S., Kaibuchi, K., Nabeshima, Y., and Hama, C. (1999). Identification of the stef gene that encodes a novel guanine nucleotide exchange factor specific for Rac1. *J. Biol. Chem.* **274,** 17837–17844.

Iden, S., and Collard, J. G. (2008). Crosstalk between small GTPases and polarity proteins in cell polarization. *Nat. Rev. Mol. Cell Biol.* **9,** 846–859.

Lakowicz, J. R. (1999). Principles of Fluorescence Spectroscopy. Kluwer Academic, New York.

Lee, H. J., and Zheng, J. J. (2010). PDZ domains and their binding partners: Structure, specificity, and modification. *Cell Commun. Signal* **8,** 1–18.

Masuda, M., Maruyama, T., Ohta, T., Ito, A., Hayashi, T., Tsukasaki, K., Kamihira, S., Yamaoka, S., Hoshino, H., Yoshida, T., Watanabe, T., Stanbridge, E. J., *et al.* (2010). CADM1 interacts with Tiam1 and promotes invasive phenotype of human T-cell leukemia virus type I-transformed cells and adult T-cell leukemia cells. *J. Biol. Chem.* **285,** 15511–15522.

Matsuo, N., Hoshino, M., Yoshizawa, M., and Nabeshima, Y. (2002). Characterization of STEF, a guanine nucleotide exchange factor for Rac1, required for neurite growth. *J. Biol. Chem.* **277,** 2860–2868.

Matsuo, N., Terao, M., Nabeshima, Y., and Hoshino, M. (2003). Roles of STEF/Tiam1, guanine nucleotide exchange factors for Rac1, in regulation of growth cone morphology. *Mol. Cell. Neurosci.* **24,** 69–81.

Mertens, A. E., Roovers, R. C., and Collard, J. G. (2003). Regulation of Tiam1-Rac signalling. *FEBS Lett.* **546,** 11–16.

Mertens, A. E., Pegtel, D. M., and Collard, J. G. (2006). Tiam1 takes PARt in cell polarity. *Trends Cell Biol.* **16,** 308–316.

Mocz, G., Helms, M. K., Jameson, D. M., and Gibbons, I. R. (1998). Probing the nucleotide binding sites of axonemal dynein with the fluorescent nucleotide analogue 2'(3')-*O*-(-*N*-Methylanthraniloyl)-adenosine 5'-triphosphate. *Biochemistry* **37,** 9862–9869.

Owicki, J. C. (2000). Fluorescence polarization and anisotropy in high throughput screening: Perspectives and primer. *J. Biomol. Screen.* **5,** 297–306.

Perrin, F. (1926). Polarisation de la lumière de fluoresence: Vie moyenne des molécules dans l'état excité. *J Phys. Radium V Ser.* **6,** 390–401.

Ponting, C. P., Phillips, C., Davies, K. E., and Blake, D. J. (1997). PDZ domains: Targeting signalling molecules to sub-membranous sites. *Bioessays* **19,** 469–479.

Royer, C. A., and Scarlata, S. F. (2008). Fluorescence approaches to quantifying biomolecular interactions. *Methods Enzymol.* **450,** 79–106.

Saro, D., Li, T., Rupasinghe, C., Paredes, A., Caspers, N., and Spaller, M. R. (2007). A thermodynamic ligand binding study of the third PDZ domain (PDZ3) from the mammalian neuronal protein PSD-95. *Biochemistry* **46,** 6340–6352.

Schultz, J., Copley, R. R., Doerks, T., Ponting, C. P., and Bork, P. (2000). SMART: A web-based tool for the study of genetically mobile domains. *Nucleic Acids Res.* **28,** 231–234.

Shepherd, T. R., Klaus, S. M., Liu, X., Ramaswamy, S., DeMali, K. A., and Fuentes, E. J. (2010). The Tiam1 PDZ domain couples to Syndecan1 and promotes cell-matrix adhesion. *J. Mol. Biol.* **398,** 730–746.

Songyang, Z., Fanning, A. S., Fu, C., Xu, J., Marfatia, S. M., Chishti, A. H., Crompton, A., Chan, A. C., Anderson, J. M., and Cantley, L. C. (1997). Recognition of unique carboxyl-terminal motifs by distinct PDZ domains. *Science* **275,** 73–77.

Stiffler, M. A., Chen, J. R., Grantcharova, V. P., Lei, Y., Fuchs, D., Allen, J. E., Zaslavskaia, L. A., and MacBeath, G. (2007). PDZ domain binding selectivity is optimized across the mouse proteome. *Science* **317,** 364–369.

Tonikian, R., Zhang, Y., Sazinsky, S. L., Currell, B., Yeh, J. H., Reva, B., Held, H. A., Appleton, B. A., Evangelista, M., Wu, Y., Xin, X., Chan, A. C., *et al.* (2008). A specificity map for the PDZ domain family. *PLoS Biol.* **6,** e239.

Tyler, R. C., Peterson, F. C., and Volkman, B. F. (2010). Distal interactions within the par3-VE-cadherin complex. *Biochemistry* **49,** 951–957.

CHAPTER FIVE

Thermodynamic Analysis of Metal Ion-Induced Protein Assembly

Andrew B. Herr *and* Deborah G. Conrady

Contents

Abstract

A large number of biological systems are regulated by metal ion-induced protein assembly. This phenomenon can play a critical role in governing protein function and triggering downstream biological responses. We discuss the basic thermodynamic principles of linked equilibria that pertain to metal ion-induced dimerization and describe experimental approaches useful for studying such systems. The most informative techniques for studying these systems are sedimentation velocity and sedimentation equilibrium analytical ultracentrifugation, although a wide range of other spectroscopic, chromatographic, or qualitative approaches can provide a wealth of useful information. These experimental procedures are illustrated with examples from two systems currently under study: zinc-induced assembly of a staphylococcal protein responsible for intercellular adhesion in bacterial biofilms and calcium-induced dimerization of a human nucleotidase.

Department of Molecular Genetics, Biochemistry & Microbiology, University of Cincinnati College of Medicine, Cincinnati, Ohio, USA

Methods in Enzymology, Volume 488
ISSN 0076-6879, DOI: 10.1016/B978-0-12-381268-1.00005-7
© 2011 Elsevier Inc.
All rights reserved.

1. Introduction

Protein–ligand binding equilibria can often be modulated by secondary interactions with solute molecules, including salt, protons, various ions, or other components. In general, any time that binding of a macromolecule M to a ligand X is altered by interaction with a solute Y, the M–X and M–Y binding events are linked equilibria. A commonly observed type of linked equilibrium involves protein dimerization that is modulated by metal cations such as calcium, magnesium, or zinc. There are many examples in biology in which metal cations influence protein self-assembly, either directly or indirectly. For example, calcium is required for cadherin dimerization (Alattia *et al.*, 1997; Nagar *et al.*, 1996); zinc and copper can induce oligomerization of the Alzheimer-related Aβ or prion peptides (Curtain *et al.*, 2001; Jobling *et al.*, 2001); either zinc or cobalt can induce self-association of KIR2D, a human natural killer cell inhibitory receptor (Fan *et al.*, 2000); zinc induces dimerization of the streptococcal superantigen SPE-C (Roussel *et al.*, 1997); sodium and calcium modulate dimerization of glutamate receptors (Chaudhry *et al.*, 2009); and magnesium induces dimerization of the inositol monophosphatase enzyme SuhB (Brown *et al.*, 2007). The mechanism by which metal ions induce protein assembly varies; in some cases, the metal induces a conformational transition in the protein that allows self-assembly, such when calcium binds to E-cadherin and stabilizes the interdomain interface (Nagar *et al.*, 1996). Alternately, protein dimerization can occur via ligation of the metal ion by side chain residues of different protomers, as seen for the superantigen SPE-C (Roussel *et al.*, 1997). Finally, in several cases, the precise mechanism has not yet been resolved, but it is still imperative to understand the relationship between metal binding and assembly. Designing experiments with a basic understanding of linked equilibria can generate a substantial amount of information about the system of interest, as described below. In this chapter, we will focus on two systems under study in our laboratory: a nucleotidase enzyme whose activity is modulated by calcium-induced dimerization (Yang *et al.*, 2006, 2008), and a bacterial cell-surface protein that self-assembles in the presence of zinc, leading to critical intercellular adhesion in biofilms (Conrady *et al.*, 2008).

Human soluble calcium-activated nucleotidase (SCAN) is an enzyme that catalyzes the hydrolysis of nucleotide diphosphates to the monophosphate form (reviewed in (Smith and Kirley, 2006)). This reaction is also carried out by the well-known enzyme CD39, a member of the NTPDase family of enzymes. Although the two enzymes catalyze similar reactions, SCAN is structurally and functionally homologous to the apyrases from blood-sucking insects, which inhibit platelet activation through ADP hydrolysis. Human SCAN, like the insect apyrases, is significantly more thermostable than CD39 or related NTPDases and thus is of interest for therapeutic applications. However, the preferred substrates of SCAN are GDP and UDP, which lack the platelet-inhibitory affects of ADP, so developing SCAN for clinical use relies on altering its specificity to improve its specific activity for ADP hydrolysis. It was observed years ago that SCAN can undergo a calcium-dependent spectroscopic shift that coincides with increased nucleotidase activity (Murphy *et al.*, 2003); we helped to show that this calcium-dependent event was SCAN dimerization (Yang *et al.*, 2006, 2008).

Our laboratory also studies the mechanistic basis for biofilm formation in staphylococcal species. Biofilms are surface-adherent colonies of bacteria that are widely resistant to antibiotics and host immune responses (Costerton *et al.*, 1999). A critical early stage in staphylococcal biofilm formation is the clustering of bacterial cells to form microcolonies; this process in *Staphylococcus epidermidis* requires the accumulation-associated protein (Aap; Hussain *et al.*, 1997). The C-terminal half of Aap, which remains on the bacterial surface after proteolytic processing, contains between 5 and 17 nearly perfect repeats of a 128-amino acid sequence called simply the B-repeat (each repeat contains an 80-amino acid "G5 domain" Bateman *et al.*, 2005). This B-repeat region has been shown to be the essential portion of Aap required for microcolony formation (Corrigan *et al.*, 2007; Rohde *et al.*, 2005, 2007). We recently expressed Aap constructs containing one or two intact B-repeat domains along with a conserved C-terminal half-repeat (called Brpt1.5 or Brpt2.5, respectively) for biophysical studies. Based on both biophysical experiments and biofilm assays, we showed that the B-repeat domain of Aap can form a zinc-dependent dimer, and that this zinc-based adhesion event is necessary for staphylococcal biofilms to form (Conrady *et al.*, 2008).

2. Linked Equilibria—General Concepts

Any time binding of M (the macromolecule) to X (any particular ligand) is influenced by the binding of M to Y (another ligand or solute molecule), the two equilibria are linked. In other words, if the M–X equilibrium *changes* upon a change in the free ligand concentration of Y, the two equilibria are linked; if Y can bind to M but does not change the M–X equilibrium, then the two are not linked equilibria. The components X and Y can be any ligand or solute molecule. A general feature of such linked equilibria is that if binding of Y promotes M binding to X (i.e., MY binds to X more tightly than M alone does; or $K_3 > K_1$ in Fig. 5.1A), then by definition, the converse is true: binding of X also promotes M binding

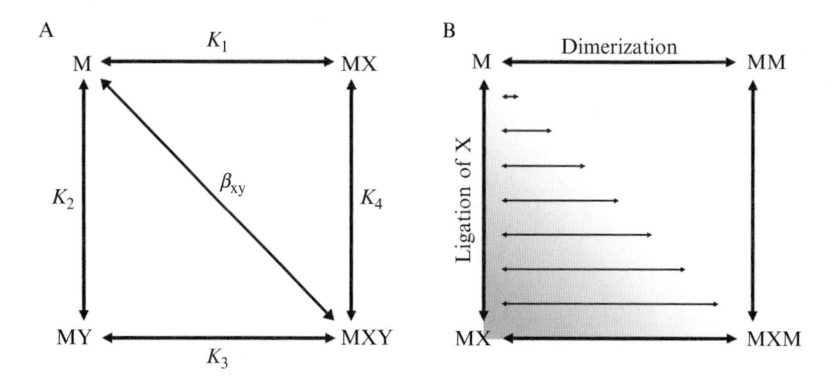

Figure 5.1 Linked equilibria. (A) The thermodynamic cycle for a generalized linked equilibrium in which the macromolecule M can bind both X and Y. Stepwise equilibrium association constants are labeled K_1 through K_4, and the overall constant that governs the equilibrium between unligated M and fully ligated MXY is labeled β_{xy}. If binding of X and Y to M are *not* linked, $K_1 = K_3$ and $K_2 = K_4$, and binding of Y to M has no effect on the binding of M and X. The equilibria are linked if $K_3 > K_1$ (and therefore $K_4 > K_2$) or if $K_3 < K_1$ (and $K_4 < K_2$). (B) The linked equilibrium between protein dimerization and metal ligation. The horizontal arrows of varying length within the square illustrate the differing degrees of protein dimerization that will be observed at different discrete metal concentrations.

to Y (i.e., $K_4 > K_2$). This can be seen by looking at the equilibrium constants of the thermodynamic cycle; since this is a closed system, $K_1K_4 = K_2K_3 = \beta_{xy}$ (Fig. 5.1A). In this chapter, we will primarily be discussing an equilibrium governing the macromolecule M binding to another M (i.e., dimerization), which is linked to an equilibrium for M binding to a metal ion (throughout the rest of the chapter, X will be used to represent the metal ion; Fig. 5.1B). We will also briefly discuss an additional layer of complexity, in which a metal ion-dependent protein dimerization event is also linked to protonation (i.e., the dimerization changes as a function of pH).

A linked equilibrium such as metal-induced dimerization can be described by a thermodynamic cycle, as shown in Fig. 5.1B. The free macromolecule M can undergo two reactions: the top horizontal arrow describes the dimerization reaction, and the left vertical arrow describes the metal-binding reaction. What often turns out to be the case with linked equilibria is that one reaction is "visible," or easily observable, whereas the second reaction is "invisible," or less easily observed. For example, protein dimerization can be followed by a wide range of experimental techniques, which will be described below; in contrast, these same techniques will typically not register the slight changes in mass, shape, or size due to metal binding. Thus, using the schematic in Fig. 5.1B as a visual aid, one can follow the visible horizontal reaction (i.e., dimerization) at different discrete locations along the vertical equilibrium (i.e., different defined free metal ion concentrations). Thus, if metal binding favors protein dimerization, one will see the monomer–dimer equilibrium shift toward dimeric species as the free concentration of the metal ion is increased. This is shown schematically as horizontal arrows that increase in length as the experimental conditions shift downward along the vertical arrow (Fig. 5.1B). If the extent of dimerization is determined over a range of metal ion concentrations, the resulting data will show a sigmoidal transition from monomer to dimer as a function of log[X] (i.e., the log of the free metal ion concentration).

The general concepts of linked equilibria have been described elegantly in the classic book *Binding and Linkage* (Wyman and Gill, 1990). There are some very useful ways in which one can analyze the thermodynamics of linked equilibria to determine important information about the system under study. One elegant type of analysis is to plot the logarithm of the association constant (i.e., $1/K_D$) governing the M–M equilibrium as a function of the logarithm of the free metal concentration (log[X]). The slope of this plot will reveal the net number of X ions bound or released upon dimerization of M (e.g., see Fig. 5.5C). Likewise, this same approach can be used to analyze the pH dependence of dimerization; the slope of a $\log K$ versus $\log[H^+]$ plot will reveal the net number of protons bound or released upon dimerization. As described in Section 3.2, we have used these approaches to establish a model for zinc-dependent adhesion in staphylococcal biofilms.

There are two simple, practical considerations that must be stressed before one begins analyzing a linked equilibrium such as those described here. First, it is important to remember that equilibrium thermodynamics are based on free ligand activities, which can typically be approximated by free ligand concentrations when working under dilute conditions. Thus, in order to obtain accurate information about the system, it is important to dialyze the protein at various concentrations of metal ion in order to achieve accurate free metal concentrations, rather than simply adding a fixed total concentration of metal to your protein sample (which will be partitioned between bound and free metal ions after addition). After dialysis, the equilibrated free metal concentration in the protein sample will be effectively equivalent to the metal concentration of the dialyzate buffer, due to the large excess volume of the dialyzate buffer compared to the protein sample. Secondly, it is always important to choose an experimental approach such that the concentration range over which protein dimerization occurs is appropriate for the technique used. This ensures that the experimental approach will be able to determine the equilibrium constants accurately. For example, analytical ultracentrifugation is an optical technique, so one typically uses sample concentrations in the micromolar range (for absorbance or interference optics). If the dimerization dissociation constant is in the nanomolar range, a fluorescence-based technique is more appropriate due to the increased sensitivity.

3. Experimental Approaches—Analytical Ultracentrifugation

There are many techniques that can be used to study metal-induced protein assembly. The technique that offers the most advantages is analytical ultracentrifugation (AUC), which will be the primary focus of this chapter. We will also discuss both high-tech and simple alternative techniques. AUC is considered to be the gold-standard technique for analyzing protein assembly, given that it allows one to directly analyze such equilibria in solution, under a wide range of physiologically relevant conditions, without the need for any standards. There are two complementary experimental approaches one can use: (1) sedimentation velocity provides information on the size and shape of the sedimenting species, and can give an overall view of the number of species in solution; (2) sedimentation equilibrium can provide molecular weights of sedimenting species and equilibrium constants for monomer–oligomer assembly reactions. Laue and Stafford have written an excellent introduction to analytical ultracentrifugation theory (Laue and Stafford, 1999), which is recommended for anyone new to the technique. The AUC data presented here have been collected in a Beckman XL-I analytical ultracentrifuge using absorbance optics with a four-hole An-Ti80 rotor.

3.1. Sedimentation velocity

In sedimentation velocity experiments, the samples are spun at relatively high speeds and the rate of sedimentation is determined; in other words, this is a kinetic experiment. The rate of sedimentation is related to the size and shape of the sedimenting protein. The centrifugal field acting on the protein causes it to sediment toward the outside of the sample cell, creating a concentration gradient. This gradient then causes diffusion of the protein "upstream" against the applied centrifugal field. Larger species will sediment more rapidly, yielding larger sedimentation coefficients, meaning that they sediment more quickly compared to smaller species of identical shape. However, two proteins of identical mass that differ notably in shape can give very different rates of sedimentation, primarily due to drag; elongated proteins will have significantly larger frictional coefficients. In particular, the shape of a protein will profoundly influence the rate of diffusion against the applied centrifugal field.

In a typical sedimentation velocity experiment, 400 μl of a protein sample and a similar volume of matched buffer are loaded into a two-sector centerpiece of the sample cell and spun for 12–16 h. The distribution of protein across the sample cell is measured every few minutes, using either absorbance or interference optics. These "snapshots" of the protein's distribution yield a series of essentially sigmoidal curves that show the boundary between protein and buffer as the protein sediments through the cell. The basic concept for analysis of these concentration–distribution curves is to follow the rate at which this boundary moves as a function of the radial position (i.e., distance from the center of the rotor). For a single species, one could mark the midpoint of each boundary and follow the radial movement of the boundary as a function of time. However, analysis is typically more complicated due to the potential presence of multiple sedimenting species in the sample and the influence of protein shape on sedimentation profiles. Specifically, proteins that are highly nonglobular, such as rod-shaped proteins, will give rise to significantly steeper sigmoidal boundary curves in the raw data. This is due to the restricted diffusion caused by the high frictional coefficient of these elongated proteins. Modern software is able to determine not only the sedimentation coefficients of the sedimenting species, but also the frictional ratio f/f_0 (i.e., the frictional coefficient of the protein compared to that of an ideal sphere of identical volume). This development allows sedimentation velocity analysis to provide approximate information on molecular weights, based on the Svedberg equation (Schuck, 2000).

There are many different software packages that are available to analyze sedimentation velocity data, and they all have different advantages depending on the needs of the user. These include programs that directly fit the raw data to solutions of the Lamm equation, such as Svedberg (Philo, 1997); those that use the van Holde–Weischet approach for analysis (van Holde and Weischet, 1978), such as UltraScan (Demeler, 2005); and several that use variations of the time-derivative approach first developed by Stafford (Philo, 2000; Schuck, 2000; Stafford, 1992) (i.e., the $g(s^*)$, dc/dt, $c(s)$, and $c(M)$ methods), such as DCDT+ (Philo, 2000), SEDANAL (Stafford, 1992), SEDFIT, and SEDPHAT (Schuck, 2000, 2003). Some of these program suites such as UltraScan, SEDFIT, and SEDPHAT offer multiple types of analysis methods, and a couple of programs, such as SEDANAL and SEDPHAT, are particularly useful for analysis of interacting systems. Finally, SEDFIT and UltraScan have recently been improved to allow analysis of velocity data with simultaneous resolution of sedimentation coefficients and f/f_0 ratios for components in solution, which greatly increases the amount of information about the sedimenting species in a sample. We should also point out one other program, SEDNTERP, which is not used directly for data analysis but is a very valuable tool that helps in the interpretation of sedimentation data (Laue *et al.*, 1992). All programs mentioned here can be either obtained freely or purchased for a reasonable price. Scott and Schuck (2005) have written an excellent introductory chapter comparing the programs mentioned here and a few others.

All velocity data discussed in this chapter have been analyzed using the $c(s)$ method in SEDFIT. This analysis method deconvolutes the raw experimental data into a sedimentation coefficient distribution plot, which shows peaks for all sedimenting species, with the x-axis representing the apparent sedimentation coefficient (s^*). The value s^* reported here is the *apparent* sedimentation coefficient because it has not been corrected for temperature, protein concentration, or buffer to yield the absolute value $s^{\circ}_{20,w}$. The overall appearance of the $c(s)$ distribution plot is similar to a gel filtration trace, although in the $c(s)$ plot, the smaller components will be on the left, since they will have smaller sedimentation coefficients. The $c(s)$ approach is a variant of the time-derivative approach that specifically accounts for diffusion, which both prevents broadening of the resultant peaks in the sedimentation coefficient profile and allows estimation of molecular weights of the sedimenting species.

Analyzing sedimentation velocity data with the $c(s)$ approach is an excellent way to study metal ion-induced protein assembly. The protein of interest needs to be analyzed by itself in order to rule out any inherent self-assembly or aggregation phenomena. In the case of the Brpt1.5 construct from the staphylococcal surface protein Aap, the $c(s)$ analysis revealed that the protein in a minimal buffer was a stable monomer over a wide range of concentrations, with no evidence of self-assembly (Fig. 5.2A). In contrast, when the velocity experiment was repeated in the presence of various divalent metal cations, the peak for Brpt1.5 shifted significantly to the right only in the presence of $ZnCl_2$ (Fig. 5.2B). Such a dramatic shift in the sedimentation profile implies that either protein oligomerization or a dramatic conformational change has occurred. In this case, further analysis of the velocity data using the $c(M)$ method in SEDFIT was able to provide a molecular weight estimate, suggesting the formation of a Brpt1.5 dimer. As described below, sedimentation equilibrium experiments were then carried out to confirm that the larger species was in fact a dimer. Likewise, the human soluble calcium-activated nucleotidase (SCAN) showed a clear shift to a larger sedimentation coefficient in the presence of either 2 mM $CaCl_2$ or $SrCl_2$, but not in the presence of $MgCl_2$ (Fig. 5.2C). Panel D shows intermediate cases of partial dimerization at 0.2 mM $CaCl_2$ for wild-type and mutant SCAN enzymes. It is clear that the sedimentation peaks shift from a monomer s^* value toward a dimer s^* value as a function of metal ion concentration. To observe the linked equilibrium between metal ion binding and dimerization, the weight-averaged s^* value can be calculated by integrating the area under the entire s^* distribution range (this is easily determined in SEDFIT). All data sets should be integrated over the same range of s^* values (in the examples shown here, Brpt1.5 data was integrated from 1 to 3.5), and the weight-averaged s^* values can be plotted as a function of metal concentration to illustrate a sigmoidal transition between monomer and dimer.

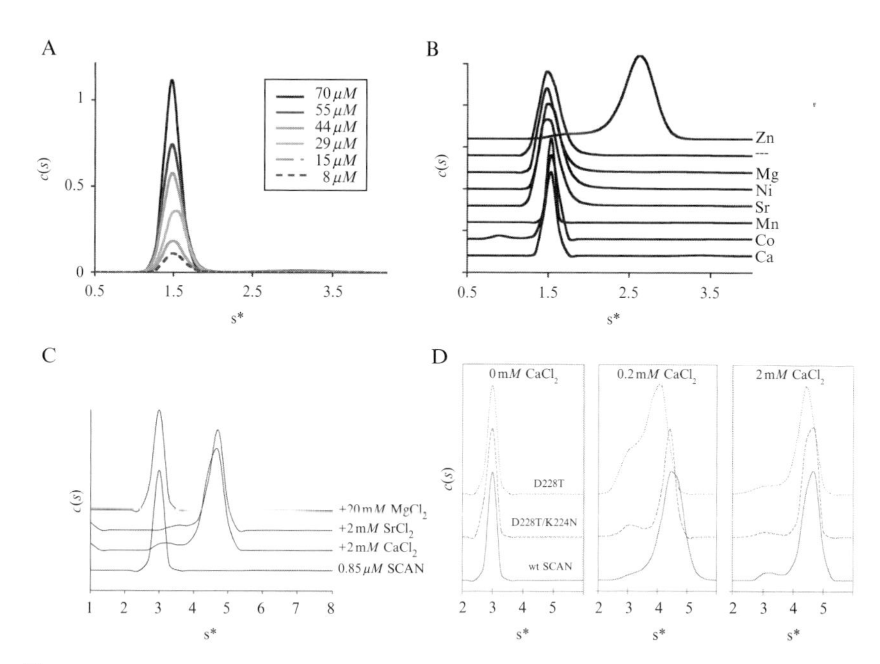

Figure 5.2 Sedimentation velocity analysis of metal ion-induced protein assembly. (A) Sedimentation coefficient distributions for the Brpt1.5 construct of Aap in the absence of metal cations, determined at multiple Brpt1.5 concentrations. All distributions are consistent with monomeric protein. The distributions were determined from a $c(s)$ analysis using SEDFIT. (B) Sedimentation coefficient distributions for the Brpt1.5 construct of Aap in the presence or absence (—) of divalent metal cations. In the presence of Zn^{2+}, Brpt1.5 shifts to a species with a larger sedimentation coefficient, consistent with a dimer. Panels A and B are reprinted from *PNAS* (Conrady *et al.*, 2008), © 2008 National Academy of Sciences, USA. (C) Sedimentation coefficient distributions for human SCAN alone or in the presence of $CaCl_2$, $SrCl_2$, or $MgCl_2$. In the presence of $CaCl_2$ or $SrCl_2$, SCAN shifts to a larger species that was shown to be a dimer. Panel C was originally published in *JBC* (Yang *et al.*, 2006), © 2006 the American Society for Biochemistry and Molecular Biology. (D) Sedimentation coefficient distributions for wild-type SCAN and two mutants, showing the behavior of SCAN variants at intermediate and saturating $CaCl_2$ concentrations. Panel D was originally published in *Biochemistry* (Yang *et al.*, 2008). Reprinted with permission, © 2008 American Chemical Society.

In many cases, the sedimentation profile for intermediate states corresponding to a mixture of monomeric and dimeric species will show a single peak between the monomer and dimer s^* values that shifts progressively toward the dimer position as the metal ion is increased. This is expected due to the fact that monomer and dimer are typically in fast equilibrium relative to the time scale of the velocity experiment, which generally takes 12–16 h (Correia *et al.*, 2009; Gilbert and Gilbert, 1978). This phenomenon can be seen more clearly in Fig. 5.3, where we

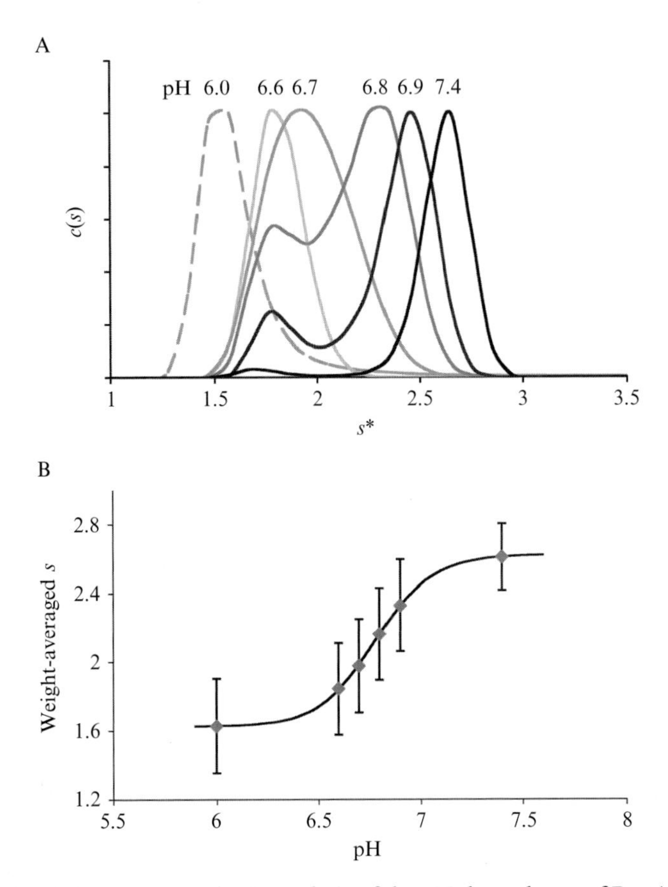

Figure 5.3 Sedimentation velocity analysis of the pH dependence of Brpt1.5 assembly. (A) Sedimentation coefficient distributions for Brpt1.5 with 10 m*M* $ZnCl_2$ as a function of pH. At pH 7.4, Brpt1.5 is predominantly dimeric, but the equilibrium shifts toward monomer as the pH is lowered to 6.0. Note that under intermediate conditions of pH, the peak in the distribution plot shifts to an intermediate position between the monomer and dimer *s** values, according to Gilbert theory for assembling systems. Panel A was reprinted from *PNAS* (Conrady *et al.*, 2008), © 2008 National Academy of Sciences, USA. (B) Sigmoidal plot of the weight-averaged sedimentation coefficient *s** as a function of pH, calculated from the data shown in panel A. The weight-averaged *s** values were determined using SEDFIT over the entire range of the distribution plot (integrated from 1 to 3.5 S). The sigmoidal curve was fitted to a modified Hill equation to reveal an apparent pK_a of 6.8 for the transition.

investigated another linked equilibrium in the Brpt1.5 system. We analyzed the dependence of dimer formation on pH, since we were interested in determining the potential role of histidine residues in zinc coordination. It can be seen clearly that at pH 7.4, the Brpt1.5 protein in the presence of 10 m*M* $ZnCl_2$ predominantly forms a dimer. But as the pH is lowered to 6.0, the s^* peak gradually shifts to the left until a predominantly monomeric peak is seen at pH 6.0 (Fig. 5.3A). By plotting the weight-averaged s^* values as a function of pH, we can see the sigmoidal transition between monomer and dimer as a function of pH (Fig. 5.3B). These data were fitted to a modified Hill equation to determine an apparent pK_a of the transition of 6.8, which suggests that one or more histidines are responsible for the loss of zinc coordination at lower pH. It should be noted that although this approach is valid, the error bars for the velocity analysis were significantly larger than those resulting from a global analysis of equilibrium data measuring the same phenomenon (Fig. 5.5B).

One way to quantitatively analyze a metal ion-linked dimerization equilibrium is to measure the equilibrium dimerization constant as a function of metal ion concentration. Typically, this is done using sedimentation equilibrium, as discussed in Section 3.2. However, there may be reasons why a sedimentation equilibrium experiment is not feasible. In particular, for proteins with limited stability, the length of time needed for a full equilibrium experiment (often 3–4 days for the entire run) may be impractical. In this case, one can analyze the sedimentation velocity data directly using programs such as SEDANAL or SEDPHAT to derive equilibrium constants for the assembly reaction. For example, the Correia group has used this approach to study a rather complicated system of isodesmic tubulin self-assembly in the presence of vinca alkaloids using sedimentation velocity data (Correia, 2000, 2010). Finally, although the data presented here were analyzed with the $c(s)$ method in SEDFIT, a similar approach can of course be carried out by analyzing velocity data with other analysis methods in SEDFIT or other programs, such as the dc/dt or $g(s^*)$ approach in DCDT+ or the van Holde–Weischet, dc/dt, or $c(s)$ approaches in UltraScan.

3.2. Sedimentation equilibrium

Sedimentation equilibrium is a distinct yet highly complementary approach to velocity analysis that can be conducted in the analytical ultracentrifuge. In this experiment, samples are spun at lower speeds until the opposing forces of sedimentation and diffusion acting on the protein come into equilibrium. At that point, there is no net movement of the proteins in the sample cell, and a stable equilibrium concentration gradient is formed. The equilibrium gradient of each individual sedimenting species can be described by an exponential curve of the form:

$$c(r) = c_0 e^{\sigma\left(r^2/2 - r_0^2/2\right)} \tag{5.1}$$

where $c(r)$ is the concentration at radial position r, c_0 is the concentration at reference position r_0, and σ is the reduced buoyant molecular weight of the sedimenting species. The reduced buoyant molecular weight is the main experimental parameter of interest in sedimentation equilibrium, defined as

$$\sigma = \frac{M(1 - \bar{v}\rho)\omega^2}{RT} \tag{5.2}$$

where M is the molecular weight of the sedimenting species, $\bar{v}$ the partial specific volume calculated based on protein composition (the program SEDNTERP is recommended for this), ρ the buffer density, ω the rotor's angular velocity (rpm $\cdot$ $\pi/30$), R the universal gas constant, and T is the temperature in kelvin. The important concept from Eq. (5.1) is that the value of the reduced buoyant molecular weight σ is derived directly from the shape of the exponential gradient in the raw data. At the same time, Eq. (5.2) shows that σ is directly proportional to the molecular weight of the protein, after accounting for experimental parameters such as rotor speed, temperature, buffer density, and protein composition. Thus, sedimentation equilibrium is able to directly determine the experimental molecular weight of a protein under a wide range of physiologically relevant solution conditions based on thermodynamic first principles, without the need for comparison to any standards. Because by definition these data are collected at equilibrium where there is no net movement of the protein, the shape of the sedimenting species does not affect the result, although elongated proteins will take longer to achieve equilibrium.

A second important concept is that *each* individual species in solution will form an equilibrium distribution of its own, based on its molecular weight, according to Eq. (5.1). Thus, if a monomer–dimer equilibrium exists for the protein under study, the observed signal will represent the sum of two exponential concentration gradients, that of the monomer and the dimer (Fig. 5.4A). Furthermore, because the concentrations of monomer and dimer are governed by an equilibrium constant, the equation for the observed concentration distribution can be written as the sum of two exponentials in which the dimer concentration terms are defined in terms of monomer concentration and the equilibrium constant (Laue and Stafford, 1999; Scott and Schuck, 2005). Likewise, higher order assembly systems can be described by the sum of multiple exponential gradients, each of which has a concentration term defined in terms of monomer concentration and the appropriate monomer–*N*mer equilibrium constant. Whether the system under study features a single species or an equilibrium between oligomeric species, the standard approach is to measure data at multiple concentrations and multiple rotor speeds. Typically, 100 μl samples of a protein at three concentrations (ideally spanning a 10-fold concentration range) are loaded in each sample cell, and the protein is spun at three or four rotor speeds, depending on the range of oligomeric states needing to be

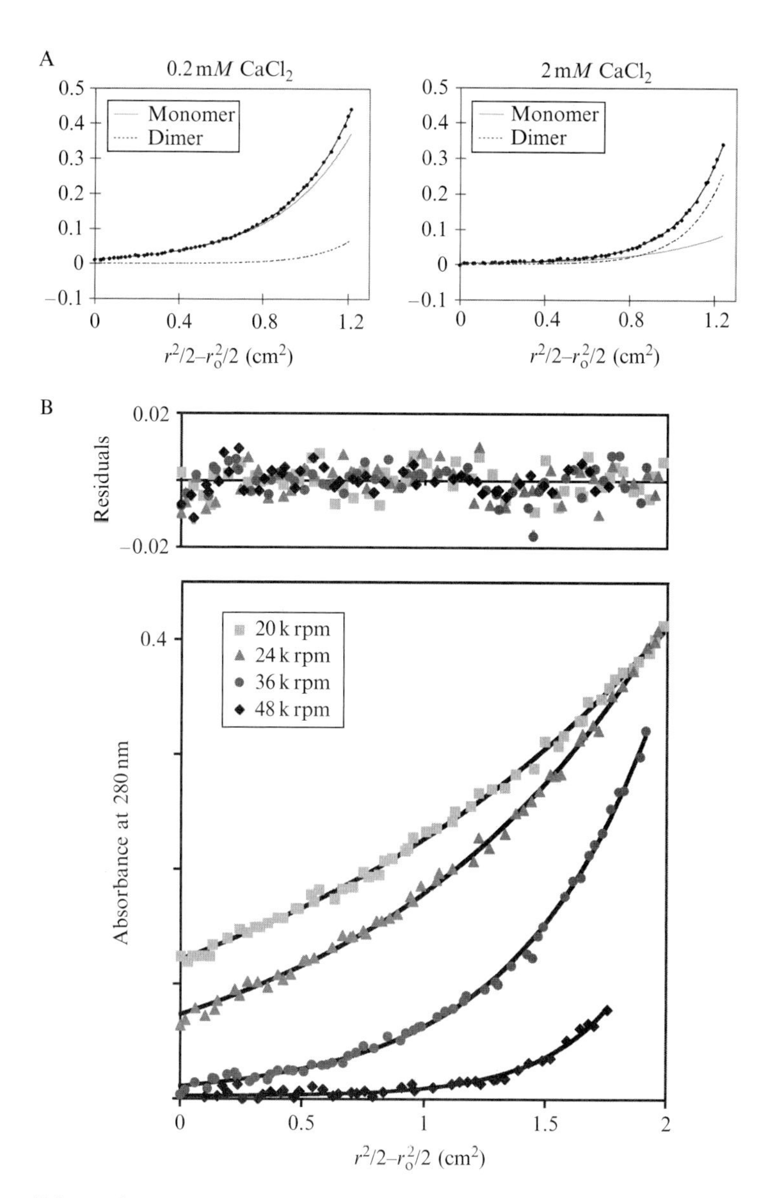

Figure 5.4 Sedimentation equilibrium analysis of metal ion-induced protein dimerization. (A) Representative sedimentation equilibrium curves for SCAN in the presence of 0.2 or 2 m*M* $CaCl_2$. The observed data are shown as circles, with the final fitted curve superimposed. The experimental concentration gradient is the sum of the component gradients for the monomer and dimer species, shown in gray solid (monomer) or dashed black (dimer) lines. Panel A was originally published in *JBC* (Yang *et al.*, 2006), © 2006 the American Society for Biochemistry and Molecular Biology. (B) Four representative sedimentation equilibrium curves for Brpt1.5 collected at different speeds, taken from a set of 12 curves that were globally fitted using the program

resolved. The raw data curves from the different experimental conditions are fitted simultaneously in a global analysis in order to extract accurate values for the reduced buoyant molecular weight σ and any relevant equilibrium constants. Figure 5.4B shows four representative curves for Brpt1.5 (out of 12 total) that were globally analyzed to determine the assembly state of Brpt1.5 in the absence of $ZnCl_2$. This experiment was repeated in the presence of $ZnCl_2$ to establish that Brpt1.5 formed a zinc-induced dimer (Conrady *et al.*, 2008). The program we use to analyze self-assembly reactions from sedimentation equilibrium data is WinNONLIN (Johnson *et al.*, 1981), although SEDPHAT and UltraScan can also be used for global analysis of sedimentation equilibrium data.

There are two approaches to analyzing a linked equilibrium using sedimentation equilibrium data. The simplest approach is to analyze the data according to Eq. (5.1), whether the data actually describe a monomer or higher-order oligomers. This yields a weight-averaged value of σ, or σ_w. For a metal-induced dimerization event, σ_w (or equivalently, the relative molecular weight, given by $\sigma_w/\sigma_{monomer}$) can be plotted versus metal ion concentration to yield a sigmoidal curve (Fig. 5.5A); likewise, this approach can be used for any linked equilibrium, including pH-dependent dimerization (Fig. 5.5B). Fitting these data to modified Hill equations allowed us to determine the EC_{50} values for zinc-induced dimerization of two different Aap constructs from the data in Fig. 5.5A, or the apparent pK_a for pH-induced dissociation of the Brpt1.5 dimer in the presence of zinc (Conrady *et al.*, 2008). The second approach is to analyze the monomer–dimer equilibrium under varying conditions to determine how the equilibrium constant changes as a function of the free metal ion concentration. As for all linked equilibria, one can plot log$K$ versus log[metal]; for example, for zinc-induced dimerization, the slope of this plot yields ΔZn^{2+}, or the net number of zinc ions bound or released upon dimer formation (Fig. 5.5C). Likewise, for pH-dependent dimerization, the slope of the plot yields $-\Delta H^+$, or the number of ionizable residues implicated in dimer dissociation (Fig. 5.5D). We used this approach to show that doubling the number of complete B-repeat domains in the protein construct approximately doubled the number of zinc ions bound within the dimer interface. Brpt1.5 dimerization required 2–3 Zn^{2+} ions, whereas Brpt2.5 needed $\sim$5 zinc ions to form the dimer. These data suggested that the B-repeat domains in Aap

WinNONLIN. Raw data points are shown with symbols, with the fits superimposed. The data were fitted to a single-species model, yielding an experimental molecular weight of 22,075 Da, consistent with a monomeric state of Brpt1.5 in the absence of Zn^{2+} (predicted monomer weight = 22,284 Da). Residuals for the fits are shown in the upper panel above the experimental data. A similar type of analysis revealed that Brpt1.5 and Brpt2.5 both formed dimers in the presence of Zn^{2+}. Panel B is reprinted from *PNAS* (Conrady *et al.*, 2008), © 2008 National Academy of Sciences, USA.

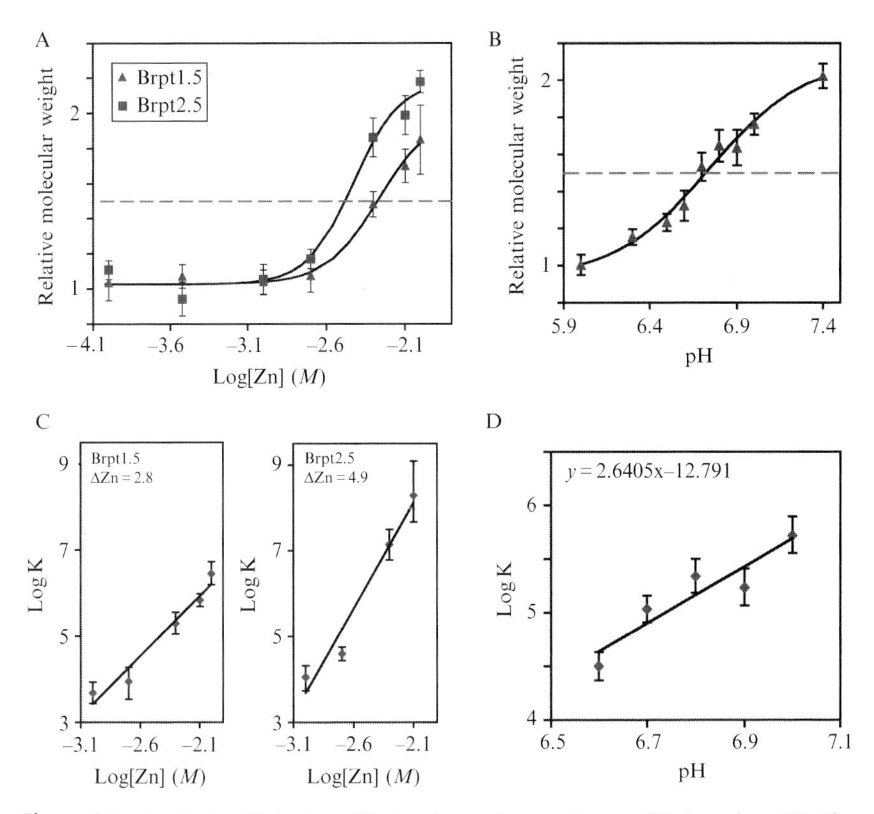

Figure 5.5 Analysis of linked equilibria using sedimentation equilibrium data. (A) Plot of the sigmoidal transition observed for the relative molecular weights of Brpt1.5 and Brpt2.5 as a function of Zn^{2+} concentration. Each data point represents the weight-averaged experimental molecular weight relative to the monomer weight, determined from σ_w, at discrete concentrations of $ZnCl_2$. The EC_{50} values for zinc-induced dimerization were 5.4 and 3.7 m*M* for Brpt1.5 and Brpt2.5, respectively. (B) Plot of the sigmoidal transition between monomer and dimer for Brpt1.5 with 10 m*M* Zn^{2+}, observed as a function of pH. The apparent pK_a from this curve was 6.7, suggesting the involvement of one or more histidine residues in the coordination of zinc. (C) Log*K* versus log[Zn^{2+}] plots for Brpt1.5 and Brpt2.5. The slope of these plots yields ΔZn^{2+}, the net number of zinc ions bound or released upon dimer formation. Note that two to three zinc ions are involved in dimerization of one full B-repeat (Brpt1.5), whereas approximately twice the number of zinc ions are implicated in dimerization of two full B-repeats (Brpt2.5). (D) Log*K* versus pH plot for Brpt1.5. The slope of this plot yields $-\Delta H^+$, indicating that an uptake of approximately three protons is linked to the disassembly of the Brpt1.5 dimer in the presence of Zn^{2+}. All panels are reprinted from *PNAS* (Conrady *et al.*, 2008), © 2008 National Academy of Sciences, USA.

self-assemble in a modular fashion, with 2–3 zincs per dimer interface. Since the full-length Aap protein contains up to 17 B-repeats, this type of modular self-assembly enables the formation of an extensive adhesive contact between bacterial cells in biofilms, which we have termed the "zinc zipper." Consistent with these biophysical data, we found that removal of free zinc through use of the chelator DTPA prevented the formation of biofilms by both *S. epidermidis* and *S. aureus* (Conrady *et al.*, 2008).

3.3. Other experimental approaches

Although analytical ultracentrifugation is the preferred method for studying metal ion-induced protein assembly, AUC may not be accessible to all investigators. There are many other approaches that can be used in lieu of AUC to analyze such linked equilibria. In general, any technique that monitors a change in shape or molecular weight could be used to analyze metal ion-induced dimerization. Examples of these techniques include dynamic or static light scattering, gel filtration, or even native gel electrophoresis. As stated previously, in order to get accurate quantitative information, it is necessary to ensure that protein dimerization is measured as a function of *free* metal ion concentration rather than total concentration. One example of such an alternative approach is shown in Fig. 5.6A. Gel filtration was used to show that in the presence of 2 m*M* $CaCl_2$, SCAN underwent a self-assembly reaction, as evidenced by the shift in retention time of the SCAN peak as a function of protein and calcium concentrations (Yang *et al.*, 2006). If AUC had not been a viable option for this project, we could have repeated the gel filtration experiment at a fixed protein concentration with varying calcium concentrations to observe the calcium-dependent monomer–dimer transition. Likewise, the experiment shown in Fig. 5.6A could have been fitted to determine a momomer–dimer equilibrium constant; this procedure could then have been repeated at multiple discrete free calcium concentrations in order to generate a log$K$ versus log $[Ca^{2+}]$ plot, similar to the plots shown in Fig. 5.5C.

Spectroscopic approaches are powerful techniques that can be used to analyze metal ion-induced protein assembly, particularly for cases in which the dimerization is high-affinity. Intrinsic fluorescence can provide information on dimerization if a fluorescent residue is in or near the dimer interface, or if dimerization induces conformational changes that affect fluorescent residues. The signal measured could be enhancement or quenching of the intrinsic fluorescence intensity or a shift in the peak of maximum emission (i.e., a red- or blue-shift), depending on the chemical environment of the fluorophore before and after dimerization. Figure 5.6B shows another example for SCAN. The first indication that SCAN exhibited a calcium-induced dimerization equilibrium was the discovery that calcium induced an increase in tryptophan fluorescence (Murphy *et al.*, 2003). After solving the crystal structure of SCAN, we identified point mutations that would disrupt the dimer interface (Yang *et al.*, 2006, 2008). As can be seen in Fig. 5.6B, the obligate monomer mutants I170K and

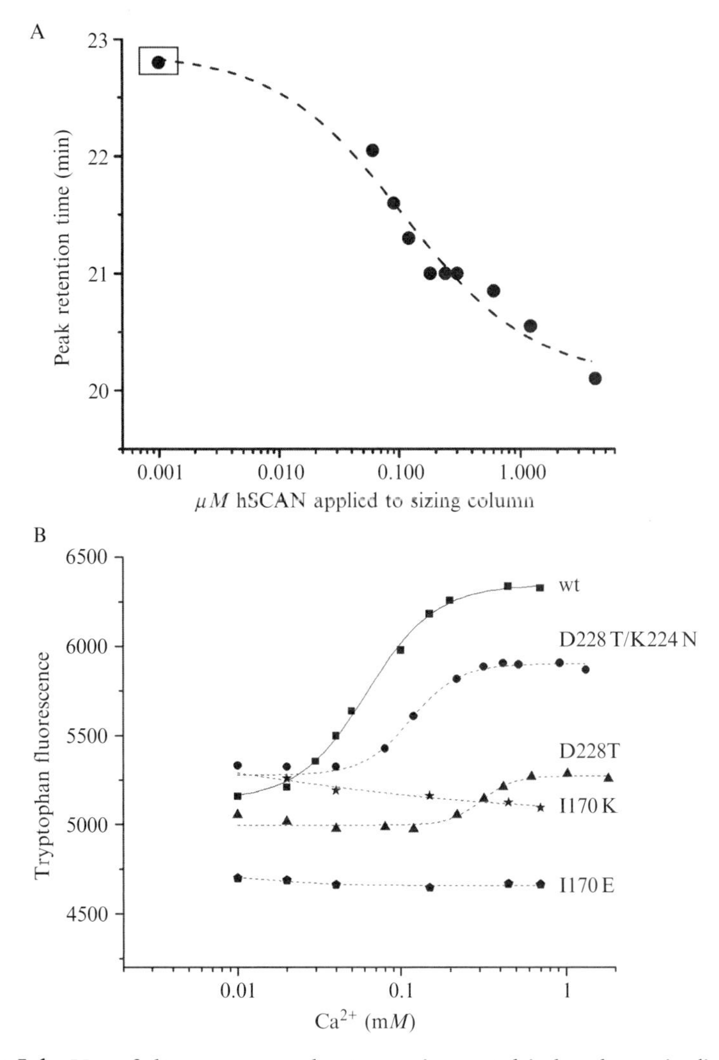

Figure 5.6 Use of alternate approaches to monitor metal-induced protein dimerization. (A) Plot of gel filtration retention times for various concentrations of SCAN in the absence of $CaCl_2$ (boxed data point on the left) or in the presence of 2 mM $CaCl_2$. This is a generally applicable approach that can be used to study metal ion-dependent protein dimerization. A similar experiment can be carried out by loading the same concentration of protein in each run and varying the metal ion concentration. Panel A was originally published in *JBC* (Yang *et al.*, 2006), © 2006 the American Society for Biochemistry and Molecular Biology. (B) Intrinsic fluorescence of SCAN and SCAN mutants as a function of $CaCl_2$ concentration. Wild-type SCAN exhibits a pronounced increase in fluorescence due to calcium-induced dimerization. Obligate monomeric mutants I170K and I170E, designed based on the crystal structure of the SCAN dimer (Yang *et al.*, 2006), do not show the sigmoidal transition in fluorescence intensity as a function of calcium. Panel B was originally published in *Biochemistry* (Yang *et al.*, 2008). Reprinted with permission, © 2008 American Chemical Society.

I170E no longer showed the sigmoidal increase in fluorescence as a function of calcium concentration, unlike the wild-type enzyme (Yang *et al.*, 2008).

Another useful spectroscopic approach for monitoring dimerization is fluorescence polarization or anisotropy. This approach reports on the rate of tumbling of a fluorescent molecule; by labeling a protein with a fluorophore, one can follow the increase in polarization upon dimerization (due to the slower tumbling of the more massive dimer relative to the monomer). This approach would be particularly useful for smaller proteins or other macromolecules.

Indirect approaches can also be used to monitor metal ion-induced dimerization. For example, dimerization can stabilize the tertiary fold of proteins, so measuring the stability of a protein by circular dichroism or differential scanning calorimetry could indirectly yield information about metal ion-induced dimerization. Of course, this assumes that one has already established that the protein is assembling in the presence of metal (as opposed to measuring stabilization of a monomeric fold by metal binding).

Finally, there are other nonequilibrium approaches that can provide qualitative information about metal-induced protein assembly, although these will not provide the full range of information possible by equilibrium-based techniques. The primary example in this category is chemical cross-linking. In this case, the intent is to accurately report on a preexisting equilibrium, even though the technique itself will alter the equilibrium. It is important to choose conditions such that transient interactions between monomers do not result in covalent attachment, but that preexisting dimers are covalently bound for subsequent analysis, such as by gel electrophoresis. A similar case exists for mass spectrometry, which can report on a preexisting equilibrium between protein oligomers. However, the process of injecting the sample for mass spectrometric analysis will dramatically perturb the chemical environment of the protein and may have unknown effects on the assembly equilibrium.

4. Summary

Analyzing metal-induced protein assembly according to the thermodynamics of linked equilibria (Wyman and Gill, 1990) is a powerful approach that can provide a wealth of insight into the biological system of interest. For example, in the case of the B-repeat region of Aap, our analysis of zinc-mediated dimerization revealed that individual B-repeats from Aap are capable of forming dimers with two to three zinc ions per dimer. This suggested that each of these repeat domains comprises an independent assembly module within the context of the intact protein, which contains

5–17 tandem copies of the B-repeat domain. Zinc-dependent self-association of the full B-repeat region of Aap would give rise to an extensive adhesive contact between staphylococcal cells within a biofilm, the so-called zinc zipper. Based on these results, we successfully predicted that zinc chelation would prove effective at inhibiting the formation of staphylococcal biofilms, which are a primary cause of recurrent hospital-acquired infections (Otto, 2008). In the case of SCAN, the approaches described here were able to resolve the basis for the calcium-dependent fluorescence change, and they revealed that dimerization is linked to higher enzymatic activity. This discovery could help lead to the design of engineered forms of SCAN that are thermostable and show high activity, characteristics that would be useful for antithrombotic applications.

ACKNOWLEDGMENTS

We acknowledge K. Horii, T.L. Kirley, and C.C. Brescia, whose work on the SCAN and Aap projects has been featured here. We thank R.A. Kovall for comments on the chapter. Several figures were reprinted courtesy of the *Proceedings of the National Academy of Sciences (USA)*, the *Journal of Biological Chemistry*, and *Biochemistry*. Funding for these projects was provided by the State of Ohio Eminent Scholar Program, NIH R01-HL72382, NIH T32-AI055406, and a pilot grant from the Midwest Center for Emerging Infectious Diseases at the University of Cincinnati (MI-CEID).

REFERENCES

Alattia, J. R., Ames, J. B., Porumb, T., Tong, K. I., Heng, Y. M., Ottensmeyer, P., Kay, C. M., and Ikura, M. (1997). Lateral self-assembly of E-cadherin directed by cooperative calcium binding. *FEBS Lett.* **417,** 405–408.

Bateman, A., Holden, M. T., and Yeats, C. (2005). The G5 domain: A potential *N*-acetylglucosamine recognition domain involved in biofilm formation. *Bioinformatics* **21,** 1301–1303.

Brown, A. K., Meng, G., Ghadbane, H., Scott, D. J., Dover, L. G., Nigou, J., Besra, G. S., and Fütterer, K. (2007). Dimerization of inositol monophosphatase *Mycobacterium tuberculosis* SuhB is not constitutive, but induced by binding of the activator Mg^{2+}. *BMC Struct. Biol.* **7,** 55.

Chaudhry, C., Plested, A. J., Schuck, P., and Mayer, M. L. (2009). Energetics of glutamate receptor ligand binding domain dimer assembly are modulated by allosteric ions. *Proc. Natl. Acad. Sci. USA* **106,** 12329–12334.

Conrady, D. G., Brescia, C. C., Horii, K., Weiss, A. A., Hassett, D. J., and Herr, A. B. (2008). A zinc-dependent adhesion module is responsible for intercellular adhesion in staphylococcal biofilms. *Proc. Natl. Acad. Sci. USA* **105,** 19456–19461.

Correia, J. J. (2000). Analysis of weight average sedimentation velocity data. *Methods Enzymol.* **321,** 81–100.

Correia, J. J. (2010). Analysis of tubulin oligomers by analytical ultracentrifugation. *Methods Cell Biol.* **95,** 275–288.

Correia, J. J., Alday, P. H., Sherwood, P., and Stafford, W. F. (2009). Effect of kinetics on sedimentation velocity profiles and the role of intermediates. *Methods Enzymol.* **467,** 135–161.

Corrigan, R. M., Rigby, D., Handley, P., and Foster, T. J. (2007). The role of *Staphylococcus aureus* surface protein SasG in adherence and biofilm formation. *Microbiology* **153,** 2435–2446.

Costerton, J. W., Stewart, P. S., and Greenberg, E. P. (1999). Bacterial biofilms: A common cause of persistent infections. *Science* **284,** 1318–1322.

Curtain, C. C., Ali, F., Volitakis, I., Cherny, R. A., Norton, R. S., Beyreuther, K., Barrow, C. J., Masters, C. L., Bush, A. I., and Barnham, K. J. (2001). Alzheimer's disease amyloid-beta binds copper and zinc to generate an allosterically ordered membrane-penetrating structure containing superoxide dismutase-like subunits. *J. Biol. Chem.* **276,** 20466–20473.

Demeler, B. (2005). UltraScan—A comprehensive data analysis software package for analytical ultracentrifugation experiments. *In* "Analytical Ultracentrifugation: Techniques and Methods," (D. J. Scott, S. E. Harding, and A. J. Rowe, eds.), pp. 210–229. Royal Society of Chemistry, Cambridge, UK.

Fan, Q. R., Long, E. O., and Wiley, D. C. (2000). Cobalt-mediated dimerization of the human natural killer cell inhibitory receptor. *J. Biol. Chem.* **275,** 23700–23706.

Gilbert, L. M., and Gilbert, G. A. (1978). Molecular transport of reversibly reacting systems: Asymptotic boundary profiles in sedimentation, electrophoresis, and chromatography. *Methods Enzymol.* **48,** 195–212.

Hussain, M., Herrmann, M., von Eiff, C., Perdreau-Remington, F., and Peters, G. (1997). A 140-kilodalton extracellular protein is essential for the accumulation of *Staphylococcus epidermidis* strains on surfaces. *Infect. Immun.* **65,** 519–524.

Jobling, M. F., Huang, X., Stewart, L. R., Barnham, K. J., Curtain, C., Volitakis, I., Perugini, M., White, A. R., Cherny, R. A., Masters, C. L., Barrow, C. J., Collins, S. J., *et al.* (2001). Copper and zinc binding modulates the aggregation and neurotoxic properties of the prion peptide PrP106-126. *Biochemistry* **40,** 8073–8084.

Johnson, M. L., Correia, J. J., Yphantis, D. A., and Halvorson, H. R. (1981). Analysis of data from the analytical ultracentrifuge by nonlinear least-squares techniques. *Biophys. J.* **36,** 575–588.

Laue, T. M., and Stafford, W. F. (1999). Modern applications of analytical ultracentrifugation. *Annu. Rev. Biophys. Biomol. Struct.* **28,** 75–100.

Laue, T. M., Shah, B. D., Ridgeway, T. M., and Pelletier, S. M. (1992). Computer-aided interpretation of sedimentation data for proteins. *In* "Analytical Ultracentrifugation in Biochemistry and Polymer Science," (S. E. Harding, A. J. Rowe, and J. C. Horton, eds.), pp. 90–125. Royal Society of Chemistry, London.

Murphy, D. M., Ivanenkov, V. V., and Kirley, T. L. (2003). Bacterial expression and characterization of a novel, soluble, calcium-binding, and calcium-activated human nucleotidase. *Biochemistry* **42,** 2412–2421.

Nagar, B., Overduin, M., Ikura, M., and Rini, J. M. (1996). Structural basis of calcium-induced E-cadherin rigidification and dimerization. *Nature* **380,** 360–364.

Otto, M. (2008). Staphylococcal biofilms. *Curr. Top. Microbiol. Immunol.* **322,** 207–228.

Philo, J. S. (1997). An improved function for fitting sedimentation velocity data for low-molecular-weight solutes. *Biophys. J.* **72,** 435–444.

Philo, J. S. (2000). A method for directly fitting the time derivative of sedimentation velocity data and an alternative algorithm for calculating sedimentation coefficient distribution functions. *Anal. Biochem.* **279,** 151–163.

Rohde, H., Burdelski, C., Bartscht, K., Hussain, M., Buck, F., Horstkotte, M. A., Knobloch, J. K., Heilmann, C., Herrmann, M., and Mack, D. (2005). Induction of *Staphylococcus epidermidis* biofilm formation via proteolytic processing of the accumulation-associated protein by staphylococcal and host proteases. *Mol. Microbiol.* **55,** 1883–1895.

Rohde, H., Burandt, E. C., Siemssen, N., Frommelt, L., Burdelski, C., Wurster, S., Scherpe, S., Davies, A. P., Harris, L. G., Horstkotte, M. A., Knobloch, J. K., Ragunath, C., *et al.* (2007). Polysaccharide intercellular adhesin or protein factors in biofilm accumulation of *Staphylococcus epidermidis* and *Staphylococcus aureus* isolated from prosthetic hip and knee joint infections. *Biomaterials* **28,** 1711–1720.

Roussel, A., Anderson, B. F., Baker, H. M., Fraser, J. D., and Baker, E. N. (1997). Crystal structure of the streptococcal superantigen SPE-C: Dimerization and zinc binding suggest a novel mode of interaction with MHC class II molecules. *Nat. Struct. Biol.* **4,** 635–643.

Schuck, P. (2000). Size-distribution analysis of macromolecules by sedimentation velocity ultracentrifugation and Lamm equation modeling. *Biophys. J.* **78,** 1606–1619.

Schuck, P. (2003). On the analysis of protein self-association by sedimentation velocity analytical ultracentrifugation. *Anal. Biochem.* **320,** 104–124.

Scott, D. J., and Schuck, P. (2005). A brief introduction to the analytical ultracentrifugation of proteins for beginners. *In* "Analytical Ultracentrifugation: Techniques and Methods," (D. J. Scott, S. E. Harding, and A. J. Rowe, eds.), pp. 1–25. Royal Society of Chemistry, Cambridge, UK.

Smith, T. M., and Kirley, T. L. (2006). The calcium activated nucleotidases: A diverse family of soluble and membrane associated nucleotide hydrolyzing enzymes. *Purinergic Signal* **2,** 327–333.

Stafford, W. F., III (1992). Boundary analysis in sedimentation transport experiments: A procedure for obtaining sedimentation coefficient distributions using the time derivative of the concentration profile. *Anal. Biochem.* **203,** 295–301.

van Holde, K. E., and Weischet, W. O. (1978). Boundary analysis of sedimentation-velocity experiments with monodisperse and paucidisperse solutes. *Biopolymers* **17,** 1387–1403.

Wyman, J., and Gill, S. J. (1990). Binding and linkage: Functional chemistry of biological macromolecules. University Science Books, Mill Valley, CA.

Yang, M., Horii, K., Herr, A. B., and Kirley, T. L. (2006). Calcium-dependent dimerization of human soluble calcium activated nucleotidase: Characterization of the dimer interface. *J. Biol. Chem.* **281,** 28307–28317.

Yang, M., Horii, K., Herr, A. B., and Kirley, T. L. (2008). Characterization and importance of the dimer interface of human calcium-activated nucleotidase. *Biochemistry* **47,** 771–778.

CHAPTER SIX

Thermodynamic Dissection of Colicin Interactions

Nicholas G. Housden *and* Colin Kleanthous

Contents

Abstract

Bacteriocins are selective protein antibiotics that bind and kill specific bacterial species, the best studied of which are the colicins that target *Escherichia coli*. Colicins tend to parasitize cell envelope systems that are important for cell viability under nutrient-limited conditions or environmental stress. In this chapter, we review how in conjunction with other biophysical methods and structural information, isothermal titration calorimetry (ITC) has been used to investigate

Department of Biology (Area 10), University of York, York, United Kingdom

Methods in Enzymology, Volume 488
ISSN 0076-6879, DOI: 10.1016/B978-0-12-381268-1.00006-9
© 2011 Elsevier Inc.
All rights reserved.

how colicins enter *E. coli* cells. In particular, we summarize current understanding of the thermodynamics of outer membrane receptor binding and how this has been linked to biological function. We also summarize thermodynamic investigations using ITC that have helped elucidate the mechanisms by which colicins bind and parasitize proteins in the periplasm, forming protein–protein interactions that ultimately trigger translocation across the outer membrane. Our review focuses on the two major cytotoxic classes of colicin that have been the subject of intense investigation, pore-forming toxins, and nonspecific endonucleases (DNases).

DNase colicin-producing *E. coli* avoid committing suicide through the production of a small antidote protein known as the immunity (Im) protein, with the Im protein only released once cell-entry is initiated. Exosite binding by Im proteins has driven an evolutionary arms race among colicin-producing bacteria whereby markedly different colicin DNase–Im protein interaction specificities have evolved without impacting on cytotoxicity. Extensive investigations have shown that homologous colicin DNase–Im protein complexes have K_ds, for cognate and noncognate complexes, that vary by 10-orders of magnitude, essentially matching the entire spectrum of binding affinities seen for protein–protein interactions in biology. Hence, this system has proved to be a powerful model for investigating the thermodynamics of specificity in protein–protein interactions, with ITC being the principal tool. We review this literature and point to how the thermodynamic information that has been generated complements various other kinetic and structural data.

1. Introduction

Outside the laboratory, bacteria live in a competitive environment where they use all means at their disposal to gain an advantage over other bacterial strains. Colicins are plasmid-encoded protein antibiotics produced by some strains of *Escherichia coli* under conditions of nutritional or environmental stress to eliminate other closely related strains, thereby enhancing the chances of survival of the colicin-producing strain (Cascales *et al.*, 2007). Colicins have been shown to promote microbial diversity within structured environments such as the mammalian colon (Kirkup and Riley, 2004). Once released from the producing cell, the colicin gains entry into its target bacterium through the parasitization of proteins within the outer membrane and periplasm (Lazdunski *et al.*, 1998). Colicins are organized into three functional domains, each one of which is responsible for a critical process required for cell killing to occur. The central receptor-binding domain locks the colicin onto the surface of the target bacterium through a high-affinity interaction with an outer membrane nutrient receptor. The N-terminal translocation domain must then interact with a translocator protein in the outer membrane and proteins within the periplasm of the target cell to mediate delivery of the C-terminal cytotoxic domain to the relevant cellular location for it to bring about cell killing. A variety of colicins exist which bring about cell killing through a number of different mechanisms, including the insertion of a depolarizing pore into the bacterial inner membrane and the delivery of a lethal degrading enzyme to either the periplasm or the cytoplasm.

Colicins are grouped according to the periplasmic proteins with which they interact to facilitate delivery of their cytotoxic domains into the target cell. Group A colicins bind components of the Tol system, while group B colicins interact with members of the Ton system. Both the Tol and Ton systems access the inner membrane proton motive force; therefore, interacting with these systems may provide energy for driving the colicin into the cell. The protein–protein interactions that occur between colicins and proteins within their target bacteria are critical to the processes that bring about cell killing. Here, we examine these interactions for two group A colicins, the pore-forming colicin N (Bourdineaud *et al.*, 1990) that brings about cell killing through destruction of the proton motive force, and the nonspecific DNase colicin E9 (James *et al.*, 1996; Fig. 6.1). The receptor-binding and translocation domains of eight enzymatic E colicins (E2–E9) are highly conserved, indicating a common mechanism for the delivery of varied toxic enzymes to the bacterial cytoplasm. The cytotoxic nuclease domain of enzymatic E colicins takes the form of a ribosomal RNase, a transfer RNA-specific RNase, or a nonspecific DNase which target the bacterial genomic DNA. Each of the enzymatic E colicins is produced as a heterodimer with its own specific immunity protein to prevent the colicin-producing cell from committing suicide (Kleanthous and Walker, 2001).

Understanding the protein–protein interactions that take place during colicin translocation is pivotal in understanding how these protein antibiotics function. Isothermal titration calorimetry (ITC) has been used extensively to investigate the binding parameters of various colicin-related complexes. Moreover, the four different DNase colicins and their associated immunity proteins are a model system for studying protein–protein interactions, displaying both specificity and promiscuity. The combination of cognate and noncognate DNase–immunity complexes spans the range of biologically relevant binding affinities seen within nature. Insight into the processes of receptor-binding and the recruitment of periplasmic proteins involved in delivery of the cytotoxic domain into the target cell is essential to developing an understanding of the mode of action of these potent antibiotics. The application of ITC to studying these complexes, mapping epitopes responsible for binding, and building a picture of the protein–protein interaction network central to colicin uptake—are all discussed in the following sections.

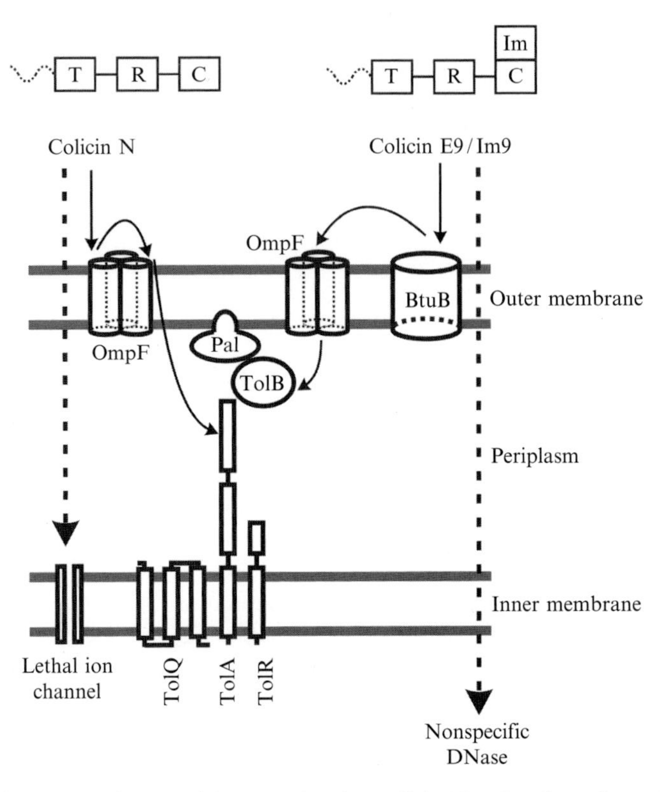

Figure 6.1 Route of entry of the pore-forming colicin N and endonuclease colicin E9 across the *E. coli* cell envelope. Colicins are modular proteins (top of figure), containing an N-terminal translocation (T-) domain, a central receptor-binding (R-) domain, and a C-terminal cytotoxic (C-) domain. T-domains have both a structured region and a natively unfolded region (shown as a dotted line). Nuclease colicins are cosynthesized with high-affinity immunity proteins that are lost as the toxin enters a susceptible cell. The receptor-binding domain of colicin N binds to OmpF in the outer membrane, followed by interaction of its T-domain with OmpF, allowing entry of the intrinsically unstructured N-terminus into the periplasm where it can bind to TolA. TolA binding initiates the transport of the C-terminal cytotoxic domain of colicin N to the *E. coli* inner membrane where it inserts and forms a lethal ion channel destroying the proton motive force across the inner membrane. The receptor-binding domain of colicin E9 locks the toxin onto the bacterial surface through interaction with the vitamin B_{12} receptor BtuB, from where the intrinsically disordered N-terminus of the T-domain recruits OmpF, allowing entry of the TolB binding epitope into the periplasm where it can bind TolB. TolB binding initiates translocation of the C-terminal nonspecific DNase into the *E. coli* cytoplasm where it randomly degrades the genomic DNA by a mechanism that is currently not understood. Both TolA and TolB are part of the transperiplasmic Tol assembly in Gram-negative bacteria (TolQ, TolR, TolA, TolB, and Pal) which is involved in stabilizing the outer membrane.

1.1. Measuring equilibrium dissociation constants

When studying any protein–ligand interaction, the ability to quantify binding affinities through the measurement of equilibrium dissociation constants (K_d) is central to our understanding of the complexes formed. Two strategies exist for determining K_d values, either the microscopic association and dissociation rate constants are measured through pre-equilibrium binding experiments or the amount of ligand present in complex at the equilibrium is assessed across a range of ligand concentrations. Both approaches have been used in colicin research. While pre-equilibrium measurements can be highly informative, complex formation is rarely a simple one-step process in which K_d can be determined from the rate constant of complex dissociation (k_{-1}) divided by the rate constant for complex formation (k_1).

$$\text{Protein} + \text{Ligand} \underset{k_{-1}}{\overset{k_1}{\rightleftharpoons}} \text{Protein}\cdot\text{Ligand} \qquad K_d = k_{-1}/k_1$$

In order to determine the K_d from pre-equilibrium rate constants, it is essential that the binding interaction is fully understood and that all rate constants can be measured and assigned correctly. Pre-equilibrium rate constants are typically measured through stopped-flow fluorescence spectroscopy or surface plasmon resonance, the former of which has successfully been used in the study of E colicin nuclease domains binding to their inhibitory immunity proteins (Keeble and Kleanthous, 2005; Li *et al.*, 2004). Stopped-flow fluorescence spectroscopy requires a usable change in the intrinsic fluorescence of the complex components upon complex formation. If there is no change in the intrinsic fluorescence properties of the complex, it may be possible to introduce a fluorescent reporter through either site-directed mutagenesis or chemical modification, both of which risk changing the very properties of the complex that are under investigation. While surface plasmon resonance does not require fluorescent reporter groups, the need to immobilize one of the components of the complex onto a surface potentially alters the properties of the complex under investigation.

Through the measurement of complex formation under equilibrium conditions across a range of ligand concentrations, the K_d can be determined without the need for a detailed dissection of the binding mechanism. At given concentrations of protein and ligand, the amount of ligand in complex with the protein is defined by the K_d of the interaction.

$$K_d = [\text{Protein}][\text{Ligand}]/[\text{Protein} \times \text{Ligand}]$$

A wide variety of biophysical parameters can be used to determine the amount of ligand present in the complex. The most fundamental measurement that can be made in the case of equilibrium dialysis experiments is the concentration of free ligand. If the total ligand concentration is known, measuring the free ligand concentration also gives the concentration of the protein–ligand complex. While equilibrium dialysis requires minimal specialist equipment, it is demanding both in terms of the duration of the experiment and the amounts of protein and ligand required, with ligands often requiring specific radiolabeling to allow their detection and quantification. Measurement of a biophysical property of the complex that differs from its components such as fluorescence, anisotropy, or molar ellipticity allows complex formation as a product of ligand titration to be monitored and dissociation constant and stoichiometry of binding to be calculated.

1.2. Isothermal titration calorimetry

Intrinsic spectroscopic signals which change significantly upon complex formation are not always available. Introduction of extrinsic reporter can be problematic as discussed above. ITC offers a label-free solution for monitoring complex formation. Binding processes may either be exothermic or endothermic, releasing or taking in heat from their surroundings. ITC monitors binding through the measurement of the amounts of heat exchanged with the surroundings upon complex formation. Not only does this allow the affinity of binding interactions to be measured, the thermodynamic basis of the binding affinity can also be determined. The amount of heat exchanged with the surroundings upon ligand binding is proportional to the enthalpy change (ΔH) and the amount of complex being formed. From the binding curve obtained, through monitoring the heat exchange with the surroundings upon titrating in ligand, it is possible to determine the K_d of the binding interaction. Once the K_d of the interaction is known, it is possible to determine the ΔH of the interaction from the observed heats and the amount of complex that is formed. The free energy change (ΔG) of the binding interaction can be derived from the measured K_d using the equation $\Delta G = RT \ln K_d$, which in combination with the measured ΔH and the equation $\Delta G = \Delta H - T\Delta S$, allows ΔS to be calculated. Therefore, a single ITC experiment, unlike other methods that measure equilibrium binding constants, gives ΔH, ΔG, and ΔS of the binding reaction.

1.3. Sample considerations

While ITC is widely regarded as demanding large amounts of protein, this is very much dependent upon the interaction being studied. The appropriate protein concentrations to be used in the ITC cell can be estimated from the C-value, which is based upon the affinity and stoichiometry (n) of ligand binding.

$$C = ([\text{protein}] \times n)/K_d$$

Accurate determination of K_d requires C values to be between 1 and 1000. As a result, to obtain binding parameters for a high-affinity interaction necessitates the use of protein at low concentrations in the cell and therefore, binding can only be observed if the ΔH is large; otherwise, the evolved heats will be below the detection limit of the calorimeter. When studying low-affinity interactions, the concentrations within the cell are higher. The concentrations of ligand used in the syringe must be sufficiently high to ensure that by the end of the titration the protein in the cell is saturated. For high-affinity interactions, this is easily achievable; however, for low-affinity binding, availability of material and solubility are the factors that limit accurate measurement of binding parameters. While the burden on amounts of material has been reduced with the introduction of commercially available ITC instruments with cell volumes reduced to ~200 μl, the solubility issues remain. Great care is needed when using high concentrations of protein, close to the limit of their solubility, as any precipitation occurring during the binding experiment will impact upon the observed heats and baseline stability.

The quality of material used in ITC experiments will affect the results obtained. When performing qualitative experiments to detect whether or not binding occurs, less than pure samples can be tolerated assuming that any observed binding is not due to contaminants that are present. Highly purified proteins with accurately determined concentrations are required for the correct determination of binding parameters. Inaccuracies in the concentration of either the cell or syringe component will result in non-integer binding stoichiometries which will impact on the other parameters determined. While protein concentrations can be determined accurately through a combination of amino acid analysis and measurements of absorbance at 280 nm, these values give no information on the proportion of material that is present in a correctly folded, active form. Therefore, even if concentrations are known with great accuracy, the observed binding parameters may not reflect the true values for the complex.

1.4. Baseline corrections

Factors unrelated to binding can impact upon the heats observed during the binding titration. For example, the high concentration of protein required for use in the injection syringe may result in the formation of oligomers; upon titration into the sample cell, these oligomers may dissociate due to the decrease in concentration. As a result, any heats observed would be a combination of the binding process and rearrangement of the monomer–oligomer equilibrium. While it is always preferable to perform an extensive dialysis of both the protein to be used in the cell and that to be used in the syringe into the same buffer in the same vessel, this is not always possible. If a small molecule ligand or a peptide is to be titrated from the syringe, the molecular weight of the ligand may prohibit dialysis. Under such conditions, the ligand should be dissolved in a dialysis buffer following dialysis of the protein, thereby minimizing the impact of buffer mismatch.

It is important to perform a control titration of ligand into buffer allowing subtraction of the observed heats of dilution from the observed binding data. If significant heats remain at the end of the titration even after subtraction of the control data, this may indicate that (i) saturation of the protein in the sample cell has not been achieved; (ii) a second weaker binding site for the ligand may exist (this could be investigated using higher ligand concentrations); (iii) dilution of the protein within the cell causes shifts in existing equilibriums (this can be verified through titration of the buffer into the protein); (iv) the control titration of the ligand into the buffer does not adequately reflect the changes that may occur during titration of the ligand into the protein.

1.5. Data fitting

The software provided by ITC manufacturers is typically capable of fitting data to a number of different binding models, including a single set of identical sites, two sets of independent sites, and sequential sites. The data discussed in the following sections were all analyzed according to the single identical sites model using Origin 7.0 (OriginLab Corporation). A detailed discussion of the mathematical basis of the single set of identical binding sites model has been given elsewhere (Indyk and Fisher, 1998).

1.6. Data presentation

Enthalpy changes are typically quoted in units of kJ/mol or kcal/mol. While the former is the appropriate SI unit for enthalpy, biologists and biochemists tend to favor the latter; therefore, kcal/mol for enthalpy and cal/mol/K for entropy will be the units used throughout this chapter. To aid the comparison of enthalpy and entropy contributions to the free energy of binding, we will consider the entropy in terms of $T\Delta S$ such that both enthalpy and entropy have units of kcal/mol.

ITC data are typically presented in a two-panel format, with one panel showing the raw data plotted as μcal/s against time and the second plotting the integrated heats for each injection in units of kcal/mol of injectant versus molar ratio. The latter of these two plots can easily be represented as a plot of total signal change upon binding versus ligand concentration, as shown in Fig. 6.2. While plotting the integrated heats/mol of injectant against molar ratio allows the binding stoichiometry to be readily observed,

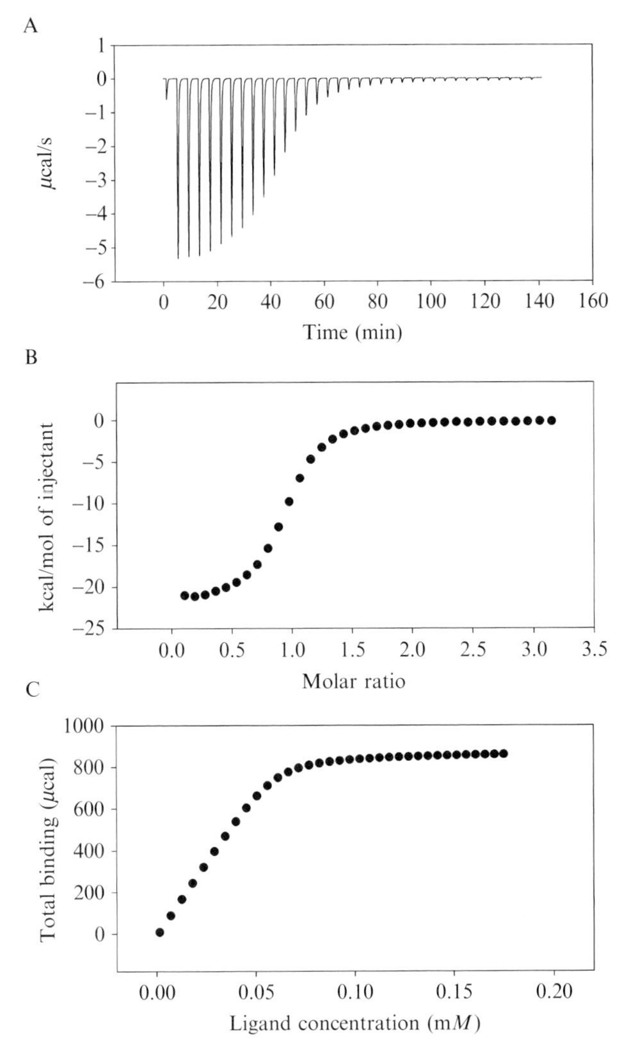

Figure 6.2 Presentation of ITC binding data. (A) Baseline corrected raw ITC data obtained on a VP-ITC (MicroCal) for the titration of 1-mM ligand into 63-μM protein for a binding interaction with the thermodynamic parameters of $K_d \sim 1.5$ μM, stoichiometry ~1 ligand/protein, and $\Delta H \sim -22$ kcal/mol. (B) Integrated injection heats from above presented in the typical kcal/mol of injectant versus Molar Ratio format. (C) Saturation curve of summated integrated injection heats giving total binding plotted against ligand concentration.

it is imperative that the figure legend accompanying any such plot clearly states the concentration of protein used in the cell and the concentration of the ligand within the syringe. Without these values, the presented data are relatively meaningless. For example, if we consider two titrations, both using 25 injections each of 10 μL, the first titration using 200 μM ligand in the syringe titrated into 10 μM protein in the cell for a binding interaction with a K_d of 1 μM and a ΔH of −2 kcal/mol, the observed kcal/mol of injectant plotted against molar ratio would be identical to a second titration in which 20 μM ligand is titrated into 1 μM protein for a 100-nM complex with a ΔH of −20 kcal/mol. When data are presented as total signal change upon binding versus ligand concentration, the differences in the affinity for the two titrations are readily apparent from the scale of the x-axis.

2. DNase Domain–Immunity Protein Interactions

E colicins that kill susceptible *E. coli* cells through the action of a nonspecific DNase domain include ColE2, ColE7, ColE8, and ColE9. Each of the four 15 kDa DNase cytotoxic domains is coproduced with its own specific 10 kDa immunity protein, namely, Im2, Im7, Im8, and Im9, which prevents the colicin-producing cell from committing suicide through degradation of its own genomic DNA. These immunity proteins do not bind to the active site of the cytotoxic domain, but instead bind to an adjacent site. Exosite binding inhibits the DNase through steric and electrostatic clashes between the immunity protein and the negatively charged phosphate backbone of the DNA substrate. As inhibition does not occur through binding at the active site, new DNase/immunity pairs have evolved without impacting on the cytotoxic DNase activity of the toxin. The different DNase/immunity pairs are the product of divergent evolution, with the DNases and their immunity proteins retaining 65 and 50% sequence identity, respectively.

The cognate complexes formed between the colicin DNases and their equivalent immunity proteins are among the highest affinity protein–protein interactions observed in nature, with K_ds of $\sim 10^{-15}$ M (Li *et al.*, 2004). Measurement of such high-affinity binding interactions presents a considerable challenge, requiring determination of the rapid association rate constants through pre-equilibrium stopped-flow fluorescence spectroscopy. Rate constants for the dissociation of these complexes are extremely slow, occurring with half-lives of several days, necessitating radioactive subunit exchange for their accurate measurement. While *in vivo* function is restricted to cognate complexes, all 16 DNase–Im complexes can be measured *in vitro*. The affinities of the noncognate complexes are between 10^6- and 10^{10}-fold weaker than those of the cognate interactions. As a result, the DNase–Im complexes are a powerful model system for studying specificity in protein–protein interactions as they span the spectrum of biologically relevant K_ds ranging from the millimolar to femtomolar range.

To complement the kinetically derived K_d values, ITC studies have been employed to investigate the thermodynamic basis of the binding affinities of cognate and noncognate DNase–Im complexes (Keeble *et al.*, 2006). While the accuracy of ITC is limited when it comes to measuring low nanomolar K_ds, the technique remains accurate in the determination of ΔH of binding for subnanomolar complexes. If the affinity of the complex is known through other techniques, a full thermodynamic profile can still be ascertained, calculating ΔS from the known ΔG and the ΔH value determined through ITC. Entropy values measured by ITC contain the highest error levels of any of the determined parameters as they combine the errors of the ΔG and ΔH measurements. When using ΔG values ascertained from other techniques in combination with the ΔH measured by ITC, the errors on the calculated ΔS are likely to be considerably higher.

2.1. Thermodynamics of cognate DNase–Im complexes

Zinc is known to bind at the active site of the colicin DNase domain where it plays an inhibitory role against nuclease activity (Ni^{2+} ions are, however, active cofactors). While the all four cognate DNase–Im complexes formed in the presence of zinc show similar femtomolar binding, dissection of these binding processes into their enthalpy and entropy components reveals underlying differences. All four complexes are enthalpically favorable, but the magnitude of the enthalpy contribution to binding varies significantly with ΔH values of -33.1, -30.3, -24.2, and -10.5 kcal/mol for the cognate complexes of E2, E7, E8, and E9 DNase, respectively. The complexes of E2, E7, and E8 DNase show decreasingly unfavorable entropy terms ($T\Delta S = -13.5$, -10.5, and -4.4 kcal/mol, respectively), while the binding of E9 DNase to Im9 is entropically favorable ($T\Delta S = +8.1$ kcal/mol).

2.2. The impact of zinc on DNase–Im thermodynamics

When investigating binding by ITC, it is important to consider that the observed enthalpies and entropies reflect the net changes in all equilibria within the system and not just the binding event of interest. This is clearly illustrated upon comparison of the binding of immunity protein to apo- and zinc-bound forms of the colicin DNases. The melting temperatures of the apo DNase domains are increased by $\sim$15–30 °C upon addition of zinc. This is particularly significant in the case of apo E7 DNase, which with a T_m of 26.3 °C, is only partially folded at 25 °C (Table 6.1). As a result, titration of Im7 into apo E7 DNase gives rise to $\Delta H = -56.7$ kcal/mol and $T\Delta S = -36.9$ kcal/mol attributable to binding and the refolding of partially unfolded E7 DNase, while titration of Im7 into the zinc-bound E7 DNase yields $\Delta H = -30.3$ kcal/mol and $T\Delta S = -10.5$ kcal/mol.

Table 6.1 The influence of Zn^{2+} on the DNase melting temperature and enthalpy change of binding

	T_m (°C)[a]		ΔH (kcal/mol)[b]	
	$+Zn^{2+}$	$-Zn^{2+}$	$+Zn^{2+}$	$-Zn^{2+}$
E2 DNase versus Im2	61.2	37.3	−33.1	−38.6
E7 DNase versus Im7	59.9	26.3	−30.3	−56.7
E8 DNase versus Im8	63.7	45.5	−24.2	−25.8
E9 DNase versus Im9	63.0	36.6	−10.5	−19.0

[a] Melting temperatures values determined by differential scanning calorimetry taken from van den Bremer *et al.* (2004).
[b] Enthalpy changes of binding determined through ITC taken from Keeble *et al.* (2006).

Interestingly, in the case of the colicin DNase, the presence or absence of zinc appears not to influence the affinity of the complex despite significantly impacting upon the observed thermodynamics (Fig 6.3).

2.3. Thermodynamics of noncognate DNase–Im complexes

The K_d values for the majority of the noncognate DNase–Im complexes have been determined through ITC, the notable exceptions being E9–Im7 where a small ΔH (~ -1 kcal/mol) and a K_d in the region of 100 μM make the interaction too weak to measure accurately and E7–Im8 and E8–Im7 where the subnanomolar K_ds are too tight to measure. Among the noncognate complexes, unfavorable entropy components appear less prevalent than for the cognate complexes, with only the E7–Im2 and the E8–Im2 showing significantly unfavorable entropies ($T\Delta S = -8.8$ and -2.9 kcal/mol, respectively; Fig 6.4).

It is difficult to rationalize the differences between cognate and noncognate complexes in terms of enthalpy and entropy when the thermodynamic basis of high-affinity binding varies significantly between the four cognate complexes. DNase–Im complexes occur through a dual recognition mechanism, where binding affinity is dominated through the interactions of conserved residues within helix III of the Im. Specificity of cognate complexes is brought about through additional favorable contacts of helix II of the Im. Recent crystallographic studies revealed the noncognate E9 DNase–Im2 complex to be nearly identical to the cognate E9 DNase–Im9 cognate complex. While the Im helix III interactions are conserved, the favorable specificity interactions of the cognate complex are replaced by unfavorable packing of side chains. Bound water molecules improving the complementarity of the protein surfaces through bridging hydrogen bonds are present in both the cognate and noncognate complexes. The increased

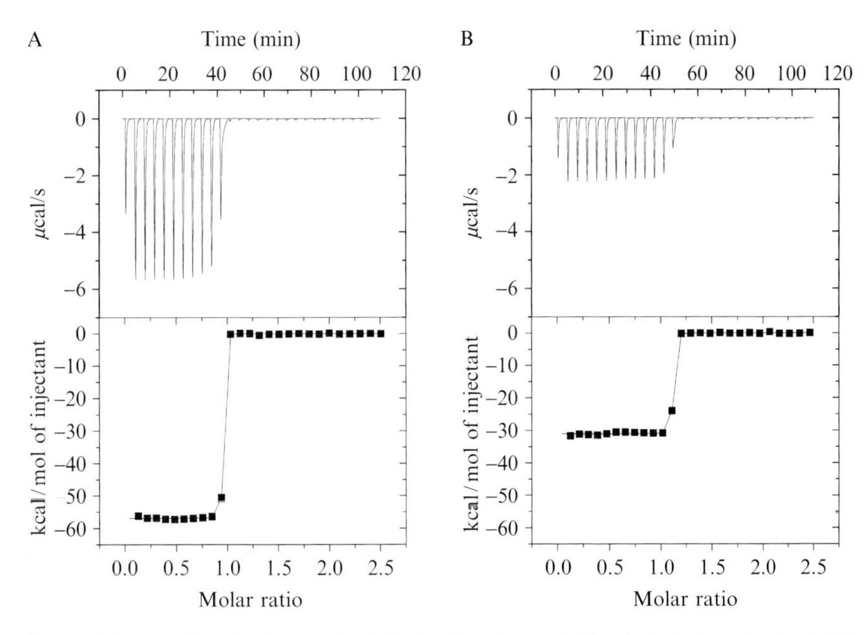

Figure 6.3 Binding isotherms for (A) the titration of 333-μ*M* Im7 into 28.5-μ*M* E7 DNase in 50-m*M* Mops, pH 7.0, 200-m*M* NaCl in the absence of Zn^{2+}, and (B) titration of 230-μ*M* Im7 into 20-μ*M* E7 DNase prebound with Zn^{2+} in 50-m*M* Mops, pH 7.0, 200-m*M* NaCl. Adapted from Keeble *et al.* (2006).

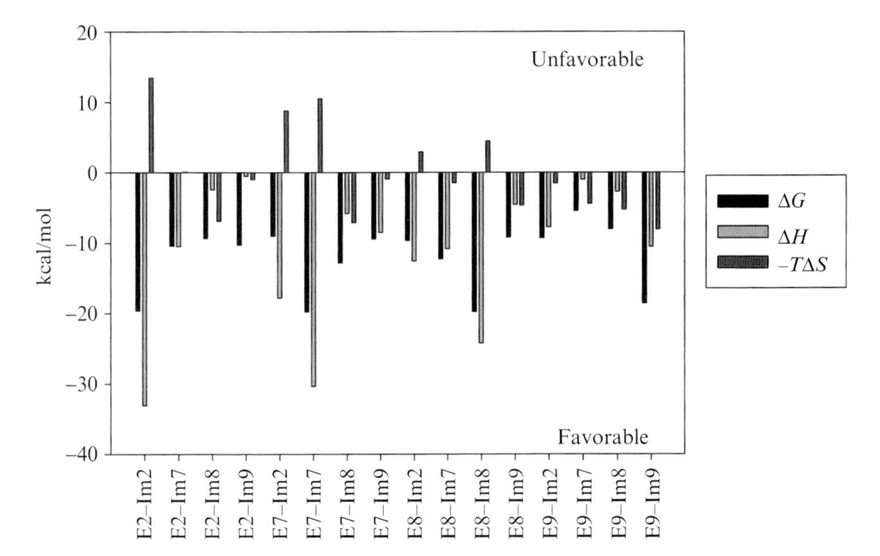

Figure 6.4 Contributions of enthalpy and entropy to the free energy of binding of colicin DNases to Im proteins. Entropy is plotted as $-T\Delta S$ such that a positive value represents an unfavorable contribution and a negative value represents a favorable contribution as is the case for ΔH. Adapted from Keeble *et al.* (2006).

Table 6.2 Kinetically determined K_d values of cognate and noncognate DNase–Im complexes[a]

	E2 DNase	E7 DNase	E8 DNase	E9 DNase
Im2	3.9×10^{-15}	3.6×10^{-10}	1.8×10^{-8}	1.0×10^{-8b}
Im7	1.4×10^{-8}	n.d.[c]	1.0×10^{-9}	$\sim 1 \times 10^{-4}$
Im8	3.8×10^{-8}	3.7×10^{-10}	n.d.	5.0×10^{-7}
Im9	1.2×10^{-8}	3.8×10^{-8}	6.4×10^{-8}	2.4×10^{-14}

[a] K_d values taken from Li *et al.* (2004).
[b] K_d value taken from Keeble and Kleanthous (2005).
[c] Although not determined, the K_d values of the E7 DNase–Im7 and E8 DNase–Im8 cognate complexes are assumed to be comparable to the values of the E2 DNase–Im2 and E9 DNase–Im9 cognate complexes.

number of bound water molecules in the noncognate complex may, in part, explain the less favorable entropy component of E9 DNase–Im2 complex formation (Table 6.2).

2.4. Measuring subnanomolar K_ds through competitive ITC

As discussed above, measuring the affinity of cognate DNase–Im complexes is far from routine, requiring a combination of stopped-flow fluorescence spectroscopy and radioactive subunit exchange experiments. While conventional ITC measurements are limited to low n*M* binding affinities or weaker, it is possible to access higher affinity interactions through competitive ITC. In such an experiment, the ligand in the ITC cell is in a preformed complex with a binding partner for which the thermodynamics of complex formation are known. Upon titration of a higher affinity competing ligand into the cell, the preformed complex dissociates and the higher affinity complex forms. The observed thermodynamic binding parameters are those for the net processes of dissociation of the preformed complex and formation of the high-affinity complex. Directed evolution has been used to select for Im7 mutants with enhanced *in vivo* stability (Foit *et al.*, 2009). The affinity with which these evolved Im7 constructs bind E7 DNase *in vitro* has been measured through competitive ITC. Displacement of Im8 from a preformed E7 DNase–Im8 noncognate complex resulting in a cognate E7 DNase–Im7 complex allowed the affinity of these mutants to be compared quantitatively with wild-type Im7. For such an experiment to work, it is imperative that the dissociation rate constant for the preformed complex is compatible with the time scale of the experiment. The higher affinity Im7 titrated into the E7 DNase–Im8 complex cannot actively dissociate Im8, it can only replace Im8 on the time scale of the natural dissociation of Im8 from the E7 DNase. While the user can adjust the spacing of injections, highly extended periods between injections are undesirable due to the occurrence of slow diffusion between the contents of the cell and the syringe. As the observed enthalpy change is that for dissociation of the preformed complex plus association of the higher affinity complex, it is essential that there is a detectable difference in the enthalpies of the two complexes. If the difference in affinity of the two complexes is caused predominantly by differences in entropy, there would be no detectable enthalpy change in the competition experiment.

3. Receptor Binding

The first step in colicin translocation is binding to an integral outer membrane protein on the surface of the target bacteria. In order to characterize these interactions by ITC, it is first necessary to purify sufficient quantities of these receptors to homogeneity in a detergent solubilized form.

3.1. Interaction of colicin N (ColN) with trimeric porins

The pore-forming toxin colicin N uses OmpF as its cell surface receptor and its translocation portal for delivering its N-terminal TolA binding epitope to the periplasm. In the absence of OmpF, colicin N-mediated cell killing can occur through interaction with the trimeric porins, OmpC and PhoE, although with impaired potency (Evans *et al.*, 1996; Table 6.3). Interestingly, the *in vitro* binding affinities of ColN with OmpF, PhoE, and OmpC are comparable (2.1 μM, 4.0 μM, and 6.7 μM, respectively), shedding little light on the disparity of the three porins in colicin N toxicity. Although the decreased affinity does follow the trend in decreasing colicin N toxicity (OmpF > PhoE > OmpC), toxicity decreases disproportionately to the modest change in affinity. When the free energy of binding is dissected into its enthalpic and entropic components, significant variations, which may account for differences in function, are seen. The observed affinity for the colicin N–OmpF complex is the net result of a large favorable enthalpy change ($\Delta H = -12.3$ kcal/mol) and an unfavorable entropy change

Table 6.3 Thermodynamic parameters of the Colicin N–trimeric porin complexes[a]

Receptor	K_d (μM)	N	ΔH (kcal/mol)	$T\Delta S$ (kcal/mol)
OmpF	2.1	2.4	−12.3	−4.6
PhoE	4.0	3.1	−6.0	+1.4
OmpC	6.7	2.8	−3.7	+3.3

[a] Values taken from Evans *et al.* (1996).

($T\Delta S = -4.6$ kcal/mol). While the enthalpy changes for the colicin N–PhoE and colicin N–OmpC complexes are less favorable ($\Delta H = -6.0$ and -3.7 kcal/mol, respectively), they are complemented by favorable entropy changes ($T\Delta S = +1.4$ and $+3.3$ kcal/mol, respectively). These data highlight the value of being able to obtain ΔG, ΔH, and ΔS from a single ITC experiment; other methods of assessing binding interactions would typically only yield information about the ΔG of binding, whereas the differences here which underlie the differences in function are within the enthalpy and entropy contributions to binding. With the observed interactions of colicin N with trimeric porins likely reflecting binding both of the receptor-binding and translocation domains to the porin, differences observed in the thermodynamics may reflect the conformational changes occurring within the colicin that facilitate translocation.

3.2. Formation of the BtuB–colicin E9 complex

The structures of BtuB in complex with the receptor-binding domains of colicin E2 (Sharma *et al.*, 2007) and colicin E3 (Kurisu *et al.*, 2003) have been solved by X-ray crystallography. No major conformational changes occur within the coiled-coil receptor-binding domain or the globular N-terminal plug domain of BtuB which blocks the lumen of its β-barrel. It is clear that the colicin cannot cross the outer membrane by passing through BtuB with the lumen of the barrel remaining plugged.

ITC measurements of the titration of colicin E9 into BtuB reveal a binding affinity of ~100 n*M* in a process characterized by an extremely favorable enthalpy change ($\Delta H \sim -40$ kcal/mol) and an unfavorable entropy component ($T\Delta S \sim -30$ kcal/mol) (Housden *et al.*, 2005; Table 6.4). Transport of vitamin B_{12} across the *E. coli* outer membrane

Table 6.4 Thermodynamic properties of Colicin E9–BtuB complexes formed in the presence and absence of Ca^{2+} ions[a]

	ColE9–Im9		$ColE9^{ss}$–Im9[b]	
	+5-m*M* EGTA	+5-m*M* $CaCl_2$	+5-m*M* EGTA	+5-m*M* $CaCl_2$
N	0.89 ± 0.01	0.94 ± 0.01	0.82 ± 0.01	0.90 ± 0.02
K_d (n*M*)	153 ± 28	2.1 ± 0.2	71.8 ± 3.3	1.0 ± 0.1
ΔH (kcal/mol)	−42.0 ± 2.4	−62.5 ± 2.3	−40.0 ± 1.0	−67.6 ± 0.1
$T\Delta S$ (kcal/mol)	−32.8 ± 2.5	−50.6 ± 2.1	−30.2 ± 1.1	−55.0 ± 0.2

[a] Values taken from Housden *et al.* (2005).
[b] $ColE9^{ss}$–Im9 is colicin E9 with a disulfide bond engineered between residues 324 and 447 at the top of the receptor-binding domain (Penfold *et al.*, 2004).

through BtuB is known to be dependent upon Ca^{2+} ions (Bradbeer *et al.*, 1986), and calcium also influences colicin E9 binding, with the affinity of the complex tightening by two orders of magnitude in its presence (Fig 6.5). Comparison of the crystal structures of apo- and calcium-bound forms of BtuB reveal ordering of the extracellular loops in the presence of Ca^{2+} ions (Chimento *et al.*, 2003). As ordering of these loops also occurs upon formation of the colicin–BtuB complex, it may be reasonable to postulate that the enhanced affinity in the presence of calcium is in part due to a reduction of the entropic penalty of binding. However, thermodynamic analysis reveals the enhanced binding affinity to be the net result of increased favorable enthalpy contributions and a lesser increase in the unfavorable entropy component. With the high-affinity nature of the BtuB–colicin E9 complex in the presence of calcium, the C-value under the experimental conditions used is approaching the limit for an accurate measurement of K_d. Under such conditions, ΔH measurements remain reliable, but the K_d should be considered to be a conservative estimate, and binding may in fact be tighter. As ΔS is calculated from the values of K_d and ΔH, inaccuracies will also exist in the measured entropy values.

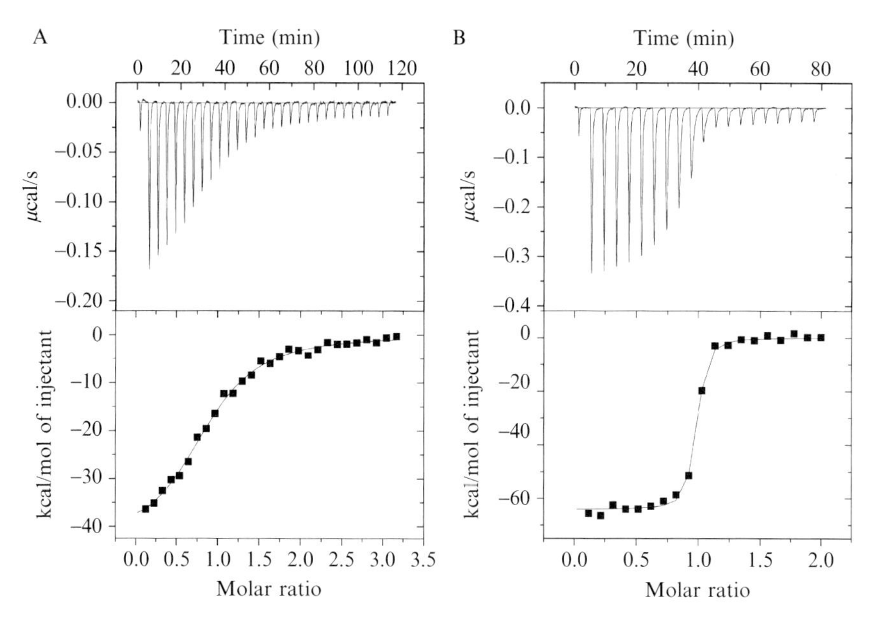

Figure 6.5 Binding isotherms for the titration of 16-μ*M* ColE9–Im9 complex into 1.2-μ*M* BtuB in 25-m*M* Tris–HCl, pH 7.5, 150-m*M* NaCl, and 1% (w/v) *n*-octyl-β-D-glucopyranoside in the presence of either 5-m*M* EGTA (A) or 5-m*M* $CaCl_2$. Figure adapted from Housden *et al.* (2005).

It has been suggested that binding to BtuB is sufficient to invoke unfurling of the colicin R-domain (Kurisu *et al.*, 2003). This, however, is not supported by the thermodynamics of binding, with the observed enthalpy and entropy changes being more consistent with an increase rather than a decrease in ordered structure. Furthermore, conformational restriction of colicin E9 through an engineered disulfide bond across the top of the receptor-binding domain has no impact on the thermodynamics of binding. Although good evidence exists in favor of unfurling of the colicin receptor-binding domain at some point during colicin uptake, *in vitro* binding of BtuB does not appear sufficient to invoke this unfurling.

4. Mapping Binding Epitopes and Signaling Networks

Once bound to BtuB on the *E. coli* outer membrane, colicin E9 must deploy its intrinsically unstructured N-terminal region across the outer membrane and into the periplasm where it can recruit TolB. Passage across the outer membrane is through interactions with OmpF.

4.1. Interaction of TolB with Pal

The Tol–Pal system is a five-component complex made up of TolA, TolQ, and TolR located within the inner membrane and Pal associated with the inner leaflet of the outer membrane (Sturgis, 2001). TolB located within the periplasm shuttles between TolA and Pal. While the *in vivo* role of the Tol–Pal system is not fully understood, it is clear that the function of this system impacts on membrane integrity. Mutation or deletion of any of the components of this system results in a phenotype characterized by blebbing of the outer membrane, leaking of periplasmic contents into the media, and increased susceptibility to antibiotics and toxic chemicals usually excluded from bacteria by the outer membrane (Webster, 1991).

The crystal structure of TolB has been solved in the presence and absence of Pal (Bonsor *et al.*, 2007; Carr *et al.*, 2000; Table 6.5). While Pal binds to the C-terminal β-propeller domain of TolB, this results in long-range conformational changes. In the structure of unbound TolB, the 12 amino acids present at its N-terminus are not seen and are assumed to be unstructured. Upon complex formation with Pal, these residues become ordered and bind to the interface between the N-terminal and C-terminal domains of TolB. The binding interaction between TolB and Pal has been characterized by ITC and is favorable both in terms of enthalpy ($\Delta H \sim -7.1$ kcal/mol) and entropy ($\Delta S \sim 2.7$ kcal/mol), giving rise to a K_d of $\sim$50 nM.

Table 6.5 Thermodynamic parameters for the complexes formed by TolB[a]

	K_d	N	ΔH (kcal/mol)	$T\Delta S$ (kcal/mol)
TolB + ColE9	125 ± 4 nM	0.93 ± 0.01	−20.6 ± 0.04	−11.3 ± 0.1
TolB + Pal	38 ± 3 nM	0.98 ± 0.00	−9.47 ± 0.10	0.8 ± 0.2
TolB + TolAIII	43 ± 2 μM	1.00 ± 0.04	2.9 ± 0.02	8.8 ± 0.03
Δ22-25TolB + ColE9	28 ± 1 nM	0.93 ± 0.01	−23.0 ± 0.07	−12.8 ± 0.1
Δ22-25TolB + Pal	313 ± 15 nM	0.91 ± 0.02	−7.2 ± 0.11	1.82 ± 0.12
Δ22-25TolB + TolAIII	No detectable binding			
TolB–ColE9 + TolAIII	13 ± 1 μM	1.16 ± 0.04	−0.58 ± 0.03	6.0 ± 0.1
TolB–Pal + TolAIII	No detectable binding			

[a] Values taken from Bonsor *et al.* (2009).

4.2. Mapping the TolB binding site within colicin E9

While site-directed mutagenesis is invaluable in identifying residues critical to biological function, these residues rarely act in isolation. Binding sites housed within intrinsically unstructured proteins are typically comprised of linear epitopes, contrasting those of globular proteins where binding residues are brought together through the three-dimensional fold of the protein, but are often distributed throughout the sequence of the protein. The binding epitope for TolB is housed within the intrinsically unstructured N-terminus of enzymatic E colicins. As a result, a precise delineation of the binding site is possible through the creation of truncated colicins and measuring binding through ITC, without incurring complications associated with disruption of the proteins' globular fold. In this way, the colicin E9 five residue TolB box (^{35}DGSGW^{39}), critical to TolB binding identified through site-directed mutagenesis, has been extended to a 16 residue peptide (^{32}GASDGSGWSSENNPWG^{47}) capable of binding to TolB, with thermodynamic parameters comparable to those of colicin E9 ($K_d \sim 1$ μM) (Loftus *et al.*, 2006). The favorable enthalpy ($\Delta H \sim -15$ kcal/mol) and unfavorable entropy ($T\Delta S \sim -7$ kcal/mol) changes are consistent with the disorder-to-order transition which occurs when the intrinsically unstructured region of colicin E9 folds on binding to TolB. The similarity of parameters obtained with full-length colicin E9 and the 16 residue TolB binding peptide indicate that this disorder to structure transition is limited to the residues in the immediate proximity of TolB.

Reducing the TolB binding epitope to its 16 residue minimal peptide was instrumental in obtaining atomic resolution detail of this complex through X-ray crystallography. The structures of the TolB–T_{32-47} peptide complex and the TolB–Pal complex show both ligands to bind to the same surface of TolB. In fact, the side chains of the T_{32-47} peptide mimic key interactions formed by Pal with TolB, including water-mediated interactions. With TolB binding Pal with a 50 nM K_d and the colicin with a 1 μM K_d, it is difficult to see how the colicin displaces Pal in order to recruit TolB. The crystal structure of TolB in complex with the colicin peptide revealed the presence of two Ca^{2+} ions bound within the narrow channel running through the β-propeller domain. Although these calcium ions do not directly contact the peptide, they do shift the electrostatics of the binding site for the peptide from negative in the absence of Ca^{2+} to more positive in its presence. In the presence of Ca^{2+}, the binding affinity for colicin increases and that for Pal decreases such that both complexes have K_d values of ~90 nM.

4.3. The interaction of TolB with TolA and the effects of colicin E9 and Pal

The affinity of the interaction between TolB and TolA is significantly weaker than the interaction of TolB with either colicin or Pal, making ITC measurements more challenging, not least in terms of the quantity and concentrations of protein required. Nevertheless, the K_d of the complex was determined as $\sim$43 μM in an entropy-driven ($T\Delta S = +\ 8.8$ kcal/mol) endothermic ($\Delta H = +\ 2.9$ kcal/mol) process (Bonsor *et al.*, 2009). The N-terminus of TolB plays a critical role in the interaction with TolA, with the deletion of four residues from the N-terminus being sufficient to eliminate the interaction. Interestingly, deletion of these same four residues impacts on the TolB–colicin and TolB–Pal complexes, causing a 4.5-fold increase and a eightfold decrease in binding affinities, respectively. While the N-terminal 12 amino acids of TolB are not seen in the crystal structure of TolB alone, NMR studies reveal this to be a result of an allosteric transition between distinct conformational states. Binding of colicin favors disorder at the TolB N-terminus, while Pal binding induces structure. Binding of Pal to TolB thereby prevents its interaction with TolA, while colicin binding reduces the dynamic movements of the TolB N-terminus, reducing the entropic penalty of TolA binding and thereby enhancing the binding affinity threefold. This is a rare example of how structural and dynamic studies can be reconciled in terms of binding thermodynamics.

5. Discussion

Titration microcalorimetry has been a core technique in the dissection of colicin-mediated protein–protein interactions, which has complemented other biophysical methods and structural information coming from NMR and X-ray crystallography. In particular, ITC has been used in the thermodynamic dissection of protein–protein interaction specificity in colicin DNase–immunity protein complexes and in the validation of their kinetic mechanisms, the latter through the agreement that is often seen between kinetically derived equilibrium dissociation constants and those stemming from ITC measurements. ITC has also played a pivotal role in our understanding of how colicins associate with bacterial receptors, recruit translocator proteins in the outer membrane, and manipulate signaling cascades in the bacterial periplasm which ultimately trigger entry of the bacteriocin into the cell. The use of ITC to investigate colicin toxicity will undoubtedly increase in the coming years as our understanding of the cellular components parasitized by these potent cytotoxins increases yet further.

REFERENCES

Bonsor, D. A., *et al.* (2007). Molecular mimicry enables competitive recruitment by a natively disordered protein. *J. Am. Chem. Soc.* **129,** 4800–4807.

Bonsor, D. A., *et al.* (2009). Allosteric beta-propeller signalling in TolB and its manipulation by translocating colicins. *EMBO J.* **28,** 2846–2857.

Bourdineaud, J. P., *et al.* (1990). Involvement of OmpF during reception and translocation steps of colicin N entry. *Mol. Microbiol.* **4,** 1737–1743.

Bradbeer, C., *et al.* (1986). A requirement for calcium in the transport of cobalamin across the outer membrane of *Escherichia coli. J. Biol. Chem.* **261,** 2520–2523.

Carr, S., *et al.* (2000). The structure of TolB, an essential component of the tol-dependent translocation system, and its protein–protein interaction with the translocation domain of colicin E9. *Structure* **8,** 57–66.

Cascales, E., *et al.* (2007). Colicin biology. *Microbiol. Mol. Biol. Rev.* **71,** 158–229.

Chimento, D. P., *et al.* (2003). The Escherichia coli outer membrane cobalamin transporter BtuB: Structural analysis of calcium and substrate binding, and identification of orthologous transporters by sequence/structure conservation. *J. Mol. Biol.* **332,** 999–1014.

Evans, L. J., *et al.* (1996). Direct measurement of the association of a protein with a family of membrane receptors. *J. Mol. Biol.* **255,** 559–563.

Foit, L., *et al.* (2009). Optimizing protein stability in vivo. *Mol. Cell* **36,** 861–871.

Housden, N. G., *et al.* (2005). Cell entry mechanism of enzymatic bacterial colicins: Porin recruitment and the thermodynamics of receptor binding. *Proc. Natl. Acad. Sci. USA* **102,** 13849–13854.

Indyk, L., and Fisher, H. F. (1998). Theoretical aspects of isothermal titration calorimetry. *Methods Enzymol.* **295,** 350–364.

James, R., *et al.* (1996). The biology of E colicins: Paradigms and paradoxes. *Microbiology* **142** (Pt. 7), 1569–1580.

Keeble, A. H., and Kleanthous, C. (2005). The kinetic basis for dual recognition in colicin endonuclease-immunity protein complexes. *J. Mol. Biol.* **352,** 656–671.

Keeble, A. H., *et al.* (2006). Calorimetric dissection of colicin DNase–immunity protein complex specificity. *Biochemistry* **45,** 3243–3254.

Kirkup, B. C., and Riley, M. A. (2004). Antibiotic-mediated antagonism leads to a bacterial game of rock-paper-scissors in vivo. *Nature* **428,** 412–414.

Kleanthous, C., and Walker, D. (2001). Immunity proteins: Enzyme inhibitors that avoid the active site. *Trends Biochem. Sci.* **26,** 624–631.

Kurisu, G., *et al.* (2003). The structure of BtuB with bound colicin E3 R-domain implies a translocon. *Nat. Struct. Biol.* **10,** 948–954.

Lazdunski, C. J., *et al.* (1998). Colicin import into *Escherichia coli* cells. *J. Bacteriol.* **180,** 4993–5002.

Li, W., *et al.* (2004). Highly discriminating protein–protein interaction specificities in the context of a conserved binding energy hotspot. *J. Mol. Biol.* **337,** 743–759.

Loftus, S. R., *et al.* (2006). Competitive recruitment of the periplasmic translocation portal TolB by a natively disordered domain of colicin E9. *Proc. Natl. Acad. Sci. USA* **103,** 12353–12358.

Penfold, C. N., *et al.* (2004). Flexibility in the receptor-binding domain of the enzymatic colicin E9 is required for toxicity against Escherichia coli cells. *J. Bacteriol.* **186,** 4520–4527.

Sharma, O., *et al.* (2007). Structure of the complex of the colicin E2 R-domain and its BtuB receptor. The outer membrane colicin translocon. *J. Biol. Chem.* **282,** 23163–23170.
Sturgis, J. N. (2001). Organisation and evolution of the tol-pal gene cluster. *J. Mol. Microbiol. Biotechnol.* **3,** 113–122.
van den Bremer, E. T., *et al.* (2004). Distinct conformational stability and functional activity of four highly homologous endonuclease colicins. *Protein Sci.* **13,** 1391–1401.
Webster, R. E. (1991). The tol gene products and the import of macromolecules into *Escherichia coli*. *Mol. Microbiol.* **5,** 1005–1011.

CHAPTER SEVEN

Energetics of Src Homology Domain Interactions in Receptor Tyrosine Kinase-Mediated Signaling

John E. Ladbury *and* Stefan T. Arold

Contents

Abstract

Intracellular signaling from receptor tyrosine kinases (RTK) on extracellular stimulation is fundamental to all cellular processes. The protein–protein interactions which form the basis of this signaling are mediated through a limited number of polypeptide domains. For signal transduction without corruption, based on a model where signaling pathways are considered as linear bimolecular relays, these interactions have to be highly specific. This is particularly the case when one considers that any cell may have copies of similar binding domains found in numerous proteins. In this work, an overview of the thermodynamics of binding of two of the most common of these domains (SH2 and SH3 domains) is given. This, coupled with insight from high-resolution structural detail, provides a comprehensive survey of how recognition of cognate binding sites for these domains occurs. Based on the data presented, we conclude that specificity offered by these interactions of SH2 and SH3 domains is limited and

Department of Biochemistry and Molecular Biology, University of Texas MD Anderson Cancer Center, Houston, Texas, USA

Methods in Enzymology, Volume 488
ISSN 0076-6879, DOI: 10.1016/B978-0-12-381268-1.00007-0
© 2011 Elsevier Inc.
All rights reserved.

not sufficient to enforce mutual exclusivity in RTK-mediated signaling. This may explain the current lack of success in pharmaceutical intervention to inhibit the interactions of these domains when they are responsible for aberrant signaling and the resulting disease states such as cancer.

1. Introduction

Most fundamental cellular processes are dependent on tyrosine kinase-mediated signaling. On extracellular stimulation, membrane-bound receptor tyrosine kinases (RTKs) are activated. This promotes both receptor autophosphorylation and phosphorylation of other binding ligands. Post-translational modification of this type provides binding sites for a defined group of proteins which are transiently recruited to the intracellular region of the receptor (Pawson and Nash, 2000; Schlessinger, 2000) forming an early signaling complex (ESC; O'Rourke and Ladbury, 2003). The formation of this ESC is thought to be, at least in part, responsible for ensuring the integrity of the downstream response and the ultimate cellular outcome (O'Rourke and Ladbury, 2003). Signaling downstream from the ESC is generally thought of as a series of protein–protein interactions in the cytoplasm which ultimately lead to a cellular response. This can be in the form of translocation of a protein into the nucleus and subsequent initiation of a transcriptional event. It can also involve the intracellular stimulation of an enzyme leading to the release of secondary messengers or metabolites.

Many of the proteins involved in the transduction of an intracellular signal consist of distinct domains linked, like beads on a string, through flexible linker polypeptides. These domains are identifiable by their homologous amino acid sequences and defined secondary and tertiary structures. The requirement for signaling pathways to maintain a high level of mutual exclusivity to ensure transduction without corruption suggests that the domains have to recognize their protein targets with a high level of specificity. There are many disease states associated with aberrant signaling emanating from transduction failures in RTK-mediated pathways. As a result, the interactions of many of these domains have been validated as targets for pharmaceutical intervention and inhibitor development.

Thus, understanding of the specificity (or the difference in equilibrium dissociation constant (K_D) between a cognate and a nonspecific ligand) is fundamental to understanding the basis of intracellular signal transduction. Measurement of the thermodynamic parameters coupled with structural detail provides an insight into the source of the energetics driving an interaction. The measurement of the K_D and determination of the derived change in Gibbs free energy (ΔG^o) at a given temperature ($\Delta G^o = -RT \ln K_D$, where R is the gas constant and T is the temperature) provide a value of the respective affinity of an interaction. Comparison of these terms allows the specificity of an interaction to be determined. The change in enthalpy (ΔH^{o}) and entropy (ΔS^{o}—or more usually the product with experimental temperature $T\Delta S^{o}$) can provide further insight into the source of the underlying source of affinity since they are related to the Gibbs free energy term ($\Delta G^{o} = \Delta H^{o} - T\Delta S^{o}$).

In this chapter, we highlight how the quantification of the thermodynamic parameters associated with interactions of Src homology domains involved in RTK-mediated signaling can be correlated with the available structural information to understand how recognition occurs. This reveals the level of specificity attributable to protein–protein recognition and provides a rationale for challenges faced in the development of therapeutics to inhibit these interactions.

2. Interactions of Src Homology 2 Domains

Src (or Sarcoma) homology 2 (SH2) domains were originally identified in the Src oncoprotein. They are polypeptides containing ~100 amino acids which are fundamental in RTK-mediated signaling proteins (Pawson and Nash, 2003; Schlessinger and Lemmon, 2003). These domains interact with proteins at sites with exposed phosphorylated tyrosine (pY) residues. SH2 domains are thus able to recognize and bind to sites on activated receptors and other posttranslationally modified proteins involved in RTK signaling. They have highly homologous secondary and tertiary structures (Fig. 7.1) and consist generally of a β-sheet sandwiched by two α-helices. There are ~120 genes encoding SH2 domains in the human genome. These are expressed in around 110 different proteins. The key role of these domains in recognition of the consequence of kinase activity on receptors and other signaling proteins, and their impact on pathologic conditions such as cancer, autoimmune disease, and asthma has resulted in a great deal of interest in understanding their mode of ligand recognition (Pawson, 2004; Schlessinger and Lemmon, 2003).

Specificity of the binding of individual SH2 domains is a major issue since there is likely to be competition for a given pY-containing site from other highly homologous domains in the cytoplasm. If one follows the dogma that RTK signaling is, at least in part, based on the linear processing of a signal based on a bimolecular relay of protein–protein interactions mediated by tyrosine phosphorylation, each individual interaction in this chain would have to provide sufficient specificity to preclude the competition from other SH2 domains which are capable of being present in the same cellular location (Ladbury and Arold, 2000). Based on the number of

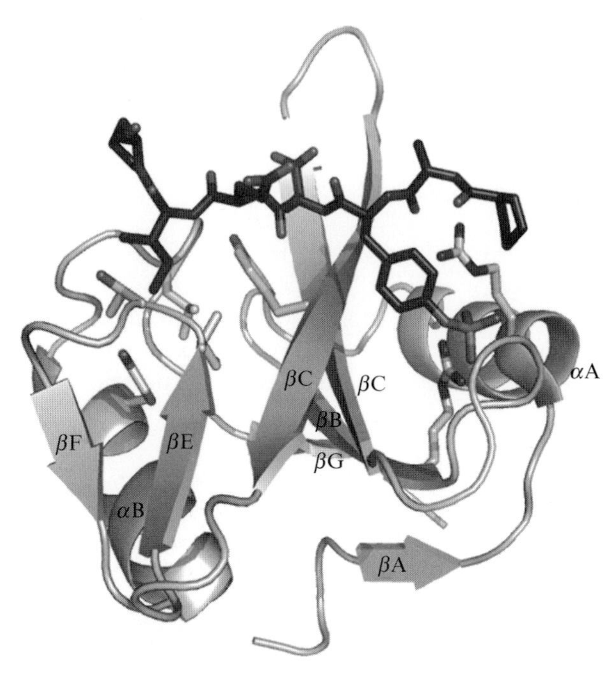

Figure 7.1 Src SH2 domain (light gray) in a complex with the peptide PQ(pY)EEIP (dark gray; PDB entry 1SPS). Key SH2 domain residues that coordinate binding to the ligand phosphotyrosine and isoleucine are shown as stick models. Labeling of secondary structure from Eck *et al.* (1993).

SH2 domain expressing genes in the human genome, a rough idea of the difference in affinity required for a specific versus nonspecific interaction can be calculated. Thus, if one assumes an SH2 domain interaction has to be specific for a particular cognate site, it has to compete against a background of ~100 other nonspecific SH2 domain-containing proteins. Assuming that these competing SH2 domains have equal affinity for the site then, because there is an effective 100-fold higher concentration of nonspecific versus specific ligands, the specific interaction would have to be approximately of two orders of magnitude higher affinity to have an equal chance of binding. Thus, in the context of reported potential fluctuations of protein concentration in the cytosol, to ensure the appropriate ligand binds, one would expect differences in binding of specific ligands of at least three orders of magnitude. This provides a useful rule of thumb in assessing specificity.

Specificity in recognition of a given SH2 domain for a cognate ligand was originally proposed to be imposed by residues proximal, and usually C-terminal of the pY residue (Songyang *et al.*, 1993). Thus, for example, the SH2 domain from the Src kinase was proposed to bind preferentially to a sequence containing pYEEI, whereas the SH2 domain from Abl kinase recognized the sequence pYENP. Available structural detail of SH2 domain interactions enhanced this view of the pY-proximal residues being important in providing sufficient mode of recognition (see, e.g., Fig. 7.2).

This reconciliation of specificity is appealing considering that if three residues proximal to the pY are required and each of the naturally occurring amino acids can be individually recognized by an SH2 domain, this provides 8000 different sequence combinations from which specificity can be derived. Thus, even allowing for some level of redundancy in the sequences (i.e., similarity in affinity between tyrosyl phosphopeptides with various residues in a given site), this should provide sufficient opportunity for specific sequences for each SH2 domain. Initial studies to address the issue of specificity were largely based on screening of a library of tyrosyl phosphopeptides against a selection of SH2 domains. Thus, they were inclined to lead to selection of a preferred sequence but did not provide any quantification of the difference in affinity between the interactions of "so-called" specific versus nonspecific. Indeed even within these exhaustive studies, high levels of degeneracy of sequence were observed. Extensive investigation of the level of specificity offered by these pY-proximal sequences has been performed using a range of techniques. As is described below, it is not

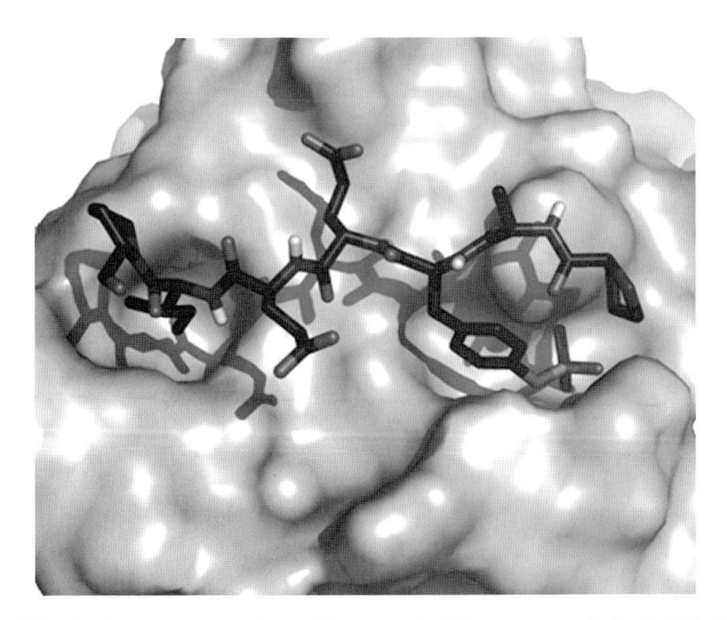

Figure 7.2 Surface representation of the canonical "two-pronged plug" SH2 domain binding mode, showing the two pockets that accommodate the isoleucine (left) and phosphotyrosine (right) of the ligand (pY)XXI motif (dark gray; PDB entry 1SPS).

clear that the range of affinities exhibited by given SH2 domains for "specific" sequences is sufficient to guarantee that aberrant interaction would not be possible.

A plethora of high-resolution structural detail has been accumulated on the interactions of SH2 domains (Bradshaw and Waksman, 2002; Kuriyan and Cowburn, 1993, 1997). There are three prevalent modes of recognition of the ligand by SH2 domains. The first has been likened to a "two-pronged plug" in which the pY residue is usually accommodated in a deep, polar pocket and a residue C-terminal of this, usually in the pY + 3 position, delves into a second pocket (Figs. 7.2 and 7.3). Thus, the SH2 domain forms the socket for the peptide plug (Bradshaw *et al.*, 1998; Waksman *et al.*, 1992, 1993). The second mode of recognition is represented by the binding of the SH2 domain from Grb2 (Fig. 7.3). In this case, the cognate ligand is forced to adopt a β-turn by the positioning of a bulky tryptophan residue in the SH2 binding site. This enhances the binding affinity of ligands which have an asparagine residue in the pY + 2 position (Bradshaw and Waksman, 2002; Rahuel *et al.*, 1996, 1998). A third variation on this theme of recognition is where the SH2 domain presents the pY pocket juxtaposed with a narrow, apolar groove which accommodates the several residues C-terminal of the pY, as in the case of the C-terminal SH2 domain of PLCγ (Fig. 7.3). We will focus primarily on these three modes of binding in the energetic analysis of SH2 domain recognition below.

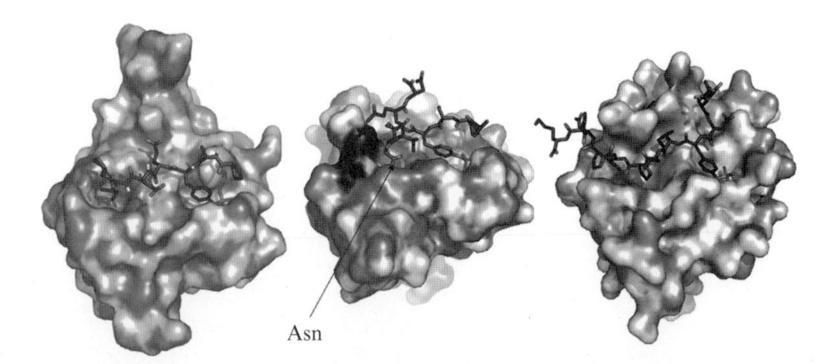

Figure 7.3 Representations of three modes of recognition to the SH2 domain binding site surface (light gray) and bound ligands (dark gray). Left: Src SH2–PQ(pY)EEIP (PDB entry 1SPS). Middle: Grb2 SH2–APS(pY)VNVQN (PDB entry 1JYR). The specificity-enhancing Grb2 SH2 tryptophan 121 is indicated in dark gray. The asparagine of the (pY)XN consensus is indicated by an arrow. Right: PLCγ SH2–DND(pY)IIPLPDPK (PDB entry 2PLD). Dark gray areas on surface correspond to hydrophobic atoms (carbons and sulfates). The orientation of SH2 domains is similar to Fig. 7.1, but has been modified to enhance visibility.

3. Recognition by the "Two-Pinned Plug"

Along with high-resolution structural detail, a large amount of thermodynamic data has helped to define the idiosyncrasies of SH2 domain recognition. The interactions of the SH2 domain from the protein Src kinase and family members (such as Lck, Fyn, and Hck kinases) are probably the most intensively investigated of all protein–protein interactions (Bradshaw and Waksman, 1999; Bradshaw *et al.*, 1999, 2000; Eck *et al.*, 1993; Ladbury, 2005; Ladbury *et al.*, 1995, 1996; Lemmon and Ladbury, 1994; Lubman and Waksman, 2002; Lubman *et al.*, 2005). The Src SH2 domain reportedly preferentially recognizes the sequence pYEEI. The crystal structure of the interaction of the SH2 domain and a tyrosyl phosphopeptide provides some topological reconciliation of how this specificity may arise since it shows the pY moiety delving into a deep polar pocket and the isoleucine probing a second apolar pocket in the canonical two-pronged plug binding mode (Bradshaw *et al.*, 1998; Waksman *et al.*, 1993). The peptide also incorporates a network of at least four water molecules which are sustained by interactions with the Glu in the pY + 2 position (Fig. 7.2). These water molecules hydrogen bond to one another and contribute ~11 hydrogen bonds to the complex (see below). The water molecules mediate all the interactions of all the atoms situated between the Glu1 and the Ile3 side chains of the tyrosyl phosphopeptide.

Based solely on data from isothermal titration calorimetry (ITC), the mean dissociation constant K_D at 25 °C for the interaction with tyrosyl phosphopeptides containing this sequence and consisting of at least eight residues is 0.26 ± 0.16 μ*M* giving a mean change in ΔG for these interactions of -38.1 ± 1.3 kJ/mol (Ladbury, 2005; see Table 7.1). This is made up from a large favorable change in enthalpy, ΔH, and a smaller but also favorable change in entropy, ΔS (or $T\Delta S$). Interestingly the ΔG for binding of the pY moiety to the SH2 domain alone is approximately -20 kJ/mol (Taylor *et al.*, 2007; see Table 7.1). Although the change in free energy terms cannot strictly be considered to be additive, it can be seen that the value for this pY group makes up more than half the total value for the intact peptide. This raises an important point regarding SH2 binding and recognition, that is, because the pY accounts for a large amount of the total free energy of binding, since this is required for all SH2 domain interactions, this leaves little room for specificity from the interactions of the other residues in the tyrosyl phosphopeptide sequence.

The rather weak binding of the isolated pY moiety is entropy driven having a small, unfavorable, ΔH, and a dominant $T\Delta S$ (Table 7.1). This is somewhat surprising based on the knowledge that the residue and the pocket are highly polar and offer the opportunity for ~11 hydrogen

Table 7.1 Binding of peptides and peptide mimetics to the Src SH2 domain[a]

Peptide/ligand	K_D (μM)	ΔH (kJ/mol)	$T\Delta S$ (kJ/mol)	ΔG (kJ/mol)	Reference
KGGQ**pYEEI**PIP	0.55	−35.4	0.6	−36.0	Ladbury (2005)
EPQ**pYEEI**PIYL	0.09	−38.7	1.4	−40.1	Chung *et al.* (1998)
PQ**pYEEI**PI	0.18	−32.3	6.3	−38.6	Bradshaw and Waksman (1999)
PQ**pYEEI**PI	0.25	−31.5	6.3	−37.8	Lubman and Waksman (2003)
PQ**pYEEI**PI	0.27	−31.5	6.3	−37.8	Lubman and Waksman (2002)
PQ**pYEEI**PI	0.2	−27.3	10.9	−38.2	Bradshaw *et al.* (1998)
pY	333	−0.4	19.3	−19.7	Bradshaw *et al.* (1999)
pY	175	−3.3	18.1	−21.4	Taylor *et al.* (2007)
PQ**pYQEI**PI	0.47	−31.9	4.2	−36.1	Bradshaw *et al.* (1999)
PQ**pYDEI**PI	0.18	−37.4	1.2	−38.6	Bradshaw *et al.* (1999)
PQ**pYAEI**PI	0.34	−32.3	4.2	−36.5	Bradshaw *et al.* (1999)
PQ**pYGEI**PI	6.25	−20.2	9.6	−29.8	Bradshaw *et al.* (1999)
PQ**pYREI**PI	8.20	−21.8	7.6	−29.4	Bradshaw *et al.* (1999)
PQ**pYEpYI**PI	0.07	−29.4	11.3	−40.7	Lubman and Waksman (2003)
PQ**pYEYI**PI	0.66	−26.8	8.4	−35.2	Lubman and Waksman (2003)
PQ**pYEQI**PI	0.53	−31.9	4.2	−36.1	Bradshaw and Waksman (1999)
PQ**pYEDI**PI	0.42	−26.0	10.5	−36.5	Bradshaw and Waksman (1999)
PQ**pYEAI**PI	1.04	−26.5	7.9	−34.4	Bradshaw and Waksman (1999)
PQ**pYEGI**PI	1.96	−25.2	7.6	−32.8	Bradshaw and Waksman (1999)
EPQ**pYEEV**PIYL	0.16	−28.6	10.2	−38.8	Chung *et al.* (1998)
EPQ**pYEEE**PIYL	0.21	−32.7	5.4	−38.1	Chung *et al.* (1998)
EPQ**pYEEW**PIYL	0.31	−32.2	4.9	−37.1	Henriques and Ladbury (2001)
EPQ**pYEED**PIYL	0.38	−27.5	9.1	−36.6	Henriques and Ladbury (2001)

Table 7.1 (*continued*)

Peptide/ligand	K_D (μ*M*)	ΔH (kJ/mol)	$T\Delta S$ (kJ/mol)	ΔG (kJ/mol)	Reference
PQ**pYEEL**PI	0.43	−23.5	13.0	−36.5	Bradshaw and Waksman (1999)
PQ**pYEEV**PI	0.46	−22.7	13.8	−36.5	Bradshaw and Waksman (1999)
PQ**pYEEA**PI	1.75	−21.4	11.3	−32.7	Bradshaw and Waksman (1999)
PQ**pYEEG**PI	3.9	−15.1	16.0	−31.1	Bradshaw and Waksman (1999)
PQ**pYAAA**PI	21.27	−16.4	10.5	−26.9	Bradshaw and Waksman (1999)
PQ**pYQPG**EN	29.4	−19.3	6.7	−26.0	Bradshaw *et al.* (1998)
EPQ**pYQPG**EN	14.3	−25.7	2.0	−27.7	Chung *et al.* (1998)
pYE-*N*-$(C_5H_{11})_2$	0.4	−18.1	19.1	−37.2	Charifson *et al.* (1997)
pYM-*N*-$(C_5H_{11})_2$	4.2	−16.4	14.6	−31.0	Charifson *et al.* (1997)
pYC-*N*-$(C_5H_{11})_2$	1.4	−18.1	15.9	−34.0	Charifson *et al.* (1997)
pYE-*N*-C_3H_6–C_5H_9	0.4	−29.8	7.2	−37.0	Charifson *et al.* (1997)
pYE-*N*-(C_5H_9) (C_4H_9)	1.0	−22.7	12.5	−35.2	Charifson *et al.* (1997)
pYE-*N*-hexanol	3.4	−13.4	18.4	−31.8	Charifson *et al.* (1997)
pYE-*N*-heptanol	2.3	−19.3	13.4	−32.7	Charifson *et al.* (1997)

[a] All data derived from isothermal titration calorimetry.

bonds to form in the complex (Henriques and Ladbury, 2001). To reconcile this, it can be assumed that before binding to pY side chain, the pocket provides a site for several water molecules which can form stable, hydrogen-bonded interactions within the pocket. Thus removing these into bulk solvent and replacing them with the pY moiety is entropically favorable but results in a net loss in the contribution to ΔH from hydrogen bonding.

Despite the fact that the thermodynamic parameters should not be considered as additive, it is interesting to see that if the quantities associated with binding of just the pY moiety are subtracted from the peptide data, all of the $T\Delta S$ terms become unfavorable (at 25 °C). This overall unfavorable entropic effect from the juxta-pY region can be accounted for, at least in part, since the X-ray structure of the peptide-bound complex shows a network of water molecules sustained primarily by Glu + 2. The pinning down of these water molecules is entropically unfavorable. However, since this network sustains ~11 hydrogen bonds (Waksman *et al.*, 1993), inclusion of these interfacial water molecules has a big enthalpic effect as is observed in the dominant ΔH for these interactions. Indeed, these bound water molecules have been shown in Src family SH2 domain interactions to stabilize the complex since they occupy sites of low potential energy and are found to enable the complex to prevail in nanoflow ESI mass spectral experiments at high cone voltages where water-free complexes are destroyed (Chung *et al.*, 1998, 1999). The energetic contribution from these water molecules and their robustness in the protein interface in the presence of potentially perturbing mutations is somewhat conjectural (see below).

Substituting different amino acids C-terminal of the pY position shows insensitivity to conservative changes (Bradshaw and Waksman, 1999). For example, substitutions in the pY + 1 position have a moderate effect on affinity. Substituting the Glu1 for a Gln, Asp, or Ala has essentially no effect on the affinity. However, the insertion of a Gly residue reduces the affinity by an order of magnitude. A similar effect is also observed in substituting the Glu1 for an Arg residue (within experimental error; Table 7.1). The effect of this charge reversal can be structurally reconciled since in the complex structure, the Glu1 is involved in direct hydrogen bonding with the Lys (βD3) from the SH2 domain (for residue positions within secondary structure notation, see Fig. 7.1). Thus, the Arg residue would be repelled by the like-charged Lys. Nonetheless, the observation that binding is still possible in the presence of this charge conflict suggests that the tyrosyl phosphopeptide interaction is tolerant to non-"specific" sequences.

Substitutions of the pY + 2 Glu would be expected to have a dramatic effect since this residue appears to sustain the network of water molecules described above. Interestingly, substituting this residue for any of the amino acids adopted has a negligible effect on ΔG (Table 7.1). From a thermodynamic view point, it would be expected that the loss of the network of water molecules would result in a less favorable ΔH and a favorable $T\Delta S$. This is not demonstrable from the data. For example the substitution of the Gly residue would not be able to mimic the Glu in the pYEEI peptide and thus the water network would be expected to collapse resulting in a loss of hydrogen bonds in the interface (and an accompanying less favorable ΔH) and a gain in $T\Delta S$ (from liberation of these water molecules). A less favorable ΔH is observed compared to the pYEEI interaction ($\Delta\Delta H \sim 10$ kJ/mol) but the $T\Delta S$ change is negligible. In the absence of an X-ray structure, it is hard to explain this; however, the presence of water molecules is robust (see, e.g., De Fabritiis *et al.*, 2008) and could be retained through other interactions. Indeed, it should be pointed out that the apo-Src SH2 domain structure itself shows clear positioning of waters with low B-factors in positions of low potential energy in the binding site (Waksman *et al.*, 1992, 1993).

Substitutions in the pY + 3 position are equally insensitive in thermodynamic terms to peptide binding (Table 7.1). Replacing the aliphatic isoleucine residue from the "specific" sequence with a range of different functional groups leads to very little change in affinity. The largest change in thermodynamic parameters observed is with the Gly substitution. The binding of pYEEG is about an order of magnitude weaker than the pYEEI interaction. From the structural information (Bradshaw and Waksman, 1999; Waksman *et al.*, 1993), the residue in the pY + 3 position makes discrete interactions with the pY + 3 pocket. This pocket is highly apolar and the isoleucine from the peptide interacts with the aromatic rings of Tyr βD5 and Tyr αB9; the methyl groups of Leu βG4, Ile βE4, and Thr EF1; and the α-carbon of Gly βG3. Thus, the substitution of this residue for Gly can be thermodynamically reconciled by the potential loss of all of these hydrophobic contacts which reduces the favorable ΔH, but increases the net $T\Delta S$ due to the previously bound atoms being liberated in the empty pocket. The residues in this pocket show significant level of dynamic flexibility (Taylor *et al.*, 2008). This flexibility of the pocket makes it accommodating to a range of amino acid side chains.

Further investigation of the effect of amino acid substitutions in the tyrosyl phosphopeptide replaced two of the three key residues C-terminal to the pY moiety. Even peptides with these double substitutions reduced affinities by only one to two orders of magnitude. In fact, the pYAAI-containing peptide only reduced binding by approximately sixfold compared to the "specific" pYEEI sequence (Table 7.1). Even completely replacing the "specific" sequence to unrelated sequences like pYAAA (ΔG = 26.9 kJ/mol) or pYQPG (ΔG = 26.0 kJ/mol) provides ligands that bind with affinities reduced by 50- to 100-fold.

Studies using thermodynamic cycles combining the data from the binding of several sequentially modified residues suggested no coupling change in free energy and hence no observable communication between the residues in the peptide (Bradshaw and Waksman, 1999; Lubman and Waksman, 2002, 2003). This is perhaps not surprising, given that the peptide binds in an extended conformation across the face of the SH2 domain.

Based on the data shown in Table 7.1 and that described above, it is apparent that the binding of the Src SH2 domain rather than being specific is actually highly promiscuous. The maximum differences in affinity do not reach the expected levels required to ensure mutual exclusivity in signal transduction pathways. Indeed, it would appear that as long as a pY moiety is present in the peptide, one can achieve binding of reasonable affinity and no less than two orders of magnitude weaker than the so-called specific interaction. The promiscuity offered in binding to the Src SH2 domain has been recently structurally reconciled by NMR studies in terms of the dynamics of the molecule. The pY pocket is tightly structured and not adaptable to exogenous ligands, whereas the pY + 3 pocket presents a large dynamic binding surface (Taylor *et al.*, 2008). There may be limitations in peptide-based studies since they offer a limited molecular surface which thus may not include additional interactions offered by protein molecules with secondary and tertiary structures (see below). However, it has been found that posttranslationally modified tyrosine residues often appear on unstructured loops in signaling proteins, as a result of needing to be accessible to kinase active sites. Thus, peptides are likely to serve as reasonable mimics of physiological interactions.

Along with tyrosyl phosphopeptides based on naturally occurring amino acid sequences, there have been extensive investigations of the Src family SH2 domains binding to peptidomimetic compounds with a view to producing lead compounds for inhibition of specific pathways. In one study, seven modified peptides which contained an intact pY moiety and modifications C-terminal of this all bound with affinities within the range observed with naturally occurring amino acid sequences present (Table 7.1; Charifson *et al.*, 1997). Thus, despite replacing the EEI sequence with largely apolar functional groups, binding was achievable. As with the tyrosyl phosphopeptides described above, all interactions were also accompanied by favorable contributions from ΔH and $T\Delta S$. However, generally the ΔH term is less favorable and the $T\Delta S$ more favorable than the pYEEI peptides. These thermodynamic effects can partially be explained since none of these compounds contain the water network-sustaining polar group in the pY + 2 position, thus potentially releasing this hydrogen-bonded group of molecules to bulk solvent. The acceptance of the variety of C-terminal functional groups on these peptidomimetic compounds into the pY + 3 pocket shown by the crystallographic detail in this study further demonstrates the promiscuity of this site on the SH2 domain.

In an exploration of modified peptides as a possible route to inhibitor development, the pY residue was constrained by inserting a 1,2,3, trisubstituted cyclopropyl group replacing two of the atoms in the peptide backbone the (α-carbon, and NH) as well as the β-carbon. This resulted in only a limited increase in affinity but significant changes in the ΔH and $T\Delta S$ terms (Davidson *et al.*, 2002). The enthalpy and entropy values adhered to the previously observed rule of both being favorable. Inhibition of the rotational degrees of freedom of the peptide, as expected, increases the $T\Delta S$. However, decreases in the observed ΔH are less easy to explain based on the fact that crystal structures reveal that the modified peptides bind in a similar way to the natural ligands. The modified pY seems to dehydrate the binding interface which will also have an effect on the thermodynamics of this interaction. It would be expected that the release of the previously held network of water molecules into bulk solvent will give the observed entropic effect, and the accompanying loss of a significant number of hydrogen bonds would reduce the favorable ΔH. Superposition of the structures of the constrained with the nonconstrained peptides shows only limited structural perturbation, and thus it is difficult to explain how the water binding site is sufficiently perturbed so as to cause this disruption.

In a recent study based on using a semisynthetic Src SH2 domain derived from three fused fragments, the specificity of the interaction with the "specific" sequence was changed (fivefold increase in affinity) to pYDEI by insertion of progressively reduced side chain of nonnatural amino acids substituting for LysβD3 (Virdee *et al.*, 2010). The enthalpic and entropic components of the free energy of binding to this modified domain were of similar in sign and magnitude to the binding of other tyrosyl phosphopeptides described above.

4. Recognition by the β-Turn Motif

As described above, the canonical two-pronged plug mode of binding can be replaced by other forms of recognition in SH2 domain interactions. In the case of the SH2 from the protein Grb2, the peptide is forced into a β-turn conformation by the presence of a Trp residue on the EF loop which closes the binding cleft C-terminal of the pY (Fig. 7.3). This requires that ligands have the consensus sequence pYXNX (where X is any residue) where the pY + 2 N is the "specificity"-determining residue. The binding of the peptide in the form of a turn is supported by a hydrogen bond between the carbonyl oxygen of the pY residue and the main chain N of the residue in the +3 position. The SH2 domain is grossly similar to other domains, but has a CD loop shortened by five residues compared to Src, and the BG loop is two residues longer (Rahuel *et al.*, 1996).

Binding studies by ITC of tyrosyl phosphopeptides to the isolated SH2 domain revealed a similar thermodynamic signature to that of the Src SH2 domain above; both the ΔH and $T\Delta S$ were favorable (Lemmon *et al.*, 1994; McNemar *et al.*, 1997). Again exploration of specificity was achieved by modifying the sequence of the "specific" sequence (Table 7.2). Alanine substitution of the interacting residues C-terminal of the pY in the +1, +3, and +4 positions resulted in relatively similar affinities. However, substitution of the "specificity"-determining residue (pY + 2 Asn to Ala) resulted in a dramatic loss in affinity from 190 nm to 0.36 m*M*. This loss of over three orders of magnitude reflects the structural constraint for peptide recognition imposed by inability of the alanine-substituted peptide to adopt a β-turn required to avoid clash with the obstructing presence of the EF loop Trp (Fig. 7.3). The Ala substitution study shows that binding of the "specific" sequence is favored by enthalpic contributions arising primarily from the pY and Asn residues. The large changes in both affinity and ΔH can be

Table 7.2 Binding of peptides to Grb2 SH2 domains[a]

Peptide/ligand	K_D (μ*M*)	ΔH (kJ/mol)	$T\Delta S$ (kJ/mol)	ΔG (kJ/mol)	Reference
SpYVNVQ[b]	0.19	−31.6	6.1	−37.7	McNemar *et al.* (1997)
ApYVNVQ[b]	0.2	−30.1	7.4	−37.5	McNemar *et al.* (1997)
SpY**A**NVQ[b]	1.63	−35.7	−3.4	−32.3	McNemar *et al.* (1997)
SpYV**A**VQ[c]	359	−8.2	11.1	−19.3	McNemar *et al.* (1997)
SpYVN**A**Q[b]	0.77	−25.0	9.2	−34.2	McNemar *et al.* (1997)
SpYVNV**A**[b]	0.3	−29.7	6.8	−36.5	McNemar *et al.* (1997)
DpYVNVPE[c]	0.27	−34.9	2.4	−37.3	Houtman *et al.* (2004)
EpYVNVSQ[c]	0.08	−33.9	6.6	−40.5	Houtman *et al.* (2004)
DpYENLQE[c]	0.17	−50.3	−11.8	−38.5	Houtman *et al.* (2004)

[a] All data derived from isothermal titration calorimetry.
[b] 50 m*M* HEPES, 150 m*M* NaCl, pH 7.5 20 °C.
[c] PBS pH 7.4 2.5 m*M* DTT. Peptide sequences 19mers (only binding site residues shown).

attributed, at least in part, to the loss of two hydrogen bonds between the asparagine's side chain and the Lys βD6 amide proton and Leu βE4 backbone carbonyl. Substitution of the pY + 1 residue for Ile, Leu, Gln, Glu, or Lys in peptides had little effect on the affinity (Table 7.2) and the ΔH was favorable and dominant in all cases. The entropy term, however, became unfavorable and very small ($T\Delta S$ ranging from −3.3 to −7.5 kJ/mol) in the case of the latter three peptides (DeLorbe *et al.*, 2009). Thus, as with the Src SH2 domain, there is only limited specificity in binding of the tyrosyl phosphopeptides, except in this case in the pY + 2 position, which requires an asparagine residue.

5. Selectivity Versus Specificity for SH2 Domain Interactions

PLCγ has two SH2 domains, known as the N-SH2 and C-SH2. The binding of ligands to the C-SH2 domain from PLCγ corresponds to the third common type of recognition by these domains (Ji *et al.*, 1999). This involves the pY C-terminal residues binding to an extended groove which is relatively hydrophobic in nature (Pascal *et al.*, 1995). The "specific" sequence for this domain is pYIIPLPD; thus, the sequence C-terminal of pY is suitably hydrophobic to interface with the SH2 domain surface (Fig. 7.3). The N-SH2 domain binds ligands in a way more like the canonical two-pronged plug seen for Src family SH2 domains and recognizes the sequence pYLDL. Some level of specificity is apparent between identical ligands binding to these two domains and recent structural information provides an intriguing twist on the mode of selective binding of the latter of these domains.

Binding of a peptide containing the pYLDL sequence to the two SH2 domains from PLCγ resulted in a fourfold difference in affinity (Table 7.3). Again, as appears to be the signature for SH2 domain interactions, the K_D affinities are in the micromolar range and both the ΔH and $T\Delta S$ are favorable with the former being the dominant contribution. This fourfold difference might be considered rather limited based on the significant differences in the topology of the SH2 binding sites offered to the tyrosyl phosphopeptide.

High-resolution X-ray crystallographic structural detail revealed that the binding of the physiological fibroblast growth factor receptor 1 (FGFR1) protein (from which the pYLDL sequence was derived) to the N-terminal SH2 domain also included a secondary binding site outside the canonical pY site (Bae *et al.*, 2009). This site comprises residues from β-strand D, the BC and DE loops from the N-SH2 domain. The binding of FGFR1 to the N-SH2 which includes this secondary interaction increases the affinity by about an order of magnitude (Table 7.3). Despite the additional bonds made by this secondary binding site, the increase in affinity is only moderate. Thus, the interaction outside the canonical pY binding site does not dramatically enhance specificity (at least not to the levels expected to ensure mutual exclusivity in signaling, see above) but it does increase selectivity, such that interaction of this additional site with FGFR1 will not be competed for by another SH2 domain.

Table 7.3 Binding of peptides and FGFR protein to PLCγ SH2 domains

Peptide/domain	K_D (μM)	ΔH (kJ/mol)	$T\Delta S$ (kJ/mol)	ΔG (kJ/mol)	Reference
TSNQEpYLDLSM + N-SH2	0.43	−28.7	7.7	−36.4	Bae *et al.* (2009)
TSNQEpYLDLSM + C-SH2	1.80	−15.8	17.2	−33.0	Bae *et al.* (2009)
FGFR + N-SH2	0.03	−33.1	9.6	−42.7	Bae *et al.* (2009)
FGFR + C-SH2	0.49	−10.7	25.4	−36.1	Bae *et al.* (2009)

6. Proline Sequence-Recognition Domains

A large number of cellular signals derived from tyrosine kinase-mediated pathways rely on the recognition of proline-rich motifs. Proline residues show a unique closure of their side chain into a five-member pyrrolidine ring. The resulting restriction of the dihedral φ angle to about $-60°$ imposes severe limits on the number of conformations that prolines and proline-rich sequences can adapt. Consequently, prolines tend to break secondary structure elements such as α-helices and β-strands, which commonly constitute the folded core region of proteins. This causes proline residues to often be exposed at the surface of proteins, rather than being buried in the core (Holt and Koffer, 2001). Proline-rich sequences promote the formation of a polyproline type II (PPII) helix, a left-handed helix with three residues per turn. This structure is very robust to substitutions of prolines with other residues. Unlike α-helices and β-strands, the PPII structure does not require stabilizing hydrogen bonds. The side chains and backbone carbonyls of PPII helices which protrude at regular intervals from the helical axis are therefore available for intermolecular hydrogen bond interactions. The restriction in conformational space of the proline-rich PPII helices also lowers the entropic cost arising from fixing PPII structures onto the surfaces of ligands. These characteristics make proline-rich sequences, and the PPII helices they form, a very distinct and recognizable motif well suited to serve in protein–protein interactions.

A number of proline-recognition domains (PRDs) have been discovered so far. Among those are the Src homology 3 (SH3) domain (Kaneko *et al.*, 2008; Li, 2005; Zarrinpar *et al.*, 2003) and the structurally very similar cytoskeleton-associated protein (CAP)-Gly domain (Saito *et al.*, 2004), the WW domain (named after two conserved tryptophan residues; Ilsley *et al.*, 2002; Kato *et al.*, 2004), the glycine-tyrosine-phenylalanine (GYF) domain (Freund *et al.*, 1999; Nishizawa *et al.*, 1998), the ubiquitin E2 variant (UEV) domain (Pornillos *et al.*, 2002), and the enabled/VASP homology (EVH1) domain (Peterson and Volkman, 2009). All these PRDs are small independently folding protein modules containing between 35 (WW domain) and 150 (UEV domain) amino acids. The topological diversity of these PRDs shows that proline recognition has evolved several times, independently, from different protein folds. For example, the structure and sequence homology of the EVH1 domain suggest that this PRD has evolved from a member of the phosphotyrosine/phosphotidylinositide-recognizing PTB/PH domain superfamily. The evolutionary success of PRD-proline-rich recognition is apparent from the great diversity of these interactions. As a group, PRDs are in fact the most abundant protein–protein recognition modules found in metazoan proteomes (Castagnoli *et al.*, 2004). About

300–400 different types of SH3 domains, the most abundant of the PRDs, exist in the human genome, as well as over 100 WW domains (Castagnoli *et al.*, 2004; Karkkainen *et al.*, 2006). All these PRDs have the daunting task of recognizing their cognate ligand within an immense pool of candidate sequences. About 25% of all human proteins harbor proline-rich sequences (Li, 2005), and proline-rich sequences constitute the most common and second most common sequence motif in the genomes of *Drosophila* and *Caernorhabditis elegans*, respectively (Rubin *et al.*, 2000). Moreover, due to the symmetry of the PPII helix, proline-rich sequences can potentially be recognized in two opposite directions (Lim *et al.*, 1994; Yu *et al.*, 1994), doubling the possible number of potential recognition motifs. The challenge of achieving fidelity in ligand recognition is particularly important since PRDs are key players in virtually all essential cellular processes, such as cell growth, differentiation, apoptosis, postsynaptic signaling, cytoskeletal remodeling, motility, and transcription (Kaneko *et al.*, 2008; Li, 2005; Zarrinpar *et al.*, 2003).

7. Interactions of SH3 Domains

The difficulty faced in selecting ligands from a pool of homologous sequences is probably best exemplified by SH3 domains, for which a plethora of structural and energetic data has been reported. The roughly 60-residue SH3 domains form a five-stranded antiparallel β-sheet and a short C-terminal 3_{10} helix (Fig. 7.4; Musacchio *et al.*, 1992; Yu *et al.*, 1992, 1994). In most cases, a shallow hydrophobic surface patch accommodates a PXXP core motif (where X is any amino acid) of the ligand sequence (for a more detailed recent structural review, see Kaneko *et al.*, 2008; Li, 2005; Zarrinpar *et al.*, 2003).

The question of SH3 domain specificity has initially, and most frequently, been addressed by screening of proline-rich peptide libraries (Rickles *et al.*, 1994; Sparks *et al.*, 1996, 1998; Tong *et al.*, 2002; Yu *et al.*, 1994). Such studies established the class I ([R/K]XXPXXP) and class II (PXXPX[R/K]) sequences as major SH3 binding motifs (Fig. 7.5). However a number of less usual motifs were also discovered, such as the PPXVXPY motif (binding to the Bem-1 SH3 domain; Tong *et al.*, 2002), the PXXDY motif (binding to SH3 domains from Eps8; Aitio *et al.*, 2008; Mongiovi *et al.*, 1999), the PXXPR motifs (identified in a number of effector molecules for CIN85; Ababou, *et al.*, 2008, 2009; Kowanetz *et al.*, 2004), or the RXXPXXXP motif (which binds to the cortactin SH3 domain; Tian *et al.*, 2006). Some SH3 binding motifs lack prolines altogether. For example, the consensus sequence (R/K)XX(K/R) was found central to mediating binding in many SH3 interactions, so much so

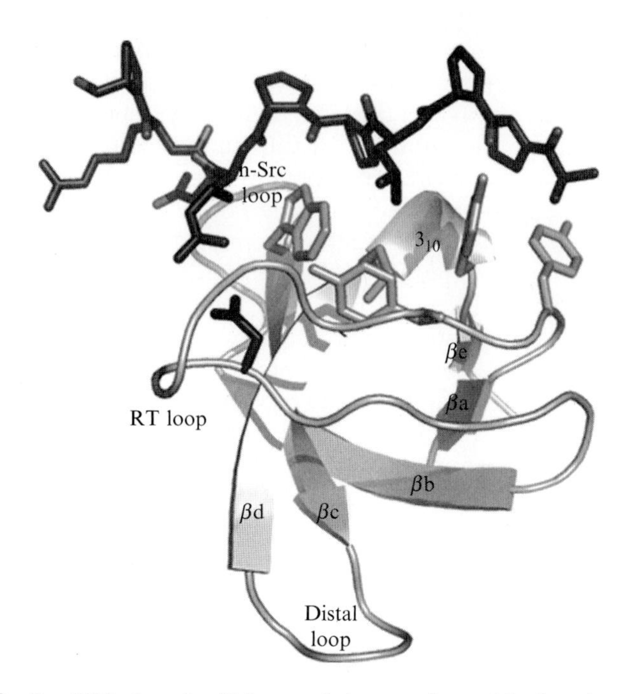

Figure 7.4 Src SH3 domain (light gray) in complex with the class II peptide APPLPPRNRP (dark gray; PDB entry 1QWE). On Src SH3, the residues that compose the hydrophobic binding patch are shown as light gray stick models. The negatively charged residue of the RT loop (an aspartic acid in Src SH3) is shown as dark gray stick model. Its interaction with an opposite charge of the binding motif determines binding orientation.

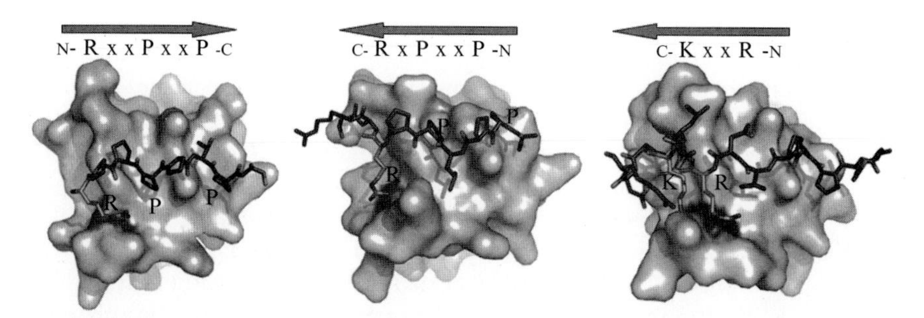

Figure 7.5 Left: β-PIX SH3 domain bound to a class I peptide (PDB entry 2P4R). Middle: Src SH3 bound to class II peptide (PDB entry 1QWE). Right: Grb2 C-terminal SH3 domain bound to a RXXK peptide (PDB entry 2VWF). This motif is sometimes referred to as class III consensus. Key arginines and lysines of the ligands are colored in light gray. Key interacting aspartic acids and glutamic acids from the SH3 domains are colored in dark gray.

that it was proposed as a class III consensus (Fig. 7.5). Sequences containing this motif were found in the scaffolding proteins Gab1 and Gab2, the T cell adaptor protein SH2 domain-containing leucocyte protein of 76 kDa (SLP76; Berry *et al.*, 2002), the SLP76 associated protein (SLAP130)/Fyn-binding protein (Fyb), the B-cell signaling protein BLNK (B-cell linker protein), and the hematopoietic progenitor kinase (HPK1; Berry *et al.*, 2002; Lewitzky *et al.*, 2004). Another motif, RKXXYXXY, present in the Src kinase associated protein of 55 kDa (SKAP55), was found to interact with the SH3 domains of Fyn and SLAP130/Fyb (Kang *et al.*, 2000).

Structural analyses showed that when bound to SH3 domains, the canonical proline-rich class I and II ligands adopt a PPII helix (Lim *et al.*, 1994; Terasawa *et al.*, 1994; Yu *et al.*, 1994). The two prolines of the PXXP motif insert into two narrow hydrophobic grooves on the SH3 domain, sculpted by almost parallel aromatic residues. This SH3 surface groove is highly selective for the proline rings which stick out parallel to the PPII helix, at positions 0 and +3 (Fig. 7.4). However, given the prevalence of proline-rich sequences, this selectivity does not allow for specificity in SH3–ligand recognition. Through screening of peptide libraries and analysis of SH3–ligand complexes, it became apparent that the charged residues included in the proline-rich sequences not only dictate the binding orientation (which is opposite for class I as compared to class II ligands) but also achieve some level of specificity, by ion-pairing interactions with opposite charges on the SH3 domains (Lim *et al.*, 1994; Terasawa *et al.*, 1994; Yu *et al.*, 1994). Nonetheless, the majority of SH3–ligand interactions display an affinity toward their ligands of between 1 and 200 μM (Ladbury and Arold, 2000; Landgraf *et al.*, 2004; Mayer, 2001). The thermodynamics of these interactions at around ambient temperature are typified by a dominant favorable enthalpy contribution and a small change in entropy (Table 7.4). To our knowledge, no SH3–ligand interaction has been reported to date with an unfavorable enthalpy contribution. This signature is somewhat unexpected.

The interaction of the PXXP core motif with the hydrophobic surface patch of SH3 domains might be expected to give rise to a significant, favorable entropic contribution, due to liberation of constrained, semiordered water molecules from the surface into the bulk solvent. Conversely, while important for orientation and specificity, the surface ion and hydrophobic bonding interactions might not contribute much in terms of net favorable enthalpy, due to the competition with the solvation counterions of these groups in the free state. Since, in addition, this signature is not necessarily associated with peptide–protein module interactions (e.g., the binding of helical peptides to the focal adhesion targeting (FAT) domain is typically entropically favorable; Garron *et al.*, 2008), this unexpected anomaly in thermodynamic signature of SH3 domains (rather like that of the binding of pY to SH2 domains described above) has been a matter of debate.

Table 7.4 Binding of ligands to SH3 domains

SH3 domain	Peptide/ligand	K_D (μM)	ΔH (kJ/mol)	$T\Delta S$ (kJ/mol)	ΔG (kJ/mol)	Reference
Cortactin	PPPLPPKPKF	1.4	−16	17	−33	Rubini *et al.* (2010)
Cortactin	KPPLLPPKKE	64	−9.6	15	−29	Rubini *et al.* (2010)
Cortactin	KPPLLPPKKEKL[a]K	5.7	−11	18	−29	Rubini *et al.* (2010)
Cortactin	KPPLLPPKPKL[a]KP	0.14	−9	30	−39	Rubini *et al.* (2010)
Grb2-N	PVPPPVPPRRRP	39	−39	−14	−25	McDonald *et al.* (2009)
Grb2-N	PVPPPVPPRAAP	84	−48	−25	−23	McDonald *et al.* (2009)
Grb2-N	PVPPPVPPARRP	490	−39	−20	−19	McDonald *et al.* (2009)
Grb2-N	PVPPPVPPAAAP	2600	TLQ	TLQ	−15	McDonald *et al.* (2009)
Grb2-N	IAGPPVPPRQST	82	−48	−25	−23	McDonald *et al.* (2009)
Abl	APSYSPPPPP	2.3	−92	−60	−32	Palencia *et al.* (2004)
Eps8L1	PPVPNPDYEPIR	24	−40	−14	−26	Aitio *et al.* (2008)
Gads C	APSIDRSTKPPLDR	0.02	−73	−28	−45	Seet *et al.* (2007)
Fyn	PPRPLPVAPGSSKT	16	−51	−24	−28	Renzoni *et al.* (1996)
Fyn	p85 DSH3	3	44	76	−32	Renzoni *et al.* (1996)
Hck	HIV-1 Nef R71	0.2	−53	−15	−38	Lee *et al.* (1995)
Hck	HIV-1 Nef T71	0.6	−63	−27	−36	Arold *et al.* (1997)
Src	HIV-1 Nef T71	11	−42	−14	−28	Arold *et al.* (1997)

Hck	HIV-1 Nef T71;Δ1-56	1.5	−32	1	−33	Arold *et al.* (1997)
Src	HIV-1 Nef T71;Δ1-56	14	−16	12	−28	Arold *et al.* (1997)
Fyn	HIV-1 Nef T71;Δ1-56	16	−2	25	−27	Arold *et al.* (1997)
Fyn[b,c]	FAK PxxP	68	−16	8	−24	Arold *et al.* (2001)
Fyn[b,c]	FAK pY	2.5	−45	−13	−32	Arold *et al.* (2001)
Fyn[b,c]	Fak pY/PxxP	0.2	−130	−94	−38	Arold *et al.* (2001)
Fyn[b,d]	Fak pY/PxxP	0.03	−89	−46	−42	Arold *et al.* (2001)
Lck[b,d]	Fak pY/PxxP	0.7	−46	−11	−35	Arold *et al.* (2001)
Fyn	SAP SH2	3.5	−18	13	−31	Chan *et al.* (2003)

TLQ = too low to quantify.
Experiment was performed at 25 °C, except for Renzoni *et al.* (1996), where the temperature was 30 °C.
[a] Norleucine was used as a methionine mimic.
[b] Fyn SH2-SH3 fragment.
[c] Tris pH 8.0, 200 m*M* NaCl.
[d] Tris pH 7.5, 150 m*M* NaCl.

Attempts to reconcile the above thermodynamic observations have been variously made. Currently, the lack of the expected favorable entropy contributions in SH3 domain binding events is attributed to a loss of degrees of freedom in the PPII ligand sequence upon going from the free state, where multiple, albeit similar, conformations exist, to the bound state where the ligand core sequence is pinned down to the SH3 surface. This would counteract the effect of the loss of solvent from the burial of hydrophobic surface. In the case of the Son of Sevenless (Sos)-derived nine-amino acid peptide binding to the *C. elegans* protein Sem-5 SH3 domain, it was shown that while the PPII helix conformation dominates in the free state, the restriction of the conformational ensemble into the SH3-binding competent form occurring upon going from the free to bound state produces changes in ΔH and $T\Delta S$ appropriate to counter balance the expected favorable entropy of desolvation and provide a significant total change in enthalpy (Ferreon and Hilser, 2004). A further potential contribution to the thermodynamics which would result in the observed parameterization may result from a reduction in conformational freedom of the SH3 loops involved in binding (RT loop and n-Src loop), combined with rearrangement of the intra- and intermolecular hydrogen and ionic bond network. In the case of Src family binding to HIV-1 Nef, such effects were used to explain the occurrence of an unfavorable entropy and, depending on the ligands, favorable enthalpy contributions (Arold *et al.*, 1998; Ferreon and Hilser, 2003; Ferreon *et al.*, 2003). Another type of contribution was illustrated by the interaction of the Abl SH3 domain with a proline-rich peptide derived from p41. In this case, Luque and coworkers showed the importance of the dynamics of individual water molecules that are found bound to the surface of the SH3 domain. The thermodynamic effects calculated from the modulation of a water-mediated hydrogen bond network helped to explain the extremely favorable enthalpy and highly unfavorable entropy of this interaction (Table 7.4; Palencia *et al.*, 2004, 2010). Similar water-mediated thermodynamics might be contributing, albeit to a lesser extent, to other SH3 ligand interactions. Nicholson and colleagues have demonstrated that ligand binding to SH3 domains subtly alters the internal hydrogen bond network of the Src SH3 domain and results in a significant reduction of SH3 backbone dynamics (Cordier *et al.*, 2000; Wang *et al.*, 2001). These responses of the SH3 domain are also expected to help in producing the experimentally observed favorable ΔH and unfavorable ΔS of SH3 ligand interactions. Together these analyses stress the fact that the energetics of SH3–ligand interactions are not restrained to the simple docking of the ligand onto the SH3 surface, and that energetics, and hence specificity and affinity, are also dominated by effects that are difficult to estimate by high-resolution structural analyses alone. Moreover, these and other energetic contributions are likely to not be limited to SH3 domains, but may contribute to *in vivo* binding characteristics of other modular ligand recognition domains. For example, the importance of localized water molecules and of loop dynamics has been shown for SH2 domains (see above), and rigidification of the PTB domain structure upon ligand binding was shown for Shc (Farooq *et al.*, 2003).

Since favorable enthalpy contributions are thought to arise from hydrogen and ionic bonds, of which the binding energy is much more sensitive to distance and direction of the interacting moieties then for hydrophobic interactions, and since hydrogen and ionic bonds are believed to be largely responsible for achieving specificity (Ladbury, 2010; Ladbury *et al.*, 2010), it could be anticipated that optimal binding sequences distinguish themselves from less optimal ones by a notably more favorable ΔH contribution. While this characteristic can be found in a subset of analyses, it is not sufficiently sustained to be taken as a rule. For example, comparing binding of a PXXPXR-consensus peptides derived from Sos to the N-SH3 domain of Grb2, the loss of the key arginine did not result in a less favorable ΔH, but a less favorable ΔS (Table 7.4; McDonald *et al.*, 2009). Substitution of two additional arginines into either alanines or other sequences even lead to a decrease in ΔH, and caused a more unfavorable ΔS.

It is also expected that a more stable "prestructuring" of a ligand into the binding-competent PPII conformation would decrease the entropy penalty for fixing this sequence onto the SH3 domain (as expected for the conformationally constrained ligands binding to SH2 domains above). However, analyses of the cortactin SH3 titrations with peptides derived from HPK1 and Shank2 showed that even though the Shank2 peptide alone did not adopt a PPII helix in solution, its binding was highly entropy driven (Table 7.4). It also bound with more favorable entropy to cortactin SH3 as the HPK1 peptide, even though this peptide adopted a PPII conformation in its free state in solution (Rubini *et al.*, 2010). These examples illustrate that the malleability of both, SH3 domains and peptides, and the contributions of interfacial water molecules compose a highly complex system. This makes it almost impossible to predict the thermodynamic fingerprint of a particular SH3–peptide interaction.

8. What Constitutes Specificity in SH3 Domain Interactions?

SH3-dependent organization of signaling complexes needs to be able to adjust quickly, and SH3-dependent regulation of signaling enzymes requires obtaining catalytic activity only in a very well-timed manner. The observation that the bulk of SH3–ligand interactions displays 1–100 μM affinities, and probably fast association and dissociation rates, was attributed to those interactions needing to respond rapidly to changes in cellular conditions (Demers and Mittermaier, 2009; Li, 2005; Mayer, 2001). It is therefore accepted that many, but certainly not all, SH3–ligand interactions are not of very high affinity. Nonetheless, as discussed above for SH2 domains, these interactions need to be of sufficient specificity to allow only correct links to be made, and only the right enzyme to be activated for the right period of time.

It is therefore somehow puzzling that associations between SH3-binding sequence motifs are very robust against amino acid substitutions, as shown by a number of systematic mutational analyses. Mutations outside consensus key residues often have only a marginal effect, typically leading to a variation in K_D of a factor of less than 50-fold (see, e.g., McDonald *et al.*, 2009 for a class I motif or Harkiolaki *et al.*, 2009 for the R/KXXK/R motif). Even mutations of key specificity residues may not abolish binding. For example, loss of the arginine residue of the PXXPXR motif only led to about a 10-fold decrease in affinity of a Sos peptide for Grb2 (McDonald *et al.*, 2009). Elimination of the complete cluster of arginines in the same sequence, leaving a proline-rich sequence without any specificity-conveying charges, lowered the affinity by less than two orders of magnitude (Table 7.4). In many SH3–ligand interactions, the determination of how much exactly the substitution of key residues affects an interaction is rendered difficult by the already low affinity of the biologically relevant ligand. In many cases, even a 10-fold drop in affinity leads to K_D values above 0.1 m*M*, which are impossible to detect in most pull-down, two-hybrid, or phage display assays, and are difficult to measure even with calorimetric or other biophysical methods. Nonetheless, as described for the SH2 domains above, in a cellular context where several hundreds of SH3 domains compete for a wealth of possible binding sites, the bulk of possible low-affinity associations may be, as an ensemble, over 100-fold more numerous than the specific interaction. In this case, a difference of only two orders of magnitude in binding affinity between the biologically relevant interaction and the bulk of nonspecific interactions would not be enough to block these low-affinity binding events from happening. Especially since enzymes fluctuate between open (active and ligand accessible conformation) and closed (inactive, ligand-binding incompetent conformation) states, even in absence of activating ligands, and would therefore present their promiscuous binding sites to nonspecific ligands (Henzler-Wildman and Kern, 2007). Moreover, even low-affinity interactions can be physiologically relevant, as illustrated by the 3 m*M* affinity interaction between the Nck SH3 domain and the PINCH-1 LIM4 domain (Vaynberg *et al.*, 2005). On the one hand, most physiological low-affinity SH3 interactions might remain unknown, because of the difficulties in their experimental detection. On the other hand, for reasons of feasibility, most studies addressed the question of specificity using peptide mimics rather than full-length proteins. In many cases, the length of the peptide motif used might not have been sufficient to reproduce the physiological interaction between SH3 and PXXP containing proteins, and important specificity and affinity enhancing supplementary contacts may thus have been overlooked. It is therefore unclear in most cases what binding energy is minimally needed for biological activity and, importantly, what difference in binding energy is needed to assure fidelity in signaling pathways.

9. Selectivity in SH3 Domain Interactions

Interactions outside the canonical proline-rich binding surface could help in multicellular eukaryotes to achieve the necessary discrimination in SH3–ligand interactions. A case of unusually high selectivity and affinity was reported for the interaction between the T cell signaling molecules Gads and SLP76. The Gads C-terminal SH3 domain associates with the SLP76 RXXK motif sequence an affinity of 8–20 n*M* (Table 7.4; Fig. 7.6; Seet *et al.*, 2007). The last four residues, although not part of the RXXK core, increase the affinity of this interaction 10-fold (Seet *et al.*, 2007). The high strength of this interaction is important for efficient T cell signaling by

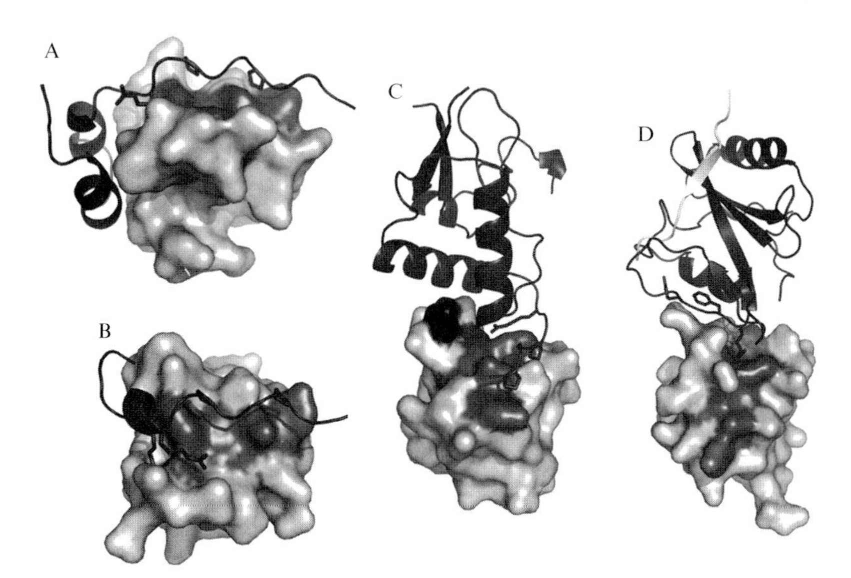

Figure 7.6 SH3 domains (light gray surfaces) engaging supplementary noncanonical interactions with ligands (dark gray ribbon). The mid-gray patch on the surfaces of the SH3 domains highlights the canonical proline-rich binding site. (A) p67^{PHOX} SH3 bound to p47^{PHOX} (PDB entry 1K4U). (B) Gads C-SH3 bound to a PXXP–RXXK motif derived from HPK1 (PDB entry 1UTI). (C) Fyn SH3 domain bound to HIV-1 Nef. The key arginine in the Fyn RT loop is indicated in dark gray. (D) Fyn SH3 bound to SAP SH2. The SLAM fragment bound to SAP SH2 is shown as light gray ribbon (PDB entry 1m27).

allowing the Grb2-family adaptor Gads to link phosphorylated LAT to SLP76. Mutations of the key residues R and K decreased the affinity to 12 and 44 μM, respectively. While still being well within the common range of SH3–ligand interactions, this interaction with the mutated RXXK motif was below the about 2 μM affinity needed for efficient T cell signaling (Seet *et al.*, 2007). Within an array of 147 unique mammalian SH3 domains, the SLP76 SH3-binding sequence associated detectably with three other SH3 domains (Grb2 C-SH3, and SH3 domains from STAM1 and STAM2), however, each time with an affinity below the threshold needed to stimulate T cell signaling (Seet *et al.*, 2007). The high affinity and selectivity of the Gads–SLP76 interaction appear to be nonetheless atypical, in that the same authors noted a much higher cross-reactivity of more conventional PXXP peptides, or even other RXXK core motifs. While efficient T cell signaling required an atypically high Gads–SLP76 affinity (and the high on and slow off rate of this interaction), it tolerated up to 400-fold reduction in strength of binding, as well as an enhanced cross-reactivity (introduced by a particular mutant of the SLP76 motif; Seet *et al.*, 2007).

Some peptidic ligand motifs were found to engage the SH3 domains with an extended binding site. An extreme example of this is found in the complex between the phagocyte oxidase (phox) proteins p47*phox* and 67*phox*. The 32-residue peptide derived from p47*phox* forms a helix-turn-helix motif that contacts the SH3 domain in addition to the PPII helix upon binding to p67*phox* SH3 (Fig. 7.6; Kami *et al.*, 2002). This extended interaction achieves a K_D of 24 nM, which is much stronger than most other peptide–SH3 interactions tested. Another example for an extended peptide–SH3 interaction is provided by the association between HPK1 and Gads. HPK1 binds the Gads C-SH3 domain using a peptide motif that contains both a PPII PxxP and a 3_{10}-helix RXXK motif (Fig. 7.6). However, despite the increased contact surface as compared to the single RXXK motif from SLP76, this sequence bound with a K_D of only 4 μM (Lewitzky *et al.*, 2004), showing that an extended recognition of SH3 domains does not necessarily attain very high affinity.

An obvious drawback of the many large-scale peptide-library based approaches is that they overlook associations between SH3 domains and ligands that are constituted by folded domains. Presenting the proline-rich sequence within the framework of a folded protein (domain) may enhance affinity and selectivity by supplementary interactions and reduce entropy loss upon binding by prestructuring the proline-rich motif. Both mechanisms might explain the approximately threefold increase in affinity, and the great gain in entropy when comparing binding of the Fyn SH3 domain to either a peptide mimic or the multidomain p85α SH3 construct (Table 7.4; Renzoni *et al.*, 1996). Unfortunately, structural information on this interaction is still missing.

One of the first tertiary SH3-domain interactions to be characterized in terms of structure and binding energetics was the association between the HIV-1 Nef protein and Src family SH3 domains (Fig. 7.6; Arold *et al.*, 1997, 1998; Lee *et al.*, 1995, 1996). HIV-1 Nef subverts cellular signaling by binding to a subset of Src family kinases (Arold and Baur, 2001). The proline-rich motif of Nef is presented to SH3 domains as part of a folded domain, and binding gives rise to tertiary interactions. These tertiary interactions increase binding affinity over 100-fold as compared to the proline-rich peptide sequence alone (Lee *et al.*, 1995), possibly by lowering entropic penalties through prestructuring of parts of the PXXPXR binding motif, and by increasing the binding surface. These supplementary contacts also increase selectivity by at least two mechanisms. First, selectivity is increased by the match or mismatch of a specificity pocket in Nef with the interacting SH3 residue from the RT loop. In HIV-1, this pocket is hydrophobic, explaining a higher affinity for Src family SH3 domains with a hydrophobic residue at this position (such as Hck and Lyn, which have an isoleucine). In HIV-2 and the simian variant of Nef, this pocket is more charged, increasing affinity for SH3 domains with a negatively charged residue at this position (such as Src and Fyn, which have an arginine; Arold *et al.*, 1997; Collette *et al.*, 2000; Lee *et al.*, 1995, 1996). Second, the tertiary nature of the interface requires the SH3 RT loop to be pinned down on the Nef surface. This is entropically more costly for SH3 domains such as Hck which have a more flexible RT loop in the free state as, for example, Src and Fyn. However, the breaking of the RT loop hydrogen bonds that exist in the free state to allow it to adopt a binding-competent conformation is more enthalpically costly for Src and Fyn (Arold *et al.*, 1998). This explains, at least in part, the resulting enthalpy–entropy compensation when comparing binding of the Nef core domain to Hck, Src, and Fyn (Table 7.4). The mostly flexible 71 N-terminal residues in full-length Nef appear to contribute to the interaction in an unknown way, leaving a thermodynamic signature that is consistent with a net gain in charged and polar interactions (more favorable ΔH as compared to Nef core interactions), concomitant with a loss in degrees of freedom upon binding (more unfavorable $T\Delta S$; Table 7.4). At least for Hck, the resulting strength of the Nef–SH3 interaction is sufficient to allow Nef to activate Hck by overriding its autoinhibition, which is otherwise maintained by SH2 and SH3 domains binding to intermolecular mimics of phosphotyrosine and proline-rich motifs (Moarefi *et al.*, 1997).

About a dozen tertiary SH3 domain interactions were identified to date (reviewed in Kaneko *et al.*, 2008; Li, 2005). Among these are also SH3 associations that are not mediated at all by a proline-rich motif in the ligand, such as the 3.5 μM strong complex between Fyn SH3 and SAP SH2 domains (Fig. 7.6; Table 7.4; Chan *et al.*, 2003). It is unclear if this relatively low number reflects the difficulties in detecting and characterizing these interactions, or if it indicates that most SH3–ligand interactions are indeed only mediated by linear proline-rich motifs. Large-scale screens that allow detecting associations between full-length proteins and SH3 domains may answer this question (Karkkainen *et al.*, 2006). Currently, tertiary interactions appear to tend to display higher affinities than linear interactions (Karkkainen *et al.*, 2006; Lee *et al.*, 1995). Depending on the dynamics of the association, these may or may not be suitable for every biological system. In the case of HIV-1 Nef, the tertiary interaction might lead to a potentially deleterious prolonged activation of Src kinases (Briggs *et al.*, 1997) which may assist viral replication. Thus, from an evolutionary point of view, it is more costly and difficult to produce a suitable and specific tertiary framework for each proline-rich interaction, especially considering that a large amount of cross-reactivity is necessary in higher eukaryotes (Kaneko *et al.*, 2008; Li, 2005).

10. Interactions Through Multiple Domains

To achieve higher affinity and selectivity, the combinatorial use of domains could provide a successful strategy. Many of the proteins involved in RTK-mediated signal transduction contain more than one recognizable domain, and in some cases, multiple domain interactions occur with a protein target. The binding of two or more domains from one protein to a given protein target would potentially provide additional affinity and thus increase the opportunity for specific interactions. Although, as mentioned above, the change in free energy terms for the interaction of more than one domain with a given cognate protein cannot strictly be considered to be additive, the binding of two domains by virtue of the increased number of possible noncovalent interactions should be of higher affinity compared to one domain. The binding of more than one domain to a protein partner has been thermodynamically characterized in a few cases for RTK-mediated signaling.

Akin to Nef, the focal adhesion kinase (FAK) activates Src kinases by out-competing the autoinhibitory intramolecular interaction between the kinase domain and the SH2–SH3 fragment. In contrast to the tertiary SH3-grip of Nef, however, FAK uses a linear 36-amino acid fragment that binds, when phosphorylated, simultaneously to the SH2 and SH3 domains of Fyn or Src (Fig. 7.6C). This combined SH2–SH3 interaction achieves an even higher affinity ($K_D = 30$ n*M*) than the tertiary association between HIV-1 Nef and Src family SH3 domains. Even though the FAK–SH3 interaction is of only moderate affinity (68 μ*M*), its addition to the FAK–SH2 interaction increases the bidentate binding affinity by one order of magnitude as compared to the single FAK-SH2 binding event (Table 7.4; Arold *et al.*,

2001). Nonetheless, the gain in free energy of the bidentate interaction is less than the sum of the isolated interactions, probably because of the high entropic cost for fixing large parts of the flexible 36-amino acid FAK fragment onto the SH2 and SH3 domains. The affinity of the FAK PXXP motif for the Src SH3 domain alone appears insufficient to activate this kinase. The association only becomes functionally relevant after autophosphorylation of the SH2-binding tyrosine in FAK. Thus, FAK provides an example where an SH3 domain interaction is controlled through a phosphorylation event. Indeed, in contrast to phosphoresidues or phospholipids which can be regulated through (de)phosphorylation, proline-rich motifs do not have an intrinsic mechanism to inactivate their binding capacity. While SH3 interactions can also be regulated directly by phosphorylation of a residue of the SH3–ligand interface (Duke-Cohan *et al.*, 2006), the SH2–SH3 association appears particularly useful in kinase signaling, since SH2 domains are more frequently found associated with SH3 domains than with any other domain (Pincus *et al.*, 2008).

Another case where the expected increase in affinity based on having two domains binding instead of one is in the interaction of the two SH2 domains from the p85 subunit of PI3-kinase. Binding of these tandem domains to a bisphosphorylated peptide with two appropriately positioned pY residues (derived from a cognate protein ligand) binds with a ΔG significantly below that expected for the addition of the individual interactions of the two domains (O'Brien *et al.*, 2000). This was hypothesized to be due to the interaction incurring an entropic penalty from the conformational change associated with having to orientate the domains appropriately for docking onto the ligand.

Perhaps the best example of the increase in specificity being enhanced by having two domains binding is that of the protein ZAP70. The SH2 domains of ZAP70 make a bidentate interaction with the bisphosphorylated ITAM peptide motifs of the ζ-subunit of the T cell receptor. This interaction is fundamental in initiating the recruitment of the ESC for signaling of the T cell receptor on presentation of an antigen (Pacini *et al.*, 2000). The specificity is ensured by the requirement to have both SH2 domains occupied before high affinity binding can occur. This is brought about by the fact that the N-terminal SH2 pY binding site requires extensive interactions between the two domains. The binding pocket has to be assembled on complex formation. The binding of individual monophosphorylated tyrosyl phosphopeptides to ZAP70 are weak $\sim$0.3 mM (O'Brien *et al.*, 2000). The stoichiometries of these interactions are consistent with only one SH2 domain being capable of binding. However, adding the bisphosphorylated peptide results in a further one-to-one interaction but of a high affinity $\sim$32 nM. The crystal structure of ZAP70 shows that binding of the bisphosphorylated ITAM motif forces the two SH2 domains to interact with each other presenting two sites for pY binding (which does not occur with binding to monophosphorylated peptides; Hatada *et al.*, 1995). The thermodynamic parameters show a large favorable enthalpic component for the bisphosphopeptide interaction, presumably consistent with a significant increase in noncovalent bonding as the SH2 domains come together for the intact binding site. The $T\Delta S$ is unexpectedly more favorable, which might result from the exclusion of solvent molecules as the SH2 domains come together and form the cognate site for the bidentate interaction.

11. Conclusions

Study of the binding thermodynamics and structural detail of interactions of proteins involved in tyrosine kinase-mediated signal transduction provides an insight as to the levels of specificity induced. A large body of thermodynamic data has been gathered on the interactions of signaling molecules. This has largely centered on the interactions of the SH2 and SH3 domains. Some data do exist on other domains; however, they are sparse and not sufficient to draw conclusions on general themes such as typical enthalpic and entropic signatures and the level of specificity between different domains and their cognate ligands.

With respect to SH2 domains, the overriding thermodynamic signature for an interaction under physiological temperature regimes is both ΔH and $T\Delta S$ are favorable, but dominated by the former. The pY residue makes a major contribution to the thermodynamics of binding providing a large favorable $T\Delta S$. SH3 domain interactions, perhaps surprisingly for molecules which offer largely hydrophobic binding sites, are also enthalpically driven. The entropy term varies significantly from system to system.

For the interactions of both domains, it can be seen that they offer little in terms of specificity. SH2 domains seem to bind to any ligand which has a pY present. The sequence surrounding this residue seems to be able to provide a maximum of two orders of magnitude difference in affinity. This is not sufficient to ensure mutual exclusivity in signal transduction. Likewise, the interactions of SH3 domains seem to have a similar range of affinity and again suggest that they are quite promiscuous in their recognition of ligands. Selectivity, in some cases, with both domains has been shown to be improved by additional sites of interaction outside the canonical pY or proline-rich binding sites; however, this is unlikely to be a ubiquitous property of interactions of these domains.

Interaction of multiple domains has been shown to provide some increase in affinity; however, these interactions often induce anticooperativity resulting from conformational changes in the protein required to produce the appropriate binding orientation of the domains. Thus, in all, it is likely that the specificity observed in transducing a signal from the cell surface to produce a cellular outcome is imposed by another way than simply through processing of signals in linear, bimolecular relays. This conclusion is supported indirectly by the limited success that the pharmaceutical industry has had in inhibiting proteins involved in RTK-mediated signaling. Despite the importance of proteins expressing these domains and the abundance of structural and biophysical information gathered over two decades, there are currently no compounds on the market place which inhibit their interactions. The promiscuity exhibited by the interactions of these domains means that specificity is hard to achieve in lead compounds leading to toxicological issues. Thus, understanding of the way in which RTK-mediated signaling is able to produce the appropriate cellular response remains a significant challenge to the biological and biophysical community.

REFERENCES

Ababou, A., Pfuhl, M., and Ladbury, J. E. (2008). The stoichiometry of binding between CIN85 SH3 domain A and a proline-rich motif from Cbl-b in solution. *Nat. Struct. Mol. Biol.* **15,** 890–891.

Ababou, A., Pfuhl, M., and Ladbury, J. E. (2009). Novel insights into binding mechanisms of CIN85 SH3 domains to Cbl proteins: Solution-based investigations and in vivo implications. *J. Mol. Biol.* **387,** 1120–1136.

Aitio, O., Hellman, M., Kesti, T., Kleino, I., Samuilova, O., Paakkonen, K., Tossavainen, H., Saksela, K., and Permi, P. (2008). Structural basis of PxxDY motif recognition in SH3 binding. *J. Mol. Biol.* **382,** 167–178.

Arold, S. T., and Baur, A. S. (2001). Dynamic Nef and Nef dynamics: How structure could explain the complex activities of this small HIV protein. *Trends Biochem. Sci.* **26,** 356–363.

Arold, S., Franken, P., Strub, M. P., Hoh, F., Benichou, S., Benarous, R., and Dumas, C. (1997). The crystal structure of HIV-1 Nef protein bound to the Fyn kinase SH3 domain suggests a role for this complex in altered T cell receptor signaling. *Structure* **5,** 1361–1372.

Arold, S., O'Brien, R., Franken, P., Strub, M. P., Hoh, F., Dumas, C., and Ladbury, J. E. (1998). RT loop flexibility enhances the specificity of Src family SH3 domains for HIV-1 Nef. *Biochemistry* **37,** 14683–14691.

Arold, S. T., Ulmer, T. S., Mulhern, T. D., Werner, J. M., Ladbury, J. E., Campbell, I. D., and Noble, M. E. (2001). The role of the Src homology 3-Src homology 2 interface in the regulation of Src kinases. *J. Biol. Chem.* **276,** 17199–17205.

Bae, J. H., Lew, E. D., Yuzawa, S., Tome, F., Lax, I., and Schlessinger, J. (2009). The selectivity of receptor tyrosine kinase signaling is controlled by a secondary SH2 domain binding site. *Cell* **138,** 514–524.

Berry, D. M., Nash, P., Liu, S. K., Pawson, T., and McGlade, C. J. (2002). A high-affinity Arg-X-X-Lys SH3 binding motif confers specificity for the interaction between Gads and SLP-76 in T cell signaling. *Curr. Biol.* **12,** 1336–1341.

Bradshaw, J. M., and Waksman, G. (1999). Calorimetric examination of high-affinity Src SH2 domain-tyrosyl phosphopeptide binding: Dissection of the phosphopeptide sequence specificity and coupling energetics. *Biochemistry* **38,** 5147–5154.

Bradshaw, J. M., and Waksman, G. (2002). Molecular recognition by SH2 domains. *Adv. Protein Chem.* **61,** 161–210.

Bradshaw, J. M., Grucza, R. A., Ladbury, J. E., and Waksman, G. (1998). Probing the "two-pronged plug two-holed socket" model for the mechanism of binding of the Src SH2 domain to phosphotyrosyl peptides: A thermodynamic study. *Biochemistry* **37,** 9083–9090.

Bradshaw, J. M., Mitaxov, V., and Waksman, G. (1999). Investigation of phosphotyrosine recognition by the SH2 domain of the Src kinase. *J. Mol. Biol.* **293,** 971–985.

Bradshaw, J. M., Mitaxov, V., and Waksman, G. (2000). Mutational investigation of the specificity determining region of the Src SH2 domain. *J. Mol. Biol.* **299,** 521–535.

Briggs, S. D., Sharkey, M., Stevenson, M., and Smithgall, T. E. (1997). SH3-mediated Hck tyrosine kinase activation and fibroblast transformation by the Nef protein of HIV-1. *J. Biol. Chem.* **272,** 17899–17902.

Castagnoli, L., Costantini, A., Dall'Armi, C., Gonfloni, S., Montecchi-Palazzi, L., Panni, S., Paoluzi, S., Santonico, E., and Cesareni, G. (2004). Selectivity and promiscuity in the interaction network mediated by protein recognition modules. *FEBS Lett.* **567,** 74–79.

Chan, B., Lanyi, A., Song, H. K., Griesbach, J., Simarro-Grande, M., Poy, F., Howie, D., Sumegi, J., Terhorst, C., and Eck, M. J. (2003). SAP couples Fyn to SLAM immune receptors. *Nat. Cell Biol.* **5,** 155–160.

Charifson, P. S., Shewchuk, L. M., Rocque, W., Hummel, C. W., Jordan, S. R., Mohr, C., Pacofsky, G. J., Peel, M. R., Rodriguez, M., Sternbach, D. D., and Consler, T. G. (1997). Peptide ligands of pp 60(c-src) SH2 domains: A thermodynamic and structural study. *Biochemistry* **36,** 6283–6293.

Chung, E., Henriques, D., Renzoni, D., Zvelebil, M., Bradshaw, J. M., Waksman, G., Robinson, C. V., and Ladbury, J. E. (1998). Mass spectrometric and thermodynamic studies reveal the role of water molecules in complexes formed between SH2 domains and tyrosyl phosphopeptides. *Structure* **6,** 1141–1151.

Chung, E. W., Henriques, D. A., Renzoni, D., Morton, C. J., Mulhern, T. D., Pitkeathly, M. C., Ladbury, J. E., and Robinson, C. V. (1999). Probing the nature of interactions in SH2 binding interfaces—Evidence from electrospray ionization mass spectrometry. *Protein Sci.* **8,** 1962–1970.

Collette, Y., Arold, S., Picard, C., Janvier, K., Benichou, S., Benarous, R., Olive, D., and Dumas, C. (2000). HIV-2 and SIV nef proteins target different Src family SH3 domains than does HIV-1 Nef because of a triple amino acid substitution. *J. Biol. Chem.* **275,** 4171–4176.

Cordier, F., Wang, C., Grzesiek, S., and Nicholson, L. K. (2000). Ligand-induced strain in hydrogen bonds of the c-Src SH3 domain detected by NMR. *J. Mol. Biol.* **304,** 497–505.

Davidson, J. P., Lubman, O., Rose, T., Waksman, G., and Martin, S. F. (2002). Calorimetric and structural studies of 1, 2, 3-trisubstituted cyclopropanes as conformationally constrained peptide inhibitors of Src SH2 domain binding. *J. Am. Chem. Soc.* **124,** 205–215.

De Fabritiis, G., Geroult, S., Coveney, P. V., and Waksman, G. (2008). Insights from the energetics of water binding at the domain-ligand interface of the Src SH2 domain. *Proteins* **72,** 1290–1297.

DeLorbe, J. E., Clements, J. H., Teresk, M. G., Benfield, A. P., Plake, H. R., Millspaugh, L. E., and Martin, S. F. (2009). Thermodynamic and structural effects of conformational constraints in protein–ligand interactions. Entropic paradoxy associated with ligand preorganization. *J. Am. Chem. Soc.* **131,** 16758–16770.

Demers, J. P., and Mittermaier, A. (2009). Binding mechanism of an SH3 domain studied by NMR and ITC. *J. Am. Chem. Soc.* **131,** 4355–4367.

Duke-Cohan, J. S., Kang, H., Liu, H., and Rudd, C. E. (2006). Regulation and function of SKAP-55 non-canonical motif binding to the SH3c domain of adhesion and degranulation-promoting adaptor protein. *J. Biol. Chem.* **281,** 13743–13750.

Eck, M. J., Shoelson, S. E., and Harrison, S. C. (1993). Recognition of a high-affinity phosphotyrosyl peptide by the Src homology-2 domain of p56lck. *Nature* **362,** 87–91.
Farooq, A., Zeng, L., Yan, K. S., Ravichandran, K. S., and Zhou, M. M. (2003). Coupling of folding and binding in the PTB domain of the signaling protein Shc. *Structure* **11,** 905–913.
Ferreon, J. C., and Hilser, V. J. (2003). Ligand-induced changes in dynamics in the RT loop of the C-terminal SH3 domain of Sem-5 indicate cooperative conformational coupling. *Protein Sci.* **12,** 982–996.
Ferreon, J. C., and Hilser, V. J. (2004). Thermodynamics of binding to SH3 domains: The energetic impact of polyproline II (PII) helix formation. *Biochemistry* **43,** 7787–7797.
Ferreon, J. C., Volk, D. E., Luxon, B. A., Gorenstein, D. G., and Hilser, V. J. (2003). Solution structure, dynamics, and thermodynamics of the native state ensemble of the Sem-5 C-terminal SH3 domain. *Biochemistry* **42,** 5582–5591.
Freund, C., Dotsch, V., Nishizawa, K., Reinherz, E. L., and Wagner, G. (1999). The GYF domain is a novel structural fold that is involved in lymphoid signaling through proline-rich sequences. *Nat. Struct. Biol.* **6,** 656–660.
Garron, M. L., Arthos, J., Guichou, J. F., McNally, J., Cicala, C., and Arold, S. T. (2008). Structural basis for the interaction between focal adhesion kinase and CD4. *J. Mol. Biol.* **375,** 1320–1328.
Harkiolaki, M., Tsirka, T., Lewitzky, M., Simister, P. C., Joshi, D., Bird, L. E., Jones, E. Y., O'Reilly, N., and Feller, S. M. (2009). Distinct binding modes of two epitopes in Gab2 that interact with the SH3C domain of Grb2. *Structure* **17,** 809–822.
Hatada, M. H., Lu, X., Laird, E. R., Green, J., Morgenstern, J. P., Lou, M., Marr, C. S., Phillips, T. B., Ram, M. K., Theriault, K., *et al.* (1995). Molecular basis for interaction of the protein tyrosine kinase ZAP-70 with the T-cell receptor. *Nature* **377,** 32–38.
Henriques, D. A., and Ladbury, J. E. (2001). Inhibitors to the Src SH2 domain: A lesson in structure–thermodynamic correlation in drug design. *Arch. Biochem. Biophys.* **390,** 158–168.
Henzler-Wildman, K., and Kern, D. (2007). Dynamic personalities of proteins. *Nature* **450,** 964–972.
Holt, M. R., and Koffer, A. (2001). Cell motility: Proline-rich proteins promote protrusions. *Trends Cell Biol.* **11,** 38–46.
Houtman, J. C., Higashimoto, Y., Dimasi, N., Cho, S., Yamaguchi, H., Bowden, B., Regan, C., Malchiodi, E. L., Mariuzza, R., Schuck, P., Appella, E., and Samelson, L. E. (2004). Binding specificity of multiprotein signaling complexes is determined by both cooperative interactions and affinity preferences. *Biochemistry* **43,** 4170–4178.
Ilsley, J. L., Sudol, M., and Winder, S. J. (2002). The WW domain: Linking cell signalling to the membrane cytoskeleton. *Cell. Signal.* **14,** 183–189.
Ji, Q. S., Chattopadhyay, A., Vecchi, M., and Carpenter, G. (1999). Physiological requirement for both SH2 domains for phospholipase C-gamma1 function and interaction with platelet-derived growth factor receptors. *Mol. Cell. Biol.* **19,** 4961–4970.
Kami, K., Takeya, R., Sumimoto, H., and Kohda, D. (2002). Diverse recognition of non-PxxP peptide ligands by the SH3 domains from p67(phox), Grb2 and Pex13p. *EMBO J.* **21,** 4268–4276.
Kaneko, T., Li, L., and Li, S. S. (2008). The SH3 domain—A family of versatile peptide- and protein-recognition module. *Front. Biosci.* **13,** 4938–4952.
Kang, H., Freund, C., Duke-Cohan, J. S., Musacchio, A., Wagner, G., and Rudd, C. E. (2000). SH3 domain recognition of a proline-independent tyrosine-based RKxxYxxY motif in immune cell adaptor SKAP55. *EMBO J.* **19,** 2889–2899.
Karkkainen, S., Hiipakka, M., Wang, J. H., Kleino, I., Vaha-Jaakkola, M., Renkema, G. H., Liss, M., Wagner, R., and Saksela, K. (2006). Identification of preferred protein interactions by phage-display of the human Src homology-3 proteome. *EMBO Rep.* **7,** 186–191.

Kato, Y., Nagata, K., Takahashi, M., Lian, L., Herrero, J. J., Sudol, M., and Tanokura, M. (2004). Common mechanism of ligand recognition by group II/III WW domains: Redefining their functional classification. *J. Biol. Chem.* **279,** 31833–31841.

Kowanetz, K., Husnjak, K., Holler, D., Kowanetz, M., Soubeyran, P., Hirsch, D., Schmidt, M. H., Pavelic, K., De Camilli, P., Randazzo, P. A., and Dikic, I. (2004). CIN85 associates with multiple effectors controlling intracellular trafficking of epidermal growth factor receptors. *Mol. Biol. Cell* **15,** 3155–3166.

Kuriyan, J., and Cowburn, D. (1993). Structures of SH2 and SH3 domains. *Curr. Opin. Struct. Biol.* **3,** 828–837.

Kuriyan, J., and Cowburn, D. (1997). Modular peptide recognition domains in eukaryotic signaling. *Annu. Rev. Biophys. Biomol. Struct.* **26,** 259–288.

Ladbury, J. E. (2005). Protein-protein recognition in phosphotyrosine-mediated intracellular signaling. *In* "Proteomics and Protein-Protein Interactions: Biology, Chemistry, Bioinformatics and Drug Design," (G. Waksman, ed.), Protein Reviews, Vol. 3, pp. 165–184. Kluwer/Plenum Publishing Corporation. New York, USA.

Ladbury, J. E. (2010). Calorimetry as a tool for understanding biomolecular interactions and an aid to drug design. *Biochem. Soc. Trans.* **38,** 888–893.

Ladbury, J. E., and Arold, S. (2000). Searching for specificity in SH domains. *Chem. Biol.* **7,** R3–R8.

Ladbury, J. E., Hensmann, M., Panayotou, G., and Campbell, I. D. (1996). Alternative modes of tyrosyl phosphopeptide binding to a Src family SH2 domain: Implications for regulation of tyrosine kinase activity. *Biochemistry* **35,** 11062–11069.

Ladbury, J. E., Klebe, G., and Freire, E. (2010). Adding calorimetric data to decision-making in lead discovery: a hot tip. *Nature Rev. Drug Disc.* **9,** 23–27.

Ladbury, J. E., Lemmon, M. A., Zhou, M., Green, J., Botfield, M. C., and Schlessinger, J. (1995). Measurement of the binding of tyrosyl phosphopeptides to SH2 domains: A reappraisal. *Proc. Natl. Acad. Sci. USA* **92,** 3199–3203.

Landgraf, C., Panni, S., Montecchi-Palazzi, L., Castagnoli, L., Schneider-Mergener, J., Volkmer-Engert, R., and Cesareni, G. (2004). Protein interaction networks by proteome peptide scanning. *PLoS Biol.* **2,** E14.

Lee, C. H., Leung, B., Lemmon, M. A., Zheng, J., Cowburn, D., Kuriyan, J., and Saksela, K. (1995). A single amino acid in the SH3 domain of Hck determines its high affinity and specificity in binding to HIV-1 Nef protein. *EMBO J.* **14,** 5006–5015.

Lee, C. H., Saksela, K., Mirza, U. A., Chait, B. T., and Kuriyan, J. (1996). Crystal structure of the conserved core of HIV-1 Nef complexed with a Src family SH3 domain. *Cell* **85,** 931–942.

Lemmon, M. A., and Ladbury, J. E. (1994). Thermodynamic studies of tyrosyl-phosphopeptide binding to the SH2 domain of p56lck. *Biochemistry* **33,** 5070–5076.

Lemmon, M. A., Ladbury, J. E., Mandiyan, V., Zhou, M., and Schlessinger, J. (1994). Independent binding of peptide ligands to the SH2 and SH3 domains of Grb2. *J. Biol. Chem.* **269,** 31653–31658.

Lewitzky, M., Harkiolaki, M., Domart, M. C., Jones, E. Y., and Feller, S. M. (2004). Mona/Gads SH3C binding to hematopoietic progenitor kinase 1 (HPK1) combines an atypical SH3 binding motif, R/KXXK, with a classical PXXP motif embedded in a polyproline type II (PPII) helix. *J. Biol. Chem.* **279,** 28724–28732.

Li, S. S. (2005). Specificity and versatility of SH3 and other proline-recognition domains: Structural basis and implications for cellular signal transduction. *Biochem. J.* **390,** 641–653.

Lim, W. A., Richards, F. M., and Fox, R. O. (1994). Structural determinants of peptide-binding orientation and of sequence specificity in SH3 domains. *Nature* **372,** 375–379.

Lubman, O. Y., and Waksman, G. (2002). Dissection of the energetic coupling across the Src SH2 domain-tyrosyl phosphopeptide interface. *J. Mol. Biol.* **316,** 291–304.

Lubman, O. Y., and Waksman, G. (2003). Structural and thermodynamic basis for the interaction of the Src SH2 domain with the activated form of the PDGF beta-receptor. *J. Mol. Biol.* **328,** 655–668.

Lubman, O. Y., Kopan, R., Waksman, G., and Korolev, S. (2005). The crystal structure of a partial mouse Notch-1 ankyrin domain: Repeats 4 through 7 preserve an ankyrin fold. *Protein Sci.* **14,** 1274–1281.

Mayer, B. J. (2001). SH3 domains: Complexity in moderation. *J. Cell Sci.* **114,** 1253–1263.

McDonald, C. B., Seldeen, K. L., Deegan, B. J., and Farooq, A. (2009). SH3 domains of Grb2 adaptor bind to PXpsiPXR motifs within the Sos1 nucleotide exchange factor in a discriminate manner. *Biochemistry* **48,** 4074–4085.

McNemar, C., Snow, M. E., Windsor, W. T., Prongay, A., Mui, P., Zhang, R., Durkin, J., Le, H. V., and Weber, P. C. (1997). Thermodynamic and structural analysis of phosphotyrosine polypeptide binding to Grb2-SH2. *Biochemistry* **36,** 10006–10014.

Moarefi, I., LaFevre-Bernt, M., Sicheri, F., Huse, M., Lee, C. H., Kuriyan, J., and Miller, W. T. (1997). Activation of the Src-family tyrosine kinase Hck by SH3 domain displacement. *Nature* **385,** 650–653.

Mongiovi, A. M., Romano, P. R., Panni, S., Mendoza, M., Wong, W. T., Musacchio, A., Cesareni, G., and Di Fiore, P. P. (1999). A novel peptide–SH3 interaction. *EMBO J.* **18,** 5300–5309.

Musacchio, A., Noble, M., Pauptit, R., Wierenga, R., and Saraste, M. (1992). Crystal structure of a Src-homology 3 (SH3) domain. *Nature* **359,** 851–855.

Nishizawa, K., Freund, C., Li, J., Wagner, G., and Reinherz, E. L. (1998). Identification of a proline-binding motif regulating CD2-triggered T lymphocyte activation. *Proc. Natl. Acad. Sci. USA* **95,** 14897–14902.

O'Brien, R., Rugman, P., Renzoni, D., Layton, M., Handa, R., Hilyard, K., Waterfield, M. D., Driscoll, P. C., and Ladbury, J. E. (2000). Alternative modes of binding of proteins with tandem SH2 domains. *Protein Sci.* **9,** 570–579.

O'Rourke, L., and Ladbury, J. E. (2003). Specificity is complex and time consuming: Mutual exclusivity in tyrosine kinase-mediated signaling. *Acc. Chem. Res.* **36,** 410–416.

Pacini, S., Valensin, S., Telford, J. L., Ladbury, J., and Baldari, C. T. (2000). Temporally regulated assembly of a dynamic signaling complex associated with the activated TCR. *Eur. J. Immunol.* **30,** 2620–2631.

Palencia, A., Cobos, E. S., Mateo, P. L., Martinez, J. C., and Luque, I. (2004). Thermodynamic dissection of the binding energetics of proline-rich peptides to the Abl-SH3 domain: Implications for rational ligand design. *J. Mol. Biol.* **336,** 527–537.

Palencia, A., Camara-Artigas, A., Pisabarro, M. T., Martinez, J. C., and Luque, I. (2010). Role of interfacial water molecules in proline-rich ligand recognition by the Src homology 3 domain of Abl. *J. Biol. Chem.* **285,** 2823–2833.

Pascal, S. M., Yamazaki, T., Singer, A. U., Kay, L. E., and Forman-Kay, J. D. (1995). Structural and dynamic characterization of the phosphotyrosine binding region of a Src homology 2 domain–phosphopeptide complex by NMR relaxation, proton exchange, and chemical shift approaches. *Biochemistry* **34,** 11353–11362.

Pawson, T. (2004). Specificity in signal transduction: From phosphotyrosine-SH2 domain interactions to complex cellular systems. *Cell* **116,** 191–203.

Pawson, T., and Nash, P. (2000). Protein–protein interactions define specificity in signal transduction. *Genes Dev.* **14,** 1027–1047.

Pawson, T., and Nash, P. (2003). Assembly of cell regulatory systems through protein interaction domains. *Science* **300,** 445–452.

Peterson, F. C., and Volkman, B. F. (2009). Diversity of polyproline recognition by EVH1 domains. *Front. Biosci.* **14,** 833–846.

Pincus, D., Letunic, I., Bork, P., and Lim, W. A. (2008). Evolution of the phospho-tyrosine signaling machinery in premetazoan lineages. *Proc. Natl. Acad. Sci. USA* **105,** 9680–9684.

Pornillos, O., Alam, S. L., Davis, D. R., and Sundquist, W. I. (2002). Structure of the Tsg101 UEV domain in complex with the PTAP motif of the HIV-1 p6 protein. *Nat. Struct. Biol.* **9,** 812–817.

Rahuel, J., Gay, B., Erdmann, D., Strauss, A., Garcia-Echeverria, C., Furet, P., Caravatti, G., Fretz, H., Schoepfer, J., and Grutter, M. G. (1996). Structural basis for specificity of Grb2-SH2 revealed by a novel ligand binding mode. *Nat. Struct. Biol.* **3,** 586–589.

Rahuel, J., Garcia-Echeverria, C., Furet, P., Strauss, A., Caravatti, G., Fretz, H., Schoepfer, J., and Gay, B. (1998). Structural basis for the high affinity of amino-aromatic SH2 phosphopeptide ligands. *J. Mol. Biol.* **279,** 1013–1022.

Renzoni, D. A., Pugh, D. J., Siligardi, G., Das, P., Morton, C. J., Rossi, C., Waterfield, M. D., Campbell, I. D., and Ladbury, J. E. (1996). Structural and thermodynamic characterization of the interaction of the SH3 domain from Fyn with the proline-rich binding site on the p85 subunit of PI3-kinase. *Biochemistry* **35,** 15646–15653.

Rickles, R. J., Botfield, M. C., Weng, Z., Taylor, J. A., Green, O. M., Brugge, J. S., and Zoller, M. J. (1994). Identification of Src, Fyn, Lyn, PI3K and Abl SH3 domain ligands using phage display libraries. *EMBO J.* **13,** 5598–5604.

Rubin, G. M., Yandell, M. D., Wortman, J. R., Gabor Miklos, G. L., Nelson, C. R., Hariharan, I. K., Fortini, M. E., Li, P. W., Apweiler, R., Fleischmann, W., Cherry, J. M., Henikoff, S., *et al.* (2000). Comparative genomics of the eukaryotes. *Science* **287,** 2204–2215.

Rubini, C., Ruzza, P., Spaller, M. R., Siligardi, G., Hussain, R., Udugamasooriya, D. G., Bellanda, M., Mammi, S., Borgogno, A., Calderan, A., Cesaro, L., Brunati, A. M., *et al.* (2010). Recognition of lysine-rich peptide ligands by murine cortactin SH3 domain: CD, ITC, and NMR studies. *Biopolymers* **94,** 298–306.

Saito, K., Kigawa, T., Koshiba, S., Sato, K., Matsuo, Y., Sakamoto, A., Takagi, T., Shirouzu, M., Yabuki, T., Nunokawa, E., Seki, E., Matsuda, T., *et al.* (2004). The CAP-Gly domain of CYLD associates with the proline-rich sequence in NEMO/IKKgamma. *Structure* **12,** 1719–1728.

Schlessinger, J. (2000). Cell signaling by receptor tyrosine kinases. *Cell* **103,** 211–225.

Schlessinger, J., and Lemmon, M. A. (2003). SH2 and PTB domains in tyrosine kinase signaling. *Sci. STKE* **2003,** RE12.

Seet, B. T., Berry, D. M., Maltzman, J. S., Shabason, J., Raina, M., Koretzky, G. A., McGlade, C. J., and Pawson, T. (2007). Efficient T-cell receptor signaling requires a high-affinity interaction between the Gads C-SH3 domain and the SLP-76 RxxK motif. *EMBO J.* **26,** 678–689.

Songyang, Z., Shoelson, S. E., Chaudhuri, M., Gish, G., Pawson, T., Haser, W. G., King, F., Roberts, T., Ratnofsky, S., Lechleider, R. J., *et al.* (1993). SH2 domains recognize specific phosphopeptide sequences. *Cell* **72,** 767–778.

Sparks, A. B., Rider, J. E., Hoffman, N. G., Fowlkes, D. M., Quillam, L. A., and Kay, B. K. (1996). Distinct ligand preferences of Src homology 3 domains from Src, Yes, Abl, Cortactin, p53bp2, PLCgamma, Crk, and Grb2. *Proc. Natl. Acad. Sci. USA* **93,** 1540–1544.

Sparks, A. B., Rider, J. E., and Kay, B. K. (1998). Mapping the specificity of SH3 domains with phage-displayed random-peptide libraries. *Methods Mol. Biol.* **84,** 87–103.

Taylor, J. D., Gilbert, P. J., Williams, M. A., Pitt, W. R., and Ladbury, J. E. (2007). Identification of novel fragment compounds targeted against the pY pocket of v-Src SH2 by computational and NMR screening and thermodynamic evaluation. *Proteins* **67,** 981–990.

Taylor, J. D., Ababou, A., Fawaz, R. R., Hobbs, C. J., Williams, M. A., and Ladbury, J. E. (2008). Structure, dynamics, and binding thermodynamics of the v-Src SH2 domain: Implications for drug design. *Proteins* **73,** 929–940.

Terasawa, H., Kohda, D., Hatanaka, H., Tsuchiya, S., Ogura, K., Nagata, K., Ishii, S., Mandiyan, V., Ullrich, A., Schlessinger, J., *et al.* (1994). Structure of the N-terminal SH3 domain of GRB2 complexed with a peptide from the guanine nucleotide releasing factor Sos. *Nat. Struct. Biol.* **1,** 891–897.

Tian, L., Chen, L., McClafferty, H., Sailer, C. A., Ruth, P., Knaus, H. G., and Shipston, M. J. (2006). A noncanonical SH3 domain binding motif links BK channels to the actin cytoskeleton via the SH3 adapter cortactin. *FASEB J.* **20,** 2588–2590.

Tong, A. H., Drees, B., Nardelli, G., Bader, G. D., Brannetti, B., Castagnoli, L., Evangelista, M., Ferracuti, S., Nelson, B., Paoluzi, S., Quondam, M., Zucconi, A., *et al.* (2002). A combined experimental and computational strategy to define protein interaction networks for peptide recognition modules. *Science* **295,** 321–324.

Vaynberg, J., Fukuda, T., Chen, K., Vinogradova, O., Velyvis, A., Tu, Y., Ng, L., Wu, C., and Qin, J. (2005). Structure of an ultraweak protein–protein complex and its crucial role in regulation of cell morphology and motility. *Mol. Cell* **17,** 513–523.

Virdee, S., Macmillan, D., and Waksman, G. (2010). Semisynthetic Src SH2 domains demonstrate altered phosphopeptide specificity induced by incorporation of unnatural lysine derivatives. *Chem. Biol.* **17,** 274–284.

Waksman, G., Kominos, D., Robertson, S. C., Pant, N., Baltimore, D., Birge, R. B., Cowburn, D., Hanafusa, H., Mayer, B. J., Overduin, M., *et al.* (1992). Crystal structure of the phosphotyrosine recognition domain SH2 of v-src complexed with tyrosine-phosphorylated peptides. *Nature* **358,** 646–653.

Waksman, G., Shoelson, S. E., Pant, N., Cowburn, D., and Kuriyan, J. (1993). Binding of a high affinity phosphotyrosyl peptide to the Src SH2 domain: Crystal structures of the complexed and peptide-free forms. *Cell* **72,** 779–790.

Wang, C., Pawley, N. H., and Nicholson, L. K. (2001). The role of backbone motions in ligand binding to the c-Src SH3 domain. *J. Mol. Biol.* **313,** 873–887.

Yu, H., Rosen, M. K., Shin, T. B., Seidel-Dugan, C., Brugge, J. S., and Schreiber, S. L. (1992). Solution structure of the SH3 domain of Src and identification of its ligand-binding site. *Science* **258,** 1665–1668.

Yu, H., Chen, J. K., Feng, S., Dalgarno, D. C., Brauer, A. W., and Schreiber, S. L. (1994). Structural basis for the binding of proline-rich peptides to SH3 domains. *Cell* **76,** 933–945.

Zarrinpar, A., Bhattacharyya, R. P., and Lim, W. A. (2003). The structure and function of proline recognition domains. *Sci. STKE* **2003,** RE8.

CHAPTER EIGHT

Structural and Functional Energetic Linkages in Allosteric Regulation of Muscle Pyruvate Kinase

J. Ching Lee* *and* Petr Herman†

Contents

* Department of Biochemistry and Molecular Biology, The University of Texas Medical Branch at Galveston, Galveston, Texas, USA
† Institute of Physics, Charles University, Ke Karlovu, Prague, Czech Republic

Methods in Enzymology, Volume 488
ISSN 0076-6879, DOI: 10.1016/B978-0-12-381268-1.00008-2
© 2011 Elsevier Inc.
All rights reserved.

Abstract

The understanding of the molecular mechanisms of allostery in rabbit muscle pyruvate kinase (RMPK) is still in its infancy. Although, there is a paucity of knowledge on the ground rules on how its functions are regulated, RMPK is an ideal system to address basic questions regarding the fundamental chemical principles governing the regulatory mechanisms about this enzyme which has a TIM $(\alpha/\beta)_8$ barrel structural motif [Copley, R. R., and Bork, P. (2000). Homology among $(\beta\alpha)8$ barrels: Implications for the evolution of metabolic pathways. *J. Mol. Biol.* **303,** 627–640; Farber, G. K., and Petsko, G. A. (1990). The evolution of α/ß barrel enzymes. *Trends Biochem.* **15,** 228–234; Gerlt, J. A., and Babbitt, P. C. (2001). Divergent evolution of enzymatic function: Mechanistically diverse superfamilies and functionally distinct superfamilies. *Annu. Rev. Biochem.* **70,** 209–246; Heggi, H., and Gerstein, M. (1999). The relationship between protein structure and function: A comprehensive survey with application to the yeast genome. *J. Mol. Biol.* **288,** 147–164; Wierenga, R. K. (2001). The TIM-barrel fold: A versatile framework for efficient enzymes. *FEB Lett.* **492,** 193–198]. RMPK is a homotetramer. Each subunit consists of 530 amino acids and multiple domains. The active site resides between the A and B domains. Besides the basic TIM-barrel motif, RMPK also exhibits looped-out regions in the α/β barrel of each monomer forming the B- and C-domains. The two isozymes of PK, namely, the kidney and muscle isozymes, exhibit very different allosteric behaviors under the same experimental condition. The only amino acid sequence differences between the mammalian kidney and muscle PK isozymes are located in the C-domain and are involved in intersubunit interactions. Thus, embedded in these two isozymes of PK are the rules involved in engineering the popular TIM $(\alpha/\beta)_8$ motif to modulate its allosteric properties. The PK system exhibits a lot of the properties that will allow mining of the ground rules governing the correlative linkages between sequence-fold-function. In this chapter, we review the approaches to acquire the fundamental functional and structural energetics that establish the linkages among this intricate network of linked multiequilibria. Results from these diverse approaches are integrated to establish a working model to represent the complex network of multiple linked reactions which ultimately leads to the observation of allosteric regulation of PK.

1. Introduction

Protein molecules are polyelectrolytes that are capable of binding ligands which would lead to biological activities or ligands which would regulate these activities, that is, proteins are involved in ligand bindings which can be expressed as linked multiequilibria (Steinhardt and Reynolds, 1969; Wyman and Gill, 1990). As a consequence of such a linkage, biological activities can be regulated exquisitely. In this chapter, we will address the applications of thermodynamics in the quest for the mechanism of the allosteric regulatory mechanism of allostery of an enzyme.

2. General Principles of Linked Multiequilibria Reactions

The following scheme is a generic expression of the thermodynamic cycle for a linkage of two equilibria of binding two ligands to protein P.

$$\begin{array}{ccc} & K_y & \\ X+P+Y & \rightleftharpoons & X+PY \\ K_x \uparrow\downarrow & & \downarrow\uparrow K_{x/y} \\ XP+Y & \rightleftharpoons & XPY \\ & K_{y/x} & \end{array}$$

Scheme 8.1 Thermodynamic cycle for linked reactions

where K_x and K_y are the binding constants of ligand X and Y to P in the absence of the second ligand, respectively. $K_{x/y}$ and $K_{y/x}$ are the binding constants of X and Y binding to P in the presence of saturating concentration of Y and X, respectively. If the presence of ligand X or Y affects the binding of Y or X, then those reactions are coupled. There are a few different ways to quantitatively express that relationship. They are:

2.1. Wyman linked function

The relation between equilibrium constants and effector concentration can be analyzed by the linked-function theory expressed by Tanford (1969)

$$\frac{\partial \ln K}{\partial \ln a_x} = \Delta v_x - \frac{n_x}{n_w} \Delta v_w = \Delta v_{x,\mathrm{pref}} \tag{8.1}$$

where K is the apparent binding constant of Y, a_x is the activity of effector ligand x, Δv_x and Δv_w are the changes in the amount of effector ligand and water, respectively, bound to P upon the change in states and n_x/n_w is the ratio of the number of moles of X to water present in the solution. Since n_x/n_w assumes a very small value, this term can be neglected. Then, measurement of binding constants of Y in the presence of varying concentrations of X can lead to an estimate of the nature of coupling between X and Y, be it positive or negative, that is, favoring of Y binding in the presence of X or *vice versa*.

2.2. Weber expression of the linked-function concept of Wyman in terms of binding energy

The term ΔG_{xy}, the free energy of interaction between ligands X and Y, is defined as (Weber, 1972)

$$\Delta G_{x/y} - \Delta G_x = \Delta G_{y/x} - \Delta G_y = \Delta G_{xy} \tag{8.2}$$

where ΔG_x and ΔG_y are the standard free energies of binding to unliganded P by X and Y, respectively, whereas $\Delta G_{x/y}$ and $\Delta G_{y/x}$ are the free energies of binding of X and Y to P which is fully liganded with the other ligand.

2.3. Reinhart derivation of the Weber expression as applied to steady-state kinetics

$$Q_{xy} = K_{x/y}/K_x = K_{y/x}/K_y, \quad \Delta G_{xy} = -RT\ln(Q_{xy}) \tag{8.3}$$

where the various terms for K have the same meaning as in Scheme 8.1. Application of this expression is valid only if it has been proven that the Michaelis constant, K_m, is equivalent to K_d, the dissociation constant of ligand binding. Readers are strongly recommended to review the elegant discussion of this application by Reinhart (2004).

ΔG_{xy} or Q_{xy} may assume either positive or negative values. This is a very valuable expression for a system that is just being investigated although these parameters do not provide information on the molecular mechanism of linkage. These coupling parameters expressed as such may include terms pertaining to conformational changes in protein P; change in protonation states of P as a result of binding of ligands, etc. For example, if P exists in an equilibrium between two conformation states, P and P′; each states can bind ligands albeit with different affinities, then in the presence of both X and Y, there are at least eight states that one needs to consider, namely, P, P′, PX, P′X, PY, P′Y, XPY and XP′Y. One is confronted with too many species to consider. Thus, it would be useful to simplify the system initially by collecting terms related to these species into ΔG_{xy} or Q_{xy}. With additional studies using different approaches one might be able to identify and monitor the two protein states, P and P′. However, it will be most difficult to find distinctive structural properties for P, P′, PX, P′X, PY, P′Y, XPY and XP′Y so as to enable one to tract the presence of each of these states. The greatest challenge is to place the appropriate functional significance of those structural states. For example, a "minor" structural change which might be interpreted as insignificant but might represent a state that is functionally and energetically different. Thus, it is most fruitful to distinguish those intermediate states by classifying their energetics such as the elegant studies of Hb by Gary Ackers (Holt and Ackers, 1995).

3. Functional Energetic Linkages in Allosteric Regulation of Rabbit Muscle Pyruvate Kinase

Rabbit muscle pyruvate kinase (RMPK) is chosen as a model system to illustrate the various strategies that we have employed to study a specific phenomenon of linked multiequilibria, the allosteric regulation of this enzyme. The focus of this chapter is on solution energetics, thus most of the high resolution atomic structural information of RMPK is not presented. For readers who are more interested in the atomic structures of RMPK or other isozymes of PK, they should consult the pioneering work by Muirhead, Reed, Rayment, Mattevi, and their coworkers (Larsen *et al.*, 1994, 1997, 1998; Mattevi *et al.*, 1995, 1996; Stuart *et al.*, 1979). More recently, the work of Fenton and Mesecar should also be consulted (Dombrauckas *et al.*, 2005; Williams *et al.*, 2006).

RMPK is an important allosteric enzyme of the glycolytic pathway catalyzing a transfer of the phosphate from phosphoenolpyruvate (PEP) to ADP (Ainsworth and MacFarlane, 1973; Consler *et al.*, 1989, 1992; Hall and Cottam, 1978; Oberfelder *et al.*, 1984a,b)

$$\text{PEP} + \text{ADP} \rightarrow \text{pyruvate} + \text{ATP}$$

RMPK has a TIM $(\alpha/\beta)_8$ barrel structural motif (Copley and Bork, 2000; Farber and Petsko, 1990; Gerlt and Babbitt, 2001; Heggi and Gerstein, 1999; Wierenga, 2001). The enzymatic reaction requires two metal ions, namely, K^+ and Mg^{2+}; it is inhibited by Phe or other amino acids and activated by phosphofructose 1,6-bisphosphate (FBP). Production of ATP is essential in the cell energetics, and therefore, it is not surprising that RMPK activity is subjected to an intriguing pattern of regulation. RMPK consists of four identical subunits (Larsen *et al.*, 1994; Stuart *et al.*, 1979) each of which is composed of three domains, A, B, and C. The binding sites of various ligands are distributed at distant regions of the molecule, as shown in Fig. 8.1. Thus, the

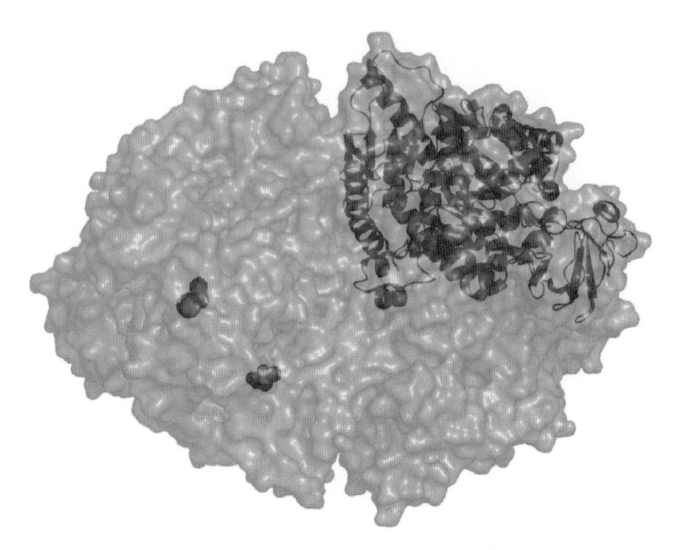

Figure 8.1 Tetrameric structure of RMPK. One of the subunits is presented as cartoon over the surface presentation. The spheres are representations of the distance between the active site and the binding site of inhibitor. The structure is generated using the coordinates of Wooll *et al.* (2001).

mechanism of communication must involve long-range communications between these distant sites. An elucidation of the regulatory mechanism must consider the linked multiequilibria interactions linking subunit–subunit communications, interfacial interactions among these domains and effects of binding of these various ligands on these macromolecular interactions. We will illustrate the various strategies that have been successfully applied to RMPK.

The minimum ligands that are important for RMPK activity include:

- Substrates: PEP and ADP
- Metal cofactors: K^+ and Mg^{2+}.
- Inhibitor: Phe and other amino acids
- Activator: FBP
- Solution environmental effects: pH and temperature.

Thus, it is immediately obvious that the basic regulatory mechanism for RMPK is complex and the investigator needs to elucidate the linkages among bindings of at least seven ligands (H^+ is also a ligand) and their effects on interfacial interaction or the reciprocal effects.

4. Functional Linkage Through Steady-State Kinetics

A useful approach to initiate a general mapping of the allosteric behavior of an enzyme system is through steady-state kinetics, which in the present case with RMPK the measurements are conducted with the coupled enzyme assay. In such a case, it is essential to ascertain that the observed kinetic behavior is a true reflection of RMPK and not influenced by the coupling enzyme, lactate dehydrogenase. Control experiments were conducted in all of the experimental conditions by monitoring the activity versus substrate concentration relation as a function of lactate dehydrogenase concentration ranging from 5 to 20 μg/mL. In all of the experimental conditions, namely, pH ranging from 6.0 to 9.0 in the absence and presence of Phe, there are no quantitative differences in the activity substrate concentration relations regardless of the concentration of lactate dehydrogenase. These results show that the coupling enzyme is not a limiting factor, and the kinetic results do reflect the intrinsic behavior of RMPK.

4.1. H^+ effect

The steady-state kinetics of RMPK was studied as a function of pH ranging from 6.0 to 9.0 in the presence of Phe concentrations ranging from 0 to 20 m*M*. In the absence of Phe, the value of $K_{m,app}$, the K_m value at a specific experimental condition, shows only a slight dependence on pH. However, the presence of Phe significantly affects the value of $K_{m,app}$, which increases with increasing pH. In earlier studies, it has been shown that $1/K_{m,app}$ can be approximated as the apparent dissociation constants of PEP, $K_{d,app}$ (Oberfelder *et al.*, 1984a,b). Thus, the dependence of $K_{m,app}$ on pH can be analyzed by the Wyman linked function (1964). The relation between equilibrium constants and effector concentration can be analyzed by the linked-function theory expressed by Tanford (1969) as shown in Eq. (8.1). In the absence of Phe, $\Delta \upsilon_H{}^+ = 0.16$, that is, RMPK activity is linked to a net absorption of H^+. However, this analysis does not provide information on the source and nature of this change.

4.2. Coupling reaction between H^+ and Phe

This set of data consists of RMPK activity as a function of H^+ in the presence of varying concentrations of Phe. The data can be analysis by the Wyman linked function. In all cases, the slope yields positive values for $\Delta \upsilon_H{}^+$ and increases with increasing Phe concentration. The values of $\Delta \upsilon_H{}^+$ range from 0.16 ± 0.03 to 0.83 ± 0.08 within the Phe concentration range of 0–20 m*M*. These results imply that in converting RMPK to the active form by PEP, a net absorption of proton is observed, although the magnitude of change depends on the amount of Phe present. An analogous analysis was applied to the data with Phe as the variable ligand. Under all pH values tested, $\Delta \upsilon_{Phe}$ assumes a negative value which becomes more negative with increasing pH until it assumes an apparently maximum value of 1.0. These results imply that conversion of RMPK to the active form in the assay mixture requires a release of Phe, the amount of which is pH dependent. Thus, the effects of H^+ and Phe are antagonistic to each other.

4.3. Coupling reaction between PEP and Phe

The relation between the apparent binding constants of PEP and Phe concentration was further analyzed by the linked function in accordance with Eq. (8.1) and a slope of -0.9 was obtained. This value suggests that the binding of an additional mole of PEP results in a net release of about 1 mol of Phe.

4.4. Coupling reactions between metal ions and other metabolites

The enzymic activity of RMPK requires the presence of both K^+ and Mg^{2+}. However, other divalent cations have been used to replace Mg^{2+}. Depending on the source of the PK under investigation, the allosteric properties may be altered when Mg^{2+} is replaced by Mn^{2+} (Larsen *et al.*, 1994, 1997; Suelter *et al.*, 1966). Mesecar and Nowak (1997a,b) tested the linkage of divalent metal ions on yeast PK activity. These investigators applied the Reinhart expression of linked function to analyze the results of their studies. Their results show that Mn^{2+} is energetically coupled to the binding of PEP and FBP. The coupling energy to PEP and FBP are -2.75 and -1.55 kcal/mol, respectively. The negative coupling energy indicates that the presence of Mn^{2+} would favor the binding of PEP and FBP. However, there is no observable coupling energy between the bindings of FBP and PEP. Thus, the coupling mechanism between observed positive cooperativity in PEP and FBP bindings is the consequence of an indirect coupling through Mn^{2+}, a novel observation, since the locations of the binding sites of these two molecules are many Ås apart.

Fenton and Alontaga (2009) conducted an extensive study on the effects of mono- and divalent cations on the binding of PEP, FBP, and inhibitor alanine to human liver PK. By applying the Reinhart expression to their steady-state kinetic data, the authors convincing demonstrated the impact of cations and anions on the allosteric behavior of liver PK. The magnitude of coupling is dependent on the identity and concentration of these ions. Thus, modulation of the complex allosteric behavior in PK is the net consequence of an intriguing and complex network of coupling among multiequilibria linked reactions. Fenton and Hutchison (2009) further studied the coupling between H^+ and other metabolites such as FBP, ATP, and Ala in human liver PK.

4.5. Dissecting the thermodynamic contribution of Phe binding

Fenton and coworkers applied the Reinhart expression to their steady-state kinetic data in the presence of various structural analogues of Phe to pin point the driving forces of various structural moieties of Phe in eliciting the inhibitory behavior (Williams *et al.*, 2006). Through this extensive and meticulous study they concluded that the L-2-aminopropanaldehyde substructure of the amino acid ligand is primarily responsible for binding to RMPK and hydrophobic interaction is the driving force of this interaction. They further determined the structure of the RMPK–Ala complex, thus is able to identify the inhibitor binding site.

5. Structural Perturbations by Ligands

The powerful linked-function analysis illustrates the usefulness of the approach in providing a quantitative indication of the nature of coupling between multiple linked reactions. In general, it is an accepted concept that protein structural changes are an integral part of allosteric mechanisms (Koshland *et al.*, 1966; Monod *et al.*, 1965). The physical identities of these states are not defined in these hypotheses, although it has been generally accepted that they are manifested as changes in secondary–tertiary structures of the enzyme. For the RMPK system, there is substantial spectroscopic evidence in the literature demonstrating that some chromophores in RMPK are perturbed by effectors and substrates (Kayne, 1973; Kayne and Price, 1972; Kwan and Davis, 1980, 1981; Suelter *et al.*, 1966). However, there have been no attempts to quantitatively relate structural changes to the enzyme kinetic observations. Since we know that kinetic and equilibrium binding data alone *do not* provide enough information to establish a molecular model for the allosteric behavior of RMPK, an extensive study was made to quantitatively correlate the change in the global RMPK structure with enzyme kinetic and ligand binding observations. The goal is to elucidate the mechanism through which allosteric regulation is elicited.

A prerequisite to developing a molecular model of allosteric regulation requires additional information to identify the structural components that are included in the general term of $\Delta v_{x,\ pref}$, ΔG_{xy}, or Q_{xy}. These structural changes may include subunit association–dissociation and domain movements. In the studies of RMPK, the specific goal is to address changes in hydrodynamic properties instead of detail atomic level information such as those revealed by spectroscopic techniques because it is difficult to dissect the spectroscopic data to distinguish changes in quaternary structures from local environmental changes around the probe. The hydrodynamic properties enable the investigator to monitor changes in solution as a function of variables such as ligand binding.

5.1. Subunit interaction by sedimentation equilibrium

Both the kinetic and ligand binding data indicate that RMPK exhibits allosteric properties in the presence of Phe; however, the physical identities of the different states of the enzyme implied by these data remained to be determined. These states might be oligomeric forms of RMPK differing from one another in their degrees of polymerization. This possibility was tested by monitoring the molecular weight of RMPK in the presence and absence of Phe by sedimentation equilibrium (Oberfelder *et al.*, 1984a). Within the concentration range of 50–750 μg/mL, the value for the weight-average molecular weight is the same as that of the number-average molecular weight, suggesting that the protein in solution is homogeneous. There is little discernible difference between the data sets obtained in the presence or absence of Phe. The average molecular weight of RMPK is shown to be 220,000 ± 5000, which is not dependent on protein concentration. These results imply that native RMPK does not undergo association–dissociation under the conditions tested. This conclusion is further substantiated by sedimentation velocity data at protein concentrations as low as 15 μg/mL. The sedimentation coefficient did not show any evidence of a dependence on protein concentration. Thus, the states of RMPK do not involve association–dissociation of this tetrameric enzyme.

5.2. Differential sedimentation velocity

Since the tetrameric RMPK does not undergo association–dissociation, the different states in the allosteric model are likely manifestations of secondary–tertiary structural changes in the enzyme. Evidence for such a change was sought through fluorescence, chemical modification, and difference sedimentation velocity experiments (Oberfelder *et al.*, 1984a). These methods were chosen to monitor different levels of structural changes which might be ligand dependent; for example, the sedimentation experiments would yield information on the global structural changes while the other experiments might indicate more localized structural changes.

The change in hydrodynamic properties of RMPK induced by ligands was monitored by difference sedimentation velocity with a Beckman analytical ultracentrifuge according to the procedure of Gerhart and Schachman (1968). For each experiment, two double-sector cells were placed in a rotor. One cell contained RMPK in the presence of the ligand being tested, while the other cell contained RMPK in the presence of tetramethylammonium chloride, the concentration of which was adjusted to an ionic strength equivalent to that of the test ligand. The difference in sedimentation coefficients, ΔS, can be computed with the following equation (Howlett and Schachman, 1977):

$$\frac{1}{\omega^2}\frac{\partial(\Delta r/\bar{r})}{\partial t} = \Delta S \qquad (8.4)$$

where ω is the angular velocity, $\Delta r = r_2 - r_1$, $\bar{r} = (r_1 + r_2)/2$, and t is time. r_2 and r_1 are the radial positions of the peaks of the sample and reference solution, respectively. The reported sedimentation coefficients were corrected to the standard conditions of 20 °C in water. The densities and viscosities of the buffer and ligand-containing solutions were measured with a precision density meter and viscometer, respectively.

Difference sedimentation experiments were performed as a function of the Phe concentration in 50 m*M* Tris at pH 7.5. These experiments were conducted to monitor the effect of Phe on RMPK in the absence of any activators. Results of these experiments show a decrease in $s_{20,w}$ and the value of $\Delta s_{20,w}$ reaches a maximum of −0.24 S at infinite concentrations of Phe. The effect of activating cations of Mg^{2+} and K^+ on the Phe-induced structural change was also monitored. The sedimentation rates of all of the Phe-containing samples were slower than those of the reference solutions. The change is dependent on Phe concentration and approaches a maximum value of −0.13 S at infinite concentrations of Phe. Thus, there are coupling between activating cations and Phe, as reflected by the difference in the change in hydrodynamic properties of RMPK, namely, −0.24 and −0.13 S without and with Mg^{2+} and K^+, respectively. The effects of a variety of other ligands on the hydrodynamic properties of PK were also tested. PEP apparently produces an increase in the sedimentation rate while the other substrate, ADP, did not appear to induce any significant changes. $MgSO_4$ and KC1 individually produce small increases in the sedimentation rate, but together they induce a more substantial increase. Hence, there seems to be coupling between Mg^{2+} and K^+ bindings. Although Ala itself does not induce any significant change in the hydrodynamic properties of RMPK, it does reverse the structural change in RMPK induced by Phe. When RMPK alone was used as a reference, 100 m*M* TMA^+Cl^- produced an increase in the sedimentation rate after correction for changes in viscosity and density. The result with TMA^+Cl^- suggests that there may be a nonspecific ionic strength effect at relatively high ionic strengths which results in an increase in the sedimentation rate.

Having established that Phe induces a global structural change in RMPK, it is *important to compare the structural and equilibrium binding data to establish a molecular model which best fits all of the data.* If the Phe-induced state change is completed prior to saturation of the binding sites by the ligand, then a model that involves a concerted structural change is the likely explanation for the observed phenomenon. On the other hand, if the state change coincides with the binding isotherm, a sequential model is suggested. In order to facilitate comparison of the data, both the state change and binding data were expressed as fractional changes. This comparison shows that the completion of state change *precedes* the saturation of Phe binding. At 0.5 m*M* Phe, the state change is greater than 80% complete while the protein is only 25% saturated with the ligand. Since there are four binding sites of Phe per RMPK tetramer, this result indicates that the binding of 1 mol of Phe results in a shift of the state equilibrium to the point where the transition is greater than three-quarters complete. On the basis of this evidence, the two-state model seems to be the most likely one to describe the observed phenomena (Oberfelder *et al.*, 1984a).

Phe binding is clearly linked to the state change in RMPK, so it is of interest to investigate the relation quantitatively. Let us assume that the fractional change in $\Delta S_{20,\mathrm{w}}$ represents the relative distribution between the two states, R and T, so that

$$\frac{\Delta S_{20,\mathrm{w}}}{\Delta S_{20,\mathrm{w,max}} - \Delta S_{20,\mathrm{w}}} = \frac{T}{R} = K_{R\leftrightarrow T} \tag{8.5}$$

where $\Delta S_{20,\mathrm{w}}$ and $\Delta S_{20,\mathrm{w,max}}$ are the observed change in the sedimentation coefficient at a particular Phe concentration and the maximum change of the sedimentation coefficient at infinite Phe concentration, respectively. T and R are the concentrations of the two proposed states of RMPK, and $K_{\mathrm{R}\leftrightarrow\mathrm{T}}$ is the equilibrium constant characterizing the transition between states. The relation between equilibrium constants and Phe concentration can be analyzed by the linked-function theory expressed by Tanford (1969), Eq. (8.1), where a_{x} is the activity of Phe, Δv is the change in the amount of Phe bound per tetrameric RMPK upon a change in states, and $n_{\mathrm{Phe}}/n_{\mathrm{w}}$ is the ratio of the number of moles of Phe to water present in the solution. Since $n_{\mathrm{Phe}}/n_{\mathrm{w}}$ assumes a very small value, this term can be neglected from Eq. (8.1). The result of such an analysis shows a slope of +0.7, thus suggesting that the occurrence of a state transition from R to T involves a net uptake of about 1 mol of Phe.

The present study provides evidence that the interactions between RMPK and ligands are closely linked to the state change.

5.3. Analytical gel filtration chromatography

The differential sedimentation velocity technique is capable of monitoring minute changes in hydrodynamic properties of RMPK, however, the concentration of protein required is in the mg/mL regime which is an order of magnitude above the estimated *in vivo* concentration of the protein. Thus, analytical gel filtration chromatography is employed (Heyduk *et al.*, 1992). This technique enables one to determine the hydrodynamic properties at μg/mL or lower concentrations. Briefly, buffer and protein solution were held in containers A and B, respectively, which are connected through a two-way tap with a thermostated chromatography column equipped with an adaptor. The eluent can be monitored by fluorescence with the aid of a flow cell with a capacity of 300 μL. Fluorescence detection allows the use of much lower protein concentrations and minimization of interference by Phe. The solution from the flow cell passes to a pump and finally to a weighing bottle for measuring flow rate. All connecting tubings were kept as short as possible. After equilibration with buffer, at least 10 ml of protein solution was then applied to the column via the two-way valve, ensuring that the conditions required for a large zone experiment were fulfilled. A large zone has to be used in order to obtain thermodynamically valid results for interacting systems (Ackers, 1970). Application of protein solution to the column has to be precisely correlated with the start of monitoring of the effluent by fluorescence. Two thousand data points were collected for each elution profile. The raw data in the form of fluorescence intensity as a function of time of elution were converted to fluorescence intensity as a function of volume of eluent, using the measured flow rate. Since a precise measurement of the flow rate is critical in these experiments, it has to be individually measured for each experiment by weighing the effluent collected from the start of recording to the appearance of a boundary. Elution volumes were determined as the centroid of the leading boundary. The centroid is defined as the elution volume at which areas for each boundary below and above solute profile are equal (Ackers, 1970). Base lines were calculated by linear regression performed on the linear portions of the elution profile. When the elution volumes of protein in the absence and presence of ligand were to be compared, the experiment with the protein in buffer was conducted first on a column equilibrated with buffer. The column was then reequilibrated with buffer in the presence of ligand.

The performance of a Pharmacia C10/40 column with a bed volume of 25 ml was monitored. For consecutive runs performed on the same sample on the same day, elution volume was reproducible to ± 7 μL (standard deviation of mean value), whereas the long term stability within a period of several weeks was ± 50 μL. The precision on determining the difference in elution volume in the presence and absence of ligand was ± 15 μL which means that for a protein of the size of 45 kDa (about 28 Å), it should be possible to detect changes of >0.1 Å, that is, $>0.4\%$ of the Stokes radius.

For RMPK Sephacryl 300 HR columns were calibrated with ribonuclease A, chymotrypsinogen A, ovalbumin, aldolase, and ferritin. *N*-Acetyltryptophan amide and blue dextran were used to determine the total volume (V_t) and the void volume (V_o) of the column, respectively. The partition coefficient, K_{av}, is calculated as

$$K_{av} = (V_e - V_o)/(V_t - V_o) \tag{8.6}$$

where V_e is the elution volume of the protein. Plotting Stokes radii of standard proteins as a function of $(-\log K_{av})^{1/2}$ (Siegel and Monty, 1966) yielded a straight line, which is used to calculate the Stokes radii. The differences in elution volume (ΔV) between RMPK in the absence and presence of Phe were monitored as a function of Phe concentration. The elution volume of the RMPK–phenylalanine complex is less than that of RMPK alone, as indicated by a positive change in ΔV. This result shows that the RMPK–Phe complex exhibits a larger Stokes radius, that is, more asymmetric or expanded, which is an observation that is in total agreement with that from a study using neutron scattering (Consler *et al.*, 1988) and differential sedimentation velocity (Oberfelder *et al.*, 1984a).

It is possible to predict the distribution between R and T states in a set of concentrations of Phe by using these constants in the following equation (Oberfelder *et al.*, 1984b).

$$\bar{T} = L\left(1 + [\mathrm{I}]/K_{\mathrm{I}}^{\mathrm{T}}\right)^4 / \left(1 + [\mathrm{I}]/K_{\mathrm{I}}^{\mathrm{R}}\right)^4 + L\left(1 + [\mathrm{I}]/K_{\mathrm{I}}^{\mathrm{T}}\right)^4 \tag{8.7}$$

where L is the intrinsic allosteric equilibrium constant, $\bar{T}$ is the fraction of molecules in the T state, [I] is Phe concentration and superscripts R and T in the equilibrium constants (K) specify the conformational state of the protein. Using the equilibrium constants derived from kinetic data (Consler *et al.*, 1989) in combination with the equation

$$\Delta V = \Delta V_{max}(\bar{T} - \bar{T}_0)/(1 - \bar{T}_0) \tag{8.8}$$

where $\bar{T}_0$ is the fraction of T in the absence of Phe, a curve was generated to express difference in elution volume ΔV as a function of Phe concentration. Values for the different constants used are $L = 0.06$, $K_{\mathrm{I}}^{\mathrm{R}} = 13$ mM, and $K_{\mathrm{I}}^{\mathrm{T}} = 0.78$ mM. The predicted data are not in good agreement with that of the experimental data, and a systematic deviation was observed. A search was undertaken to determine the values for these equilibrium constants that would best fit the experimental data. The results that lead to a perfect fit include the values $L = 0.094$, $K_{\mathrm{I}}^{\mathrm{R}} = 13$ mM, and $K_{\mathrm{I}}^{\mathrm{T}} = 0.4$ mM. These values are within the standard deviations of these constants determined by steady-state kinetics (Consler *et al.*, 1989). The coincidence of the simulated plot with the experimental data provides a strong credence to the validity of the two-state model adopted for the RMPK system, because the simulated plot is based entirely on parameters obtained from the analysis of kinetic data and the hydrodynamic experimental data are directly related to a global conformation change of the enzyme. In a separate experiment, the observed value of ΔV_{max} corresponds to a change of 1.2 Å, in Stokes radius. This correlates very well with a change of 1–2 Å in the radius of gyration observed by small angle neutron scattering (SANS) under similar experimental conditions (Consler *et al.*, 1988). In accordance with Eq. (8.3) and the equilibrium constants reported by Consler *et al.* (1989), it is possible to predict the effect of PEP upon its addition to a mixture of RMPK and Phe. PEP should shift the equilibrium toward the R state and, therefore, one should observe a shift of the elution volume of RMPK toward larger values, that is, in the presence of PEP, ΔV should be smaller than that in the presence of Phe alone and, in low concentrations of Phe, can actually become negative. Experimental results with 0.5 and 5 m*M* Phe in the presence of 0.1 m*M* PEP confirm this prediction. These results show that RMPK undergoes a concerted global structural change and the binding of Phe follows a positive cooperativity mode.

5.4. Small angle neutron scattering or small angle X-ray scattering

To characterize this change in global structure of RMPK, we studied the effects of ligands on the structure of RMPK by SANS (Consler *et al.*, 1988). The radius of gyration, R_G, decreases by about 1 Å in the presence of substrate PEP but increases by the same magnitude in the presence of inhibitor Phe.

5.4.1. The SANS data can be subjected to two separate analyses

The first method relies upon the data obtained at the lower scattering angles. Data in this region conform to the Guinier relationship:

$$\ln I(k) = -k^2 R_G^2/3 + \ln I(0) \tag{8.9}$$

where $I(k)$ and $I(0)$ are the scattering intensities at angles 2θ and 0, respectively, $k = (4\pi \sin \theta)/\lambda$, and R_G is the radius of gyration. From a Guinier plot of ln $I(k)$ versus k^2, one is able to determine R_G and $I(0)$. Data points from low angle scattering up to $k = 0.07$ Å^{-1} will be subjected to weighted linear least-squares fitting to yield values for these two parameters. The radius of gyration, R_G, decreases by about 1 Å in the presence of substrate PEP but increases by the same magnitude in the presence of inhibitor Phe.

The second method of analysis takes into account all of the SANS data up to the point where the intensity approaches 0, for example, up to $k = 0.13$ Å^{-1}. The analysis yields the length distribution function, which is defined by Eq. (8.10).

$$P(r) = (2r/\pi)\int_0^\infty kI(k)\sin(kr)\,\mathrm{d}x \tag{8.10}$$

where $P(r)$ is the frequency distribution of all the point-to-point pair distances, r, between scattering centers of the particle. This function can be approximated by the indirect transform method of Moore (1980) and yields information that describes the size, shape, and frequency distribution of all the point-to-point pair distances between scattering centers of the particle. It also yields values for R_G. Since this analysis utilizes more of the scattering data than the Guinier analysis, it contains more information. When the scattering data were analyzed as a function of $P(r)$ versus r, the results indicate that the increase in R_G is associated with a pronounced increase in the probability for interatomic distance between 80 and 110 Å.

5.4.2. Comparison of solution and crystal structures

Using the α-carbon coordinates determined by X-ray crystallography, the solution scattering behavior of the observed conformation of RMPK was predicted. The scattering curve and $P(r)$ distribution were calculated from the known α-carbon coordinates as follows. The coordinates for the symmetric tetramer were constructed by applying the appropriate symmetry operations to the crystallographic monomer coordinates. For an object assumed to consist of discrete density points, the Debye (1915) relation can be used to calculate the scattering curve directly. All coordinate pairs were used to directly calculate $P(r)$ for RMPK, in this calculation, the scattering length density was assumed to be uniformly constant throughout the molecule.

The fact that the observed scattering data contain information pertaining to interatomic distances enables one to generate length distributions from both SANS and X-ray crystallographic data. The changes in length distributions, reflecting conformational changes induced by ligands, can then be compared between the two sets of experimental observations. It must be stressed that separate comparisons were conducted for the structural parameters derived from solution and the crystal structure. Our approach thus involves generating the difference distribution, $\Delta P(r)$, which is ideal for the comparison of data sets obtained under different conditions. This function ($\Delta P(r)$) enables one to determine the (length distribution) changes that occur in solution. Once those changes are determined, they are used as a guideline for the modeling that involves manipulation of the α-carbon coordinates. The computation is deemed satisfactory when it yields a $\Delta P(r)$ function that compares favorably with that determined from solution data. A series of calculations and simulations were carried out. The common element in the analysis is the relationship of I versus k. This is obtained directly from SANS, and indirectly from X-ray crystallographic coordinates through the Debye (1915) relationship. Once the three-dimensional coordinate data are converted to the form of an I versus k distribution, the path of data analysis converges, since both solution and crystal data are treated identically. This simulation step employed computer graphics, which enabled us to conduct interactive rotation and translation, symmetry operations, and inter-α-carbon distance calculations. Having decided on a simulated conformational change, the information was then reintroduced into the path of data analysis in the form of altered α-carbon coordinates. All manipulations on the α-carbon coordinates were performed on isolated monomers. Subsequently, these newly modeled structures were used to reconstruct the tetramer by the same symmetry operations that yielded the original tetramer. The point at which the comparison between experimental and simulated data takes place is at the level of $P(r)$ distribution.

The difference between solution conformations, $\Delta P(r)_{\text{solution}}$, and crystal conformations, $\Delta P(r)_{\text{crystal}}$, was compared; when $\Delta P(r)_{\text{solution}} = \Delta P(r)_{\text{crystal}}$, the conformational change was considered adequately modeled. This approach is especially useful for the comparison of data sets obtained under different solution conditions. Hence, $P(r)$ distributions were employed to illustrate changes in molecular dimensions that are the result of these solution variations.

With the aid of computer modeling, these changes in interatomic distance is consistent with the rotation of the B domain relative to the A domain, leading to the closure or opening of the cleft between these domains as a consequence of binding to PEP and Phe, respectively. These results show that one of the dynamic motions in RMPK includes change in domain–domain interaction between A and B domains.

6. Functional Linkage Scheme of Allostery for RMPK

As a result of a series of extensive steady-state kinetic, equilibrium, and solution structural studies, a concerted, allosteric model was developed for quantitative interpretation of the kinetic and equilibrium binding data of RMPK (Consler *et al.*, 1888, 1992; Heyduk *et al.*, 1992; Oberfelder *et al.*, 1984b; Yu *et al.*, 2003). The simplest model accounting for all the experimental data involves two conformational states, inactive E^T and active E^R, shown in Fig. 8.2. E, S, P, and I are enzyme, substrate (PEP or ADP), product (pyruvate), and inhibitor (Phe), respectively.

The upper and lower faces of the cube contain all of those species assuming E^R and E^T state, respectively. The species are interconnected with equilibrium constant, $L = [E^T]/[E^R]$; K^R and K^T are the ligand binding equilibrium constants associated to the E^R and E^T state, respectively, while K_I and K_S are the equilibrium constants associated with the

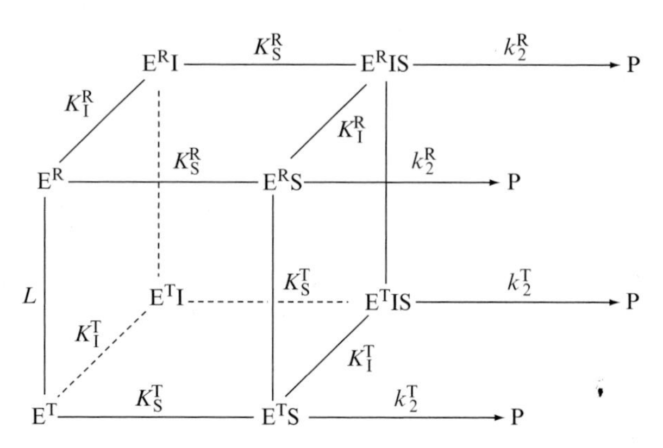

Figure 8.2 The proposed two-state model for the allosteric behavior of RMPK (Oberfelder *et al.*, 1984b). $L = [\mathrm{T}] / [\mathrm{R}]$; K^{T} and K^{R} are equilibrium constants for substrates, S, and inhibitor, I, binding to the T and R state, respectively. The model assumes that the affinities of RMPK for both S and I are dependent only on the conformational state of the enzyme, that is, the constant for the binding of S to E^{T} is the same as that of S to $\mathrm{E}^{\mathrm{T}} \cdot \mathrm{I}$. Similarly, the affinity of S for E^{R} and $\mathrm{E}^{\mathrm{R}} \cdot \mathrm{I}$ is assumed to be the same.

binding of inhibitor Phe and substrate, respectively. The conceptual significance of this model is that the two states are in a preexisting equilibrium, L. This crucial aspect of the model was independently substantiated by the literature (Harris and Winzor, 1998). The distribution of ligands bound to these two states is defined by their respective equilibrium constants to these states. The ligands shift the state changes in PK by mass action. This model differs from models which assume a mechanism consisting of a ligand induced state change not in a preexisting equilibrium.

The simple model shown in Fig. 8.2 is NOT adequate to account for the experimentally defined energy landscape of the linked equilibria that govern the enzymatic reactions. Added novel features are derived from global analysis of calorimetric and fluorescence data (see Section 7.5 for further discussion).

7. Functional Linkage Through Ligand Binding Measurements

As useful as the steady-state kinetic analysis, it is imperative to verify the quantitative parameters determined by measuring the energetic of ligand binding. Both direct (equilibrium dialysis or equivalent) and indirect (fluorescence and isothermal titration calorimetry) methods were employed in our studies.

7.1. Equilibrium binding

The binding of ligand to RMPK can be measured by the method of Hirose and Kano (1971). The advantage of this approach or its equivalent is the ability to keep the concentration of P constant. This is particularly important for proteins that may undergo association–dissociation (Na and Timasheff, 1985). This technique is based upon the principle that the protein is completely excluded from the interior of the resin, while the ligand will be distributed throughout the solution. In the presence of a protein which binds the ligand, the establishment of chemical equilibrium will result in an apparent preferential exclusion of the ligand from the interior of the resin. Having determined the concentrations of PEP and PK before and after equilibrium is reached, we can combine these parameters to obtain the amount of bound and free PEP since

$$\alpha = (P/V_o)/(P/V) \tag{8.11}$$

where α is the ratio of protein concentrations, P is the total amount of protein added to the solution, V is the total volume of solution added to the dry gel, and V_o is the volume of solution outside the resin. P/V is the concentration of the protein added, assuming that it is equally distributed throughout the solution, while P/V_o is the observed concentration of the protein. Similarly

$$\beta = (L_o/V_o)/(L/V) \tag{8.12}$$

where β is the ratio of ligand concentrations and L and L_o are the total amount of ligand added and the amount of ligand outside the gel, respectively. L/V is the concentration of ligand, assuming it is evenly distributed in the total volume added, while L_o/V_o is the concentration of ligand observed in the gel-free solution. While β is the value of the ratio in the presence of protein, β' is the value in the absence of protein. The amount of bound ligand is

$$L_0(\text{bound}) = L(\beta - \beta')/(\alpha - \beta') \tag{8.13}$$

The concentration of free ligand outside the gel is

$$L_0(\text{free}) = (L/V)\beta'(\alpha - \beta)/(\alpha - \beta') \tag{8.14}$$

The amount of ligand bound per mole of RMPK tetramer is

$$Y = L_0(\text{bound})/P = (L/P)(\beta - \beta')/(\alpha - \beta') \tag{8.15}$$

For the specific practical aspects of preparing the resin and other details, the readers are referred to Oberfelder *et al.* (1984a). The ligand solutions are made up by dilution of a concentrated stock solution of ligand that had been spiked with either an aliquot of radioactively labeled ligand unless otherwise the ligand concentration can be detected accurately by other techniques. By making the solutions by serial dilution, it is assured that the final stock solutions had the same ratio of labeled to unlabeled ligand. After the gel is swollen, RMPK and radioactive ligand are added to the appropriate tubes. The tubes are again sealed tightly with parafilm and allowed to incubate in a temperature-controlled water bath at the desired temperature for 15 min. Samples are removed from the tubes with a 100-pL Hamilton syringe; this allowed the solution outside of the gel matrix to be sampled without contamination of gel beads.

7.2. Isothermal titration calorimetry

Isothermal titration calorimetry (ITC) experiments allow direct access to fundamental thermodynamic parameters associated with ligand binding and RMPK conformational transitions (Herman and Lee, 2009a). Since ITC experiments are not based on an enzymatic activity, binding of single ligands can be assessed and interaction between ligands investigated. A detailed description of the regulatory behavior of wild-type (WT) RMPK is essential for modeling its regulation under physiological conditions. It also creates an essential baseline for understanding perturbations induced by targeted genetic modifications of the recombinant enzyme.

Concentrations of RMPK ranged from 15 to 70 μM. Stock solutions of 40 mM PEP, 100 mM Phe, and 100 mM ADP were used for injections. The titrants were prepared by dilution of the ligands in the dialyzate to prevent thermal effects caused by a buffer mismatch. The pH of the titrants was checked and adjusted, if necessary. When RMPK was titrated with Phe in the presence of a fixed concentration of ADP, the same concentration of ADP was present in the injectant. Typically, 15–25 aliquots of ligand were injected into a 1.4-mL sample. To correct data for heats of dilution of an injectant, a control experiment was performed under the same conditions except a buffer was substituted for the sample. The overall reaction heats were calculated by integration of the corrected ITC curves. The extent of irreversible processes caused by mixing of the protein at high temperatures in the calorimeter was also evaluated. Only an insignificant decrease of $3 \pm 3\%$ activity was observed after a 1-h calorimetric experiment at 40 °C.

To establish the energetic landscape of the regulatory behavior of RMPK in the presence of its ligands, we performed a series of ITC experiments at temperatures between 4 and 45 °C. To separate the contribution of the buffer ionization heat ΔH_{ion} from the overall reaction heat, we performed all titrations in two buffers with different ΔH_{ion} values.

7.3. Fluorescence

In order to rigorously test the validity of conclusions derived from the ITC data, a fluorescence approach, albeit indirect, that tracks continuous *structural* perturbations was employed (Herman and Lee, 2009b). Intrinsic Trp fluorescence of RMPK in the absence and in the presence of substrates PEP and ADP, and the allosteric inhibitor Phe was measured in the temperature range between 4 and 45 °C. For data analysis the fluorescence data were complemented by ITC experiments to obtain extended data set allowing more complete characterization of the RMPK regulatory mechanism. Twenty-one thermodynamic parameters were derived to define the network of linked interactions involved in regulating the allosteric behavior of RMPK through global analysis of the ITC and fluorescent data sets. *In this study, 27 independent curves with more than 1600 experimental points were globally analyzed.* Consequently, the consensus results not only substantiate the conclusions derived from the ITC data but also structural information characterizing the transition between the active and the inactive state of RMPK and the antagonism between ADP and Phe binding. The latter observation reveals a novel role for ADP in the allosteric regulation of RMPK.

7.4. Global fitting

Global analysis (Beechem *et al.*, 1983; Knutson *et al.*, 1983) is a powerful method for discerning between models and for accurate recovery of model parameters. The method is based on an ordinary nonlinear least-squares minimization (Marquardt, 1963) and allows for simultaneous analysis of multiple data sets. Data for the analysis can be acquired under different experimental conditions and with different techniques. Consequently, some model parameters are common for multiple curves. During global fitting, the overall sum of weighed squared deviations of measured values and values calculated from a model that encompasses all experimental data is minimized. Global analysis thus allows determination of parameter values that are consistent with all data sets. Importantly, linkage between particular parameters and different data sets sharpens the χ^2 surface with concomitant decrease of parameter correlation. Those parameters are recovered with higher accuracy (see Table 8.1; Herman and Lee, 2010). Over-determination of parameters, inherent to global analysis, helps to distinguish between alternative models by eliminating models inconsistent with data. The resolving power of global analysis is unequaled by a conventional nonlinear least-squares analysis. Global analysis has been previously used for large variety of experimental data and techniques (Ackers *et al.*, 1975; Beechem *et al.*, 1983; Eisenfeld and Ford, 1979; Johnson *et al.*, 1981; Oberfelder *et al.*, 1984a; Ucci and Cole, 2004; Verveer *et al.*, 2000). This approach of global analysis was applied to the data set obtained by ITC, fluorescence binding experiments, and fluorescence

Table 8.1 Parameters whose values are significantly improved by global fitting of the ITC and fluorescence data

Reaction	Parameter	Value		Unit[a]
		ITC	(ITC + fluorescence)	
Phe binding	$\Delta S_{Phe}{}^{R}$	7 (5)[b]	2.7 (0.9)	cal mol^{-1} K^{-1}
	$\Delta H_{Phe}{}^{R}$	0.9 (2.0)	−1.4 (0.4)	kcal mol^{-1}
	$\Delta S^{T}{}_{Phe}$	13 (7)	18.2 (0.2)	cal mol^{-1} K^{-1}
PEP binding	$\Delta S_{0,\ PEP}{}^{R}$	?	11.2 (0.2)	cal mol^{-1} K^{-1}
ADP–Phe coupling	$\Delta\Delta S^{T}{}_{ADP,\ Phe}$	−2.5 (2.0)	−4.9 (0.5)	cal mol^{-1} K^{-1}
	$R\Delta n_{ADP,\ Phe}{}^{R}$	−1.7 (1.0)[c]	−0.9 (0.5)	mol^{-1}[d]

[a] Per mol of a ligand.
[b] Standard deviations are given in parenthesis.
[c] Number of protons absorbed.
[d] Per mol of tetramer.

temperature scans for evaluation of the functional energetic landscape in the allosteric regulation of RMPK (Herman and Lee, 2009a,b). In all cases, the global analysis approach provided performance far beyond resolution of the conventional analysis. A more detail discussion on global fitting has recently been presented (Herman and Lee, 2010).

7.5. Results of global analysis of ITC and fluorescence data

7.5.1. Phe binding

The sign of heat exchange for the reaction changes from positive to negative during the titration. This is a clear indication that the reaction is complex and consists of at least two heat-generating processes with opposite signs. Actually, many reactions may contribute to the shape of ITC curves. The observation is consistent with the general mechanism of our model which includes binding to both enzyme states, a ligand induced shift of the R → T equilibrium accompanied by the change in the RMPK conformation, and linked proton reactions as described in our model. Heats resulting from reequilibration of ligands between the R and T states upon perturbation of the R → T equilibrium contribute to the detailed shape of the titration curves as well.

7.5.2. PEP binding

Since the patterns of the ITC data are quite complex, the consequence of linked multiple equilibria, a detailed interpretation of such curves is rather difficult. Therefore, we decided not to draw conclusions only on the basis of the shape of the ITC curves. Instead, at different temperatures, we titrated

RMPK to saturation, and the data are expressed as overall reaction heats. The results of this in-depth dissection of the thermodynamic signatures of RMPK interacting with metabolites provide novel insights into the mechanism of allosteric regulation of RMPK.

7.5.3. ADP binding

Binding of ADP to both states of RMPK was found to be strongly enthalpy driven with binding enthalpies to the R- and T-state about -10 and -13 kcal/mol, respectively. This is consistent with a binding mechanism that involves electrostatic interaction with the highly charged ADP substrate. Both entropy and enthalpy changes for binding of ADP were found to be the largest among all investigated ligands. ADP binding and the differential affinity toward the R- and T-states exhibit pronounced temperature dependence. The affinity to the T-state is larger at elevated temperatures. As a consequence, at high temperature ADP binds more favorably to the T-state resulting in shifting the R $\leftrightarrow$ T equilibrium toward the inactive T-state; thus, in this particular regard, ADP behaves like an inhibitor at high temperature. However, by virtue of its chemical structure, ADP is still a substrate of RMPK. In earlier studies, it was concluded that ADP behaves strictly as a substrate which plays a minor role in the allosteric regulation of RMPK because it does not show a differential affinity toward the two states of RMPK. However, as a consequence of extending the temperature range in this study, it is revealed that ADP actually plays a major role in the allosteric mechanism in RMPK under physiological relevant temperatures.

7.5.4. Energetic coupling between ADP–Phe bindings

Phe binding in the presence of ADP revealed that an antagonism exists between the two ligands, that is, Phe was found to bind more weakly in the presence of ADP. Due to the interaction enthalpy, $\Delta\Delta H^{T}{}_{\mathrm{ADP,\,Phe}}$ of about -0.5 kcal/mol, the enthalpy term $\Delta H^{T}{}_{\mathrm{ADP,\,Phe}} = \Delta H^{T}{}_{\mathrm{Phe}} + \Delta\Delta H^{T}{}_{\mathrm{ADP,\,Phe}}$ is reduced almost to zero. Concomitantly, the entropy change, $\Delta S^{T}{}_{\mathrm{ADP,\,Phe}} = \Delta S^{T}{}_{\mathrm{Phe}} + \Delta\Delta S^{T}{}_{\mathrm{ADP,\,Phe}}$ becomes smaller by 2.5 cal $\mathrm{mol}^{-1}\,\mathrm{K}^{-1}$. The net consequence of the coupling between ADP and Phe bindings is a reduction of the enthalpy term for Phe binding to essentially zero and a less favorable entropy term, that is, the presence of a bound ADP exerts a negative effect on Phe binding to the T-state. Thus, at high but physiologically relevant temperatures, the bound ADP modulates the inhibitory efficiency of Phe by reducing its affinity to the T-state.

7.5.5. Proton release or absorption

Information related to linked proton reactions associated with the ligand binding and the state transition is also available from Herman and Lee (2009a,b). The following reactions summarize these results:

Reactions with proton absorption:

$$0.2H^+ + PEP + R \rightleftarrows PEP \cdot R$$
$$0.5H^+ + R \rightleftarrows T$$
$$0.7H^+ + ADP + R \rightleftarrows ADP \cdot R$$
$$1.0H^+ + ADP + T \rightleftarrows ADP \cdot T$$

Reactions with proton release:

$$Phe + T \rightleftarrows Phe \cdot T + 0.9H^+$$
$$ADP + Phe + T \rightleftarrows ADP \cdot Phe \cdot T + 1.7H^+$$

Binding of substrates apparently always absorb protons, although the binding of ADP involves a larger amount of protons absorbed. Consequently, the binding of ADP would be more sensitive to pH perturbations. The R $\rightarrow$ T transition would also absorb protons. For the reactions which involve in proton absorption, lower pH would shift the equilibria to the right, as defined by the Le Chatelier's principle. Thus, lower pH would favor the bindings of ADP, PEP, and the T-state. The net result would be an expected change in cooperativity in substrate binding, the extent of which is pH dependent.

In contrast, binding of Phe to the T-state, in the presence or absence of ADP, leads to release of protons. Another significant observation is that the amount of proton released almost doubles in the presence of ADP. If the bindings of Phe and ADP were simply a summation of the two reactions (absorption of 0.7 and release of 0.9 H^+ for ADP and Phe binding, respectively), then the net amount of proton release is expected to be about zero; however, instead the observation is a doubling of H^+ release to 1.7. This is a clear indication of the coupling of the two binding events. Binding of Phe and ADP to the T-state seems to be more pH dependent.

These results clearly indicate that the effect of pH on the basic allosteric behavior of RMPK is a composite of the nature and magnitude of proton release or absorption linked to the various reactions. According to the Wyman's linked-function theory (1964) there should be dependence of dissociation constants on a concentration of protons and shift of the R $\leftrightarrow$ T equilibrium. For example, at low pH the T-state is favored but the binding of Phe would be weakened. Thus, a binding isotherm of Phe is pH dependent as a function of the relative values of equilibrium constants which define the distribution of the R- and T-state and the relative affinities of Phe to these states. Simultaneously, the affinity of substrates would be affected. Thus, one should expect that the allosteric behavior of RMPK is a complex phenomenon defined by the specific experimental conditions. It is most gratifying to note that such an expectation was reported by Consler *et al.* (1990). Based on steady-state kinetic studies, these authors reported the synergistic effects of proton and Phe on the regulation of RMPK.

7.5.6. Comment on the results derived from global fitting of both fluorescence and ITC data sets

Table 8.1 summarizes the results of the improvements of parameters when both sets of data were analyzed. It is clear that not only the specific values of the parameters are more accurate; the standard errors have improved significantly. These are solid proofs of the value and power of global analysis on multiple sets of data, particularly data sets collected from different techniques.

A summary of this binding study:

- The state of RMPK that ADP preferentially binds is temperature-dependent. ADP binds more favorably to the T and R states at high and low temperatures, respectively. This crossover of affinity toward the R and T states implies that ADP not only serves as a substrate but also plays an important and intricate role in regulating RMPK activity.
- The binding of Phe is negatively coupled to that of ADP in addition to the shifting of the R → T equilibrium due to the relative affinities of Phe or ADP for these two states; that is, the assumption that ligand binding to RMPK is state-dependent is only correct for PEP but not Phe in the presence of ADP.
- The release or absorption of protons linked to the various equilibria is specific to the particular reaction. As a consequence, pH will exert a complex effect on these linked equilibria, with the net effect being manifested in the regulatory behavior of RMPK.
- The R → T equilibrium is accompanied by a significant ΔCp.

Our conclusions derived from calorimetric data are in full agreement with those based on our published steady-state kinetic studies. These conclusions are further strengthened by our fluorescence data and model simulations (Herman and Lee, 2009b,c).

8. Protein Structural Dynamics—Amide Hydrogen Exchange Monitored by FT-IR (HX-FT-IR)

We further studied the structural perturbations by Fourier transform infrared (FT-IR) spectroscopy (Yu *et al.*, 2003). The experiments were designed to monitor the secondary structure of RMPK in the presence of saturating amounts of various ligands. Mg^{2+} is a divalent cation essential for activity. PEP and ADP are the two substrates while Phe is the inhibitor. In all experimental conditions, there is no significant change in the areas encompassed by the peaks assigned to either α or β structures. That implies that there is no detectable conversion of secondary structure of RMPK in the presence of all these ligands. A closer examination of the data shows that the maximum wavenumber associated with α-helix remains the same while that of the β-strand shifts as a function of ligand. In the presence of Mg^{2+}, PEP and ADP the maximum wavenumber is less than that in buffer alone or Phe. These results imply that the local environments of the β-strands are perturbed by these ligands, although the amount of β strands has not changed. The origin of the differential perturbations by ligands in the environments of secondary structures might be the modulation of structural dynamics of RMPK. Thus, the structural dynamics of RMPK in the presence of various ligands was probed by H/D exchange monitored by FT-IR. The second derivative spectra as a function of time of H/D exchange were monitored and compared. The spectra in the presence of K^+ and Mg^{2+} were compared to those of RMPK in buffer. They show that the basic pattern of exchange was retained. However, a larger change in intensity was observed even at the 1-min time point. These results imply that the activating metal ions induce an increase in the number of rapidly exchangeable amide protons. The presence of either PEP or ADP shows a pattern of exchange that is quite similar to each other, namely, a very rapid exchange was observed in both the helices and sheets. There is a clear indication of the presence of two different populations of helices. The change in the second derivative spectra reflecting the amide proton exchange in the presence of Phe showed that within the time frame of the experiment no exchangeable amide proton was detected in the β-sheets, an observation that differs from that of the helices. Thus, these H/D exchange experiments show that substrates (ADP or PEP) and activating metal ions (Mg^{2+}) lock RMPK in a more dynamic E^R structure while Phe exerts an opposite effect.

These results provide the first evidence for a differential effect of ligand binding on the dynamics of the structural elements, not major conformational changes, in RMPK. These data are consistent with our model that allosteric regulation of RMPK is the consequence of perturbation of the equilibrium of an ensemble of states the distribution of which resides mainly around the two extreme dominant end states. Sequence differences and ligands can modulate the distribution of states leading to alterations of functions.

9. Probing Interfacial Interactions

9.1. S402P mutation in the interface between the C-domains

In RMPK, there are two intersubunit interfaces—one between the C-domains while the other is between the A-domains. In order to probe the interfacial interactions, we mutated residue 402 of RMPK from S to P, in accordance to the difference in sequence between the muscle and kidney isozymes. Converting S402 to P changes neither the secondary, nor the tetrameric structure (Friesen *et al.*, 1998a). The S402P RMPK mutant exhibits steady-state kinetic behavior that indicates that it is more responsive to regulation by effectors. The sigmoidicity of the curves (activity vs. substrate concentration) is a reflection of cooperativity of substrate binding. The RMPK data show almost no sigmoidicity, as expected for an enzyme exhibiting little allosteric behavior; whereas the rabbit kidney PK shows pronounced sigmoidicity. The data for the S402P RMPK are intermediary to the muscle and kidney isozymes. The presence of 12 m*M* inhibitor Phe shifts the curve to the right, as expected, since Phe would shift the conformational state equilibrium toward E^T which has a weaker affinity for PEP. The presence of 10 n*M* activator FBP in addition to 12 m*M* Phe shifts the curve to the left. In RMPK, without the inhibitory effect of 12 m*M* Phe, it would still require millimolar concentrations of FBP to achieve the same effect. Thus, an S402P mutation confers partial restoration of allosteric behavior to the RMPK.

We have elucidated the atomic structure of the S402P RMPK variant by X-ray crystallography (Wooll *et al.*, 2001). Although the overall S402P RMPK structure is nearly identical to the WT structure within experimental error, significant differences in the conformation of the backbone are found at the site of mutation. We found that the ratio of B-factors of mutant/WT provides a good representation of the pattern of long range communications between distant sites (Wooll *et al.*, 2001). The most obvious *increase in B-factor* in the S402P RMPK is around the residue 402 indicating a significant increase in dynamics in that region. There are also significant changes in the ratio of B-factor for residues 50–200. In addition, there is an increase in the number of heterogeneity in the angle assumed by the B-domain with respect to the A-domain, that is, *increases dynamics in domain movements.* Closer examination of the X-ray data shows a disruption of a salt bridge between residues 341 and 177 of an adjacent subunit. This salt bridge and residue 402 reside in different subunit interfaces. Thus, these structural data show a communication between these two different subunit interfaces. A similar conclusion was derived from the results of our study of subunit assembly (Friesen and Lee, 1998; Friesen *et al.*, 1998b). It is evident that a mutation at residue 402 leads to *increased dynamics in distant sites through long range communication without significant changes in secondary structures.*

9.2. T340M mutation along the interface between the A-domains

In an effort to establish functional coupling among residues in RMPK, we incorporated into our studies the human genetic data (Baronciani and Beutler, 1995; Kanno *et al.*, 1991; Miwa *et al.*, 1993; Neubauer *et al.*,

1991; Valentine *et al.*, 1989). Our choice of residue 340 is based on the combined results of our structural studies and human genetic data. In our modeling study, we have further identified the residues whose inter-α carbon distances are within 15 Å (Consler *et al.*, 1994) along the axis between the adjacent A-domains. These include residues 330–350. Human genetic data identified T340M as a mutant which is observed in patients suffering from PK deficiency. The T340M RMPK and RKPK (kidney isozyme of PK) mutants are only half as active as the WT PKs. The T340M RMPK enzyme is more susceptible than RKPK to inhibition by Phe or to the activator FBP. The differences in the amino acid sequence of RMPK and RKPK are only 22 amino acids, all of which reside in the C-domain. Evidently the 22 residues in the C-domain modulate the effect of residue 340, which reside in different subunit interfaces. The resultant is a differential effect on the functional energetics of PK. This study demonstrates the linkage between distant residues in different interfacial interactions, that is, establishing the functional coupling among residues and the identities of these residues.

10. Summary Statement

The understanding of the molecular mechanisms of allostery in PK is still in its infancy. A vast amount of knowledge has been acquired; however, we have yet to identify the chemical principles underlying the observations. We do not understand the molecular mechanism that leads to the change in allosteric behavior due to mutations, the communication pathway and the thermodynamic parameters that drive the allosteric behavior. A current view of protein indicates that it is an ensemble of microstates in equilibrium (Ferreon *et al.*, 2003; Friere, 1999; Hilser *et al.*, 1998, 2006; Liu *et al.*, 2006; Luque *et al.*, 2002; Pan *et al.*, 2000; Schrank *et al.*, 2009). Mutations, change in solvent environment or binding of ligands lead to a redistribution of these microstates. The biological activity is a manifestation of the functional properties of the dominant state(s). The present knowledge of the allosteric properties of RMPK seems to favor that hypothesis, in particular, in view of the plasticity of the orientations of the B domain with respect to the A domain (Wooll *et al.*, 2001) and the small angle X-ray scattering (SAXS) results of different structures of RMPK in complex with different amino acids as inhibitors (Fenton *et al.*, 2010). The tetrameric PK is a particularly attractive model system to elucidate the ground rules of allostery. There are four isozymes and each has its own specific allosteric behavior. The sequences of some of these isozymes consist of a small number of changes, for example, only 22 amino acids out of 530 in each subunit are different between the muscle and kidney isozymes. Furthermore, these different amino acids are clustered in one domain. Thus, embedded in these two isoforms of PK are the rules involved in engineering the popular TIM $(\alpha/\beta)_8$ motif to modulate its allosteric properties.

ACKNOWLEDGMENTS

Supported by NIH GM 77551 and the Robert A. Welch Foundation (J. C. L.) and grant MSM 0021620835 of the Ministry of Education Youth and Sports of the Czech Republic (P. H.).

REFERENCES

Ackers, G. K. (1970). Analytical gel chromatography of proteins. *Adv. Protein Chem.* **24,** 343–446.

Ackers, G. K., Johnson, M. L., Mills, F. C., Halvorson, H. R., and Shapiro, S. (1975). The linkage between oxygenation and subunit dissociation in human hemoglobin. Consequences for the analysis of oxygenation curves. *Biochemistry* **14,** 5128–5134.

Ainsworth, S., and MacFarlane, N. (1973). A kinetic study of rabbit muscle pyruvate kinase. *Biochem. J.* **131,** 223–236.

Baronciani, L., and Beutler, E. (1995). Molecular study of pyruvate kinase deficient patients with hereditary nonspherocytic hemolytic anemia. *J. Clin. Invest.* **95,** 1702–1709.

Beechem, J. M., Knutson, J. R., Ross, J. B. A., Turner, B. W., and Brand, L. (1983). Global resolution of heterogeneous decay by phase modulation fluorometry—Mixtures and proteins. *Biochemistry* **22,** 6054–6058.

Consler, T. G., Uberbacher, E. C., Bunick, G. J., Liebman, M. N., and Lee, J. C. (1988). Domain interaction in rabbit muscle pyruvate kinase. II. Small angle neutron scattering and computer simulation. *J. Biol. Chem.* **263,** 2794–2801.

Consler, T. G., Woodard, S. H., and Lee, J. C. (1989). Effects of primary sequence differences on the global structure and function of an enzyme: A study of pyruvate kinase isozymes. *Biochemistry* **28,** 8756–8764.

Consler, T. G., Jennewein, M. J., Cai, G.-Z., and Lee, J. C. (1990). Synergistic effects of proton and phenylalanine on the regulation of muscle pyruvate kinase. *Biochemistry* **29,** 10765–10771.

Consler, T. G., Jennewein, M. J., Cai, G. Z., and Lee, J. C. (1992). Energetics of allosteric regulation in muscle pyruvate kinase. *Biochemistry* **31,** 7870–7878.

Consler, T. G., Liebman, M. N., and Lee, J. C. (1994). Structural elements involved in the allosteric switch in mammalian pyruvate kinase. *In* "Molecular Modeling," (M. N. Liebman and T. Kumosinski, eds.), pp. 466–485, Chapter 25.

Copley, R. R., and Bork, P. (2000). Homology among $(\beta\alpha)_8$ barrels: Implications for the evolution of metabolic pathways. *J. Mol. Biol.* **303,** 627–640.

Debye, P. (1915). Zerstreuung von Rontgenstrahlen. *Ann. Phys. (Leipzig)* **46,** 809–823.

Dombrauckas, J. D., Santarsiero, B. D., and Mesecar, A. D. (2005). Structural basis for tumor pyruvate kinase M2 allosteric regulation and catalysis. *Biochemistry* **44,** 9417–9429.

Eisenfeld, J., and Ford, C. C. (1979). A systems-theory approach to the analysis of multiexponential fluorescence decay. *Biophys. J.* **26,** 73–83.

Farber, G. K., and Petsko, G. A. (1990). The evolution of α/β barrel enzymes. *Trends Biochem. Sci.* **15,** 228–234.

Fenton, A. W., and Alontaga, A. Y. (2009). The impact of ions on allosteric functions in human liver pyruvate kinase. *Methods Enzymol.* **466,** 83–107.

Fenton, A. W., and Hutchison, M. (2009). The pH dependence of the allosteric response of human liver pyruvate kinase to fructose-1,6-bisphosphate, ATP and alanine. *Arch. Biochem. Biophys.* **484,** 16–23.

Fenton, A. W., Williams, R., and Trewhella, J. (2010). Changes in small-angle X-ray scattering parameters observed upon binding of ligand to rabbit muscle pyruvate kinase are not correlated with allosteric transitions. *Biochemistry* **49,** 7202–7209.

Ferreon, J. C., Volk, D. E., Luxon, B. A., Gorenstein, D. G., and Hilser, V. J. (2003). Solution structure, dynamics, and thermodynamics of the native state ensemble of the Sem-5 C-terminal SH3 domain. *Biochemistry* **42,** 5582–5591.

Friere, E. (1999). The propagation of binding interactions to remote sites in proteins: Analysis of the binding of the monoclonal antibody D1.3 to lysozyme. *Proc. Natl. Acad. Sci. USA* **96,** 10118–10122.

Friesen, R. H. E., and Lee, J. C. (1998). The negative dominant effects of T340M mutation on mammalian pyruvate kinase. *J. Biol. Chem.* **273,** 14772–14779.

Friesen, R. H. E., Castellani, R. J., Lee, J. C., and Braun, W. (1998a). Allostery in rabbit pyruvate kinase: Development of a strategy to elucidate the mechanism. *Biochemistry* **37,** 15266–15276.

Friesen, R. H. E., Chin, A. J., Ledman, D. W., and Lee, J. C. (1998b). Interfacial communications in recombinant rabbit kidney pyruvate kinase. *Biochemistry* **37,** 2949–2960.

Gerhart, J. C., and Schachman, H. K. (1968). Allosteric interactions in aspartate transcarbamylase. II. Evidence for different conformational states of the protein in the presence and absence of specific ligands. *Biochemistry* **7,** 538–552.

Gerlt, J. A., and Babbitt, P. C. (2001). Divergent evolution of enzymatic function: Mechanistically diverse superfamilies and functionally distinct superfamilies. *Annu. Rev. Biochem.* **70,** 209–246.

Hall, E. R., and Cottam, G. L. (1978). Isozymes of pyruvate kinase in vertebrates: Their physical, chemical, kinetic and immunological properties. *Int. J. Biochem.* **9,** 785–793.

Harris, S. J., and Winzor, D. J. (1998). Thermodynamic nonideality as a probe of allosteric mechanisms: Preexistence of the isomerization equilibrium for rabbit muscle pyruvate kinase. *Arch. Biochem. Biophys.* **265,** 458–465.

Heggi, H., and Gerstein, M. (1999). The relationship between protein structure and function: A comprehensive survey with application to the yeast genome. *J. Mol. Biol.* **288,** 147–164.

Herman, P., and Lee, J. C. (2009a). Functional energetic landscape in the allosteric regulation of muscle pyruvate kinase I. Calorimetric study. *Biochemistry* **48,** 9448–9455.

Herman, P., and Lee, J. C. (2009b). Functional energetic landscape in the allosteric regulation of muscle pyruvate kinase II. Fluorescence study. *Biochemistry* **48,** 9456–9465.

Herman, P., and Lee, J. C. (2009c). Functional energetic landscape in the allosteric regulation of muscle pyruvate kinase. III. Mechanism. *Biochemistry* **48,** 9466–9470.

Herman, P., and Lee, J. C. (2010). The advantage of global fitting of data involving complex linked reactions. *Methods Mol. Biol.* (in press).

Heyduk, E., Heyduk, T., and Lee, J. C. (1992). Global conformational changes in allosteric proteins. *J. Biol. Chem.* **267,** 3200–3204.

Hilser, V. J., Dowdy, D., Oas, T. G., and Freire, E. (1998). The structural distribution of cooperative interactions in proteins: Analysis of the native state ensemble. *Proc. Natl. Acad. Sci. USA* **95,** 9903–9908.

Hilser, V. J., Garcia-Moreno, E. B., Oas, T. G., Kapp, G., and Whitten, S. T. (2006). A statistical thermodynamic model of the protein ensemble. *Chem. Rev.* **106,** 1545–1558.

Hirose, M., and Kano, Y. (1971). Binding of ligands by proteins: A simple method with sephadex gel. *Biochim. Biophys. Acta* **251,** 376–379.

Holt, J. M., and Ackers, G. K. (1995). Pathway of allosteric control as revealed by intermediate states of hemoglobin. *Methods Enzymol.* **259,** 1–19.

Howlett, G. J., and Schachman, H. K. (1977). Allosteric regulation of aspartate transcarbamoylase. Changes in the sedimentation coefficient promoted by the bisubstrate analogue *N*-(phosphonacetyl)-L-aspartate. *Biochemistry* **16,** 5077–5083.

Johnson, M. L., Correia, J. J., Yphantis, D. A., and Halvorson, H. R. (1981). Analysis of data from the analytical ultracentrifuge by nonlinear least-squares techniques. *Biophys. J.* **36,** 575–588.

Kanno, H., Fujii, H., Hirono, A., and Miwa, S. (1991). cDNA cloning of human r-type pyruvate kinase and identification of a single amino acid substitution ($Thr^{384}\rightarrow Met$) affecting enzymatic stability in a pyruvate kinase variant (PK Tokyo) associated with hereditary hemolytic anemia. *Proc. Natl. Acad. Sci. USA* **88,** 8218–8221.

Kayne, F. (1973). Pyruvate kinase. *In* "The Enzymes," (P. D. Boyer, ed.), 3rd ed. p. 353. Academic Press, New York.

Kayne, F. J., and Price, N. C. (1972). Conformational changes in the allosteric inhibition of muscle pyruvate kinase by phenylalanine. *Biochemistry* **11,** 4415–4420.

Knutson, J. R., Beechem, J. M., and Brand, L. (1983). Simultaneous analysis of multiple fluorescence decay curves—A global approach. *Chem. Phys. Lett.* **102,** 501–507.

Koshland, D. E., Jr., Nemethy, G., and Filmer, D. (1966). Comparison of experimental binding data and theoretical models in proteins containing subunits. *Biochemistry* **5,** 365–385.

Kwan, C. Y., and Davis, R. C. (1980). pH-dependent amino acid induced conformational changes of rabbit muscle pyruvate kinase. *Can. J. Biochem.* **58,** 188–193.

Kwan, C. Y., and Davis, R. C. (1981). L-Phenylalanine induced changes of sulfhydryl reactivity in rabbit muscle pyruvate kinase. *Can. J. Biochem.* **59,** 92–99.

Larsen, T. M., Laughlin, T., Holden, H. M., Rayment, I., and Reed, G. H. (1994). Structure of rabbit muscle pyruvate kinase complexed with Mn^{2+}, K^{+}, and pyruvate. *Biochemistry* **33,** 6301–6309.

Larsen, T. M., Benning, M. M., Wesenberg, G. E., Rayment, I., and Reed, G. H. (1997). Ligand-induced domain movement in pyruvate kinase: Structure of the enzyme from rabbit muscle with Mg^{2+}, K^{+}, and I-phospholactate at 2.7 Å resolution. *Arch. Biochem. Biophys.* **345,** 199–206.

Larsen, T. M., Benning, M. M., Rayment, I., and Reed, G. H. (1998). Structure of the bis (Mg^{2+})-ATP-oxalate complex of the rabbit muscle pyruvate kinase at 2.1 Å resolution: ATP binding over a barrel. *Biochemistry* **37,** 6247–6255.

Liu, T., Whitten, S. T., and Hilser, V. J. (2006). Ensemble-based signatures of energy propagation in proteins: A new view of an old phenomenon. *Proteins Struct. Funct. Bioinform.* **62,** 728–738.

Luque, I., Leavitt, S. A., and Freire, E. (2002). The linkage between protein folding and functional cooperativity: Two sides of the same coin? *Annu. Rev. Biophys. Biomol. Struct.* **31,** 235–256.

Marquardt, D. W. (1963). An algorithm for least-squares estimation of nonlinear parameters. *J. Soc. Ind. Appl. Math.* **11,** 431–441.

Mattevi, A., Valentini, G., Rizzi, M., Speranza, M. L., Bolognesi, M., and Coda, A. (1995). Crystal structure of *E. coli* pyruvate kinase type I: Molecular basis of the allosteric transition. *Structure* **3,** 729–741.

Mattevi, A., Bolognesi, M., and Valentini, G. (1996). The allosteric regulation of pyruvate kinase. *FEBS Lett.* **289,** 15–19.

Mesecar, A. D., and Nowak, T. (1997a). Metal-ion-mediated allosteric triggering of yeast pyruvate kinase. 1. A multidimensional kinetic linked-function analysis. *Biochemistry* **36,** 6792–6802.

Mesecar, A. D., and Nowak, T. (1997b). Metal-ion-mediated allosteric triggering of yeast pyruvate kinase. 2. A multidimensional thermodynamic linked-function analysis. *Biochemistry* **36,** 6803–6813.

Miwa, S., Kanno, H., and Fujii, H. (1993). Concise review: Pyruvate kinase deficiency: Historical perspective and recent progress of molecular genetics. *Am. J. Hematol.* **42,** 31–35.

Monod, J., Wyman, J., and Changeux, J. P. (1965). On the nature of allosteric transitions: A plausible model. *J. Mol. Biol.* **12,** 88–118.

Moore, P. B. (1980). Small-angle scattering. Information content error analysis. *J. Appl. Cryst.* **13,** 168–175.

Na, G. C., and Timasheff, S. N. (1985). Measurement and analysis of ligand-binding isotherms linked to protein self-association. *Methods Enzymol.* **117,** 496–519.

Neubauer, B., Lakomek, M., Winkler, H., Paske, M., Hofferbert, S., and Schröter, W. (1991). Point mutations in the L-type pyruvate kinase gene of two children with homolytic anemia caused by pyruvate kinase deficiency. *Blood* **77,** 1871–1875.

Oberfelder, R. W., Lee, L. L., and Lee, J. C. (1984a). Thermodynamic linkages in rabbit muscle pyruvate kinase: Kinetic, equilibrium, and structural studies. *Biochemistry* **23,** 3813–3821.

Oberfelder, R. W., Barisas, B. G., and Lee, J. C. (1984b). Thermodynamic linkages in rabbit muscle pyruvate kinase: Analysis of experimental data by a two-state model. *Biochemistry* **23,** 3822–3826.

Pan, H., Lee, J. C., and Hilser, V. J. (2000). Binding sites in *Escherichia coli* dihydrofolate reductase communicate by modulating the conformational ensemble. *Proc. Natl. Acad. Sci. USA* **97,** 12020–12025.

Reinhart, G. D. (2004). Quantitative analysis and interpretation of allosteric behavior. *Methods Enzymol.* **380,** 187–203.

Schrank, T. P., Bolen, D. W., and Hilser, V. J. (2009). Rational modulation of conformational fluctuations in adenylate kinase reveals a local unfolding mechanism for allostery and functional adaptation in proteins. *Proc. Natl. Acad. Sci. USA* **106,** 16984–16989.

Siegel, L., and Monty, K. J. (1966). Determination of molecular weights and frictional ratios of proteins in impure systems by use of gel filtration and density gradient centrifugation. Application to crude preparations of sulfite and hydroxylamine reductases. *Biochim. Biophys. Acta* **112,** 346–362.

Steinhardt, J., and Reynolds, J. A. (1969). Multiple Equilibria in Proteins. Academic Press, New York, NY.

Stuart, H., Levine, M., Muirhead, H., and Stammers, D. K. (1979). Crystal structure of cat muscle pyruvate kinase at a resolution of 2·6 Å. *J. Mol. Biol.* **134,** 109–142.

Suelter, C. H., Singleton, R., Jr., Kayne, F. J., Arrington, S., Glass, J., and Mildvan, A. S. (1966). Studies on the interaction of substrate and monovalent and divalent cations with pyruvate kinase. *Biochemistry* **5,** 131–139.

Tanford, C. (1969). Extension of the theory of linked function to incorporate the effects of protein hydration. *J. Mol. Biol.* **39,** 539–544.

Ucci, J. W., and Cole, J. L. (2004). Global analysis of non-specific protein–nucleic interactions by sedimentation equilibrium. *Biophys. Chem.* **108,** 127–140.

Valentine, W. N., Tanaka, K. R., and Paglia, D. E. (1989). Pyruvate kinase and other enzyme deficiency disorders of the erythrocyte. *In* "The Metabolic Basis of Inherited Disease, Vol. 6," (C. R. Seriver, A. L. Beaudet, W. S. Sly, and D. Valle, eds.). McGrans-Hill, New York.

Verveer, P. J., Squire, A., and Bastiaens, P. I. (2000). Global analysis of fluorescence lifetime imaging microscopy data. *Biophys. J.* **78,** 2127–2137.

Weber, G. (1972). Ligand binding and internal equilibria in proteins. *Biochemistry* **11,** 864–878.

Wierenga, R. K. (2001). The TIM-barrel fold: A versatile framework for efficient enzymes. *FEBS Lett.* **492,** 193–198.

Williams, R., Holyoak, T., McDonald, G., Gui, C., and Fenton, A. W. (2006). Differentiating a ligand's chemical requirements for allosteric interactions from those for protein binding. Phenylalanine inhibition of pyruvate kinase. *Biochemistry* **45,** 5421–5429.

Wooll, J. O., Friesen, R. H. E., White, M. A., Watowich, S. J., Fox, R. O., Lee, J. C., and Czerwinski, E. W. (2001). Structural and functional linkages between subunit interfaces in mammalian pyruvate kinase. *J. Mol. Biol.* **312,** 525–540.

Wyman, J. (1964). Linked functions and reciprocal effects in hemoglobin: A second look. *Adv. Prot. Chem.* **19,** 223–286.

Wyman, J., and Gill, S. J. (1990). Bing and Linkage: Functional Chemistry of Biological Macromolecules. University Science Books, Mill Valley, CA.

Yu, S., Lee, L., and Lee, J. C. (2003). Effects of metabolites on the structural dynamics of rabbit muscle pyruvate kinase. *Biophys. Chem.* **103,** 1–11.

CHAPTER NINE

Analysis of Free Energy Versus Temperature Curves in Protein Folding and Macromolecular Interactions

Vince J. LiCata *and* Chin-Chi Liu

Contents

Abstract

Plots of free energy versus temperature are commonly called stability curves or Gibbs–Helmholtz curves, and they have proven to be extremely useful in protein folding and ligand-binding studies. Curvature in a Gibbs–Helmholtz or stability plot is indicative of a heat capacity change, and some of their primary uses in biochemistry over the past few decades have included determining ΔCp values and comparing ΔCp values between two related processes. This chapter describes basic approaches for analyzing curved Gibbs–Helmholtz plots, along with two specific extensions of standard Gibbs–Helmholtz plot analysis: (1) translating ΔG of folding versus temperature into ΔH and ΔS versus temperature for comparing mesophilic–thermophilic protein pairs, and (2) fitting Gibbs–Helmholtz plots to determine if ΔCp changes with temperature or not. Neither of these extensions is new, but they are infrequently used, and their use is particularly germane to certain molecular interpretations of thermodynamic

Department of Biological Sciences, Louisiana State University, Baton Rouge, Louisiana, USA

Methods in Enzymology, Volume 488
ISSN 0076-6879, DOI: 10.1016/B978-0-12-381268-1.00009-4
© 2011 Elsevier Inc.
All rights reserved.

information from ΔG versus temperature curves. It is shown that translating ΔG of folding into ΔH and ΔS of folding versus temperature for a mesophilic–thermophilic protein pair can immediately influence possible structural hypotheses for thermal stabilization of thermophilic proteins. It is also shown that very small temperature-dependent heat capacity changes ($\Delta\Delta$Cp values) can be obtained from extended fits to ΔG versus temperature plots, and that these very small $\Delta\Delta$Cp values can have serious consequences for any attempt to correlate ΔCp with ΔASA for some reactions.

1. Stability Curves = Gibbs–Helmholtz Curves = ΔG Versus Temperature

For about a century, the van't Hoff plot has been one of the most common methods for analyzing the temperature dependence of an equilibrium reaction. A plot of ln K_{eq} versus $1/T$ gives a line with a slope that yields the enthalpy of the reaction and an intercept that yields the entropy:

$$\ln K = \frac{-\Delta H}{RT} + \frac{\Delta S}{R} \tag{9.1}$$

The original (differential) form of the van't Hoff equation relates K_{eq} to ΔH:

$$\frac{\partial \ln K}{\partial T} = \frac{\Delta H}{RT^2} \tag{9.2}$$

Here, the change in ln K with respect to temperature is at constant pressure, but because most biochemistry is performed at constant pressure, this denotation is usually omitted. This and the fact that one is frequently not interested in any associated change in volume with temperature also account for the frequent substitution of the partial versus total derivative forms of these same equations (e.g., $\mathrm{d}\ln K/\mathrm{d}T$), since in practice for biochemists, they are effectively equivalent. Note that the more empirically useful integrated form in Eq. (9.1) is also cast in the format of $\Delta G = \Delta H - T\Delta S$. Equation (9.2) can also be rearranged other ways, however. By multiplying by R and substituting $\Delta G = -RT\ln K$, Eq. (9.2) goes from being the van't Hoff equation to being a version of the Gibbs–Helmholtz equation:

$$\frac{\partial(\Delta G/T)}{\partial T} = \frac{-\Delta H}{T^2} \tag{9.3}$$

$\Delta G = \Delta H - T\Delta S$, the definition of the Gibbs free energy, is also an integrated form of the Gibbs–Helmholtz equation (ΔS being the constant of integration). Both the Gibbs–Helmholtz equation and the van't Hoff equation can be used to obtain the ΔH of a reaction from the exact same data, but for whatever reason, the van't Hoff equation has long been the method of choice. Like the van't Hoff equation, the Gibbs–Helmholtz equation is linear only if the ΔH is constant with temperature. About four decades ago, however, a number of biophysicists started noticing that interactions between biological macromolecules were frequently associated with heat capacity changes: that is, the van' Hoff plots for many biochemical equilibria are not always linear. Curvature in a van't Hoff plot or a Gibbs–Helmholtz plot means that the ΔH changes with temperature, and the change in enthalpy with temperature is one of the most frequently empirically utilized definitions of heat capacity:

$$\frac{\partial \Delta II}{\partial T} = \Delta \mathrm{Cp} \tag{9.4}$$

Figure 9.1 illustrates the Gibbs–Helmholtz and van't Hoff plots for equivalent reactions with and without heat capacity changes. Panels A and B illustrate the linear Gibbs–Helmholtz (A) and van't Hoff (B) plots one obtains when there is no ΔCp for the reaction. The slopes of either Panel A or B will yield the ΔH of the reaction. These example plots illustrate a reaction that has a negative ΔG and a positive ΔH. Whether the slope is positive or negative depends both on the direction of reaction plotted (e.g., negative or positive ΔG direction) and on the sign of the enthalpy itself.

When a negative heat capacity is added to the equilibrium depicted in Fig. 9.1, plots C and D result. Plots C and D are simulated by using the 25 °C ΔH and ΔS values from Panels A and B, by using the ΔG value at 25 °C from Panels A and B, and by adding a ΔCp of -1.3 kcal/mol K. One convenient aspect of nonlinear Gibbs–Helmholtz plots (Panel C) is that a concave up plot always reflects a negative ΔCp and a concave down plot always reflects a positive ΔCp *for the direction of the reaction that is plotted.* This holds whether the ΔG values plotted are in the favorable or unfavorable reaction direction. Although the van't Hoff plot (Panel A) is the method of choice for analyzing the temperature dependence of reactions without a ΔCp, the Gibbs–Helmholtz plot (Panel C) is the more typical approach for reactions with a ΔCp.

To obtain the value of the heat capacity (along with enthalpy and entropy values) from curved Gibbs–Helmholtz plots like those shown in Panel C of Fig. 9.1, modified forms of the Gibbs–Helmholtz equation are used, where heat capacity has been incorporated. There are a number of different so-called "modified forms" and "integrated forms" of the Gibbs–Helmholtz equation in use. Like most thermodynamic equations, it can be

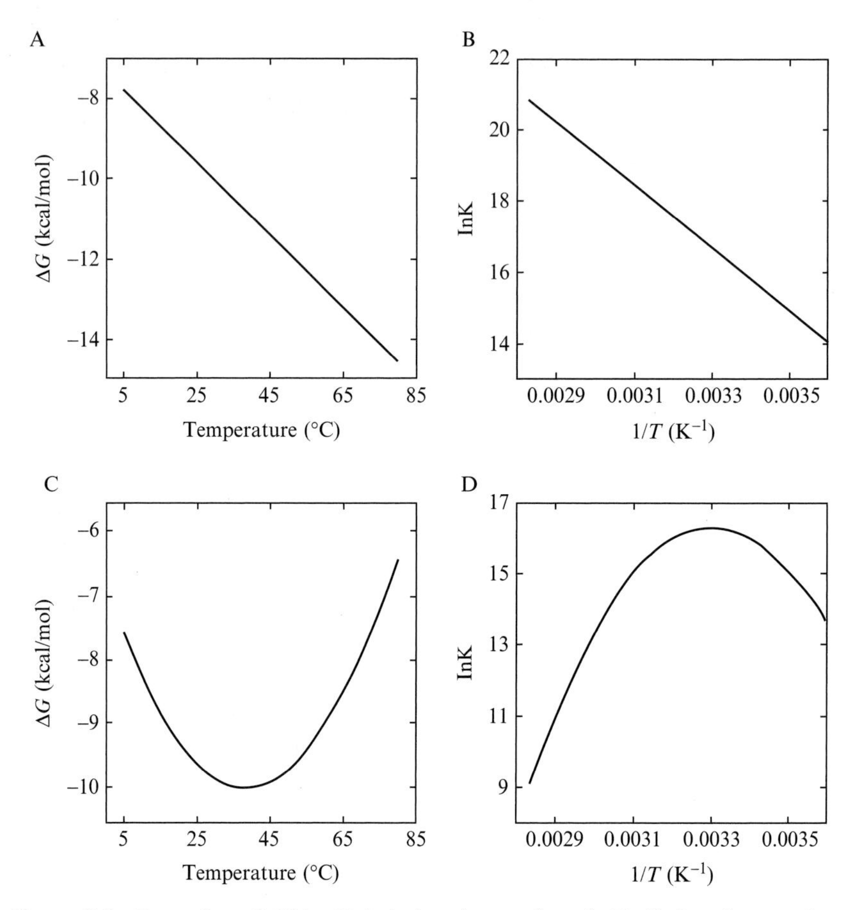

Figure 9.1 Examples of Gibbs–Helmholtz plots and van't Hoff plots for reactions without (A and B) and with (C and D) a ΔCp. In all four plots, the ΔH, $T\Delta S$, and ΔG values at 25 °C are the same (17.5, 27, and −9.5 kcal/mol, respectively). ΔH and ΔS do not vary with temperature in plots A and B, and so both the Gibbs–Helmholtz and van't Hoff plots are linear. Panels C and D show the effect of including a ΔCp of −1.3 kcal/mol K.

rearranged, integrated, expanded by substitution, etc., to produce many more (or less) useful forms. Most forms in use for analyzing ligand binding and protein folding have the following commonalities: (1) they use integrated versions of the relationships between ΔCp, ΔH, and ΔS (which are also rearranged versions of the definitions of heat capacity):

$$\Delta H(T) = \Delta H_r + \Delta \mathrm{Cp}(T - T_r) \tag{9.5}$$

$$\Delta S(T) = \Delta S_r + \Delta \mathrm{Cp} \ln\left(\frac{T}{T_r}\right) \tag{9.6}$$

and (2) they incorporate them into the basic definition of free energy ($\Delta G = \Delta H - T\Delta S$), which, as noted above, is also an integrated form of the Gibbs–Helmholtz equation. This yields forms such as

$$\Delta G(T) = \Delta H_r - T\Delta S_r + \Delta Cp\left[T - T_r - T\ln\left(\frac{T}{T_r}\right)\right] \tag{9.7}$$

which is commonly used to fit macromolecular interaction or ligand-binding data. In all the three Eqs. (9.5)–(9.7), the subscript *r* denotes a single selected reference temperature, in Kelvin (and the associated ΔH_r and ΔS_r at that temperature). Equation (9.5) is also often called "Kirchoff's law."

One starts with values of ΔG for binding at numerous temperatures. To perform the fit, any single temperature (in Kelvin) is selected as T_r and that temperature is fixed in the equation. T_r is usually selected within the range of the data, but not necessarily (as discussed later). The fitted parameters are then ΔH_r, ΔS_r, and ΔCp. The fitted value of ΔCp will remain the same regardless of the T_r chosen by the investigator. The fitted values of ΔH_r and ΔS_r will correspond to, and thus vary with, the selected T_r. (Please note that familiarity with standard nonlinear regression methods/programs is assumed in this chapter. For methodological information on nonlinear regression itself, a suggested start includes the Numerical Computer Methods volumes in this *Methods in Enzymology* series, especially Volumes 210 and 240.) We will return to macromolecular analysis using Eq. (9.7) in the section on resolving a $\Delta\Delta Cp$ value.

2. Analysis of ΔG Versus Temperature in Protein Folding

For protein folding, Eq. (9.7) is often simplified slightly more by using the melting temperature (T_m) as the reference temperature (T_r). At the T_m, $\Delta G = 0$, and so ΔS at the T_m (ΔS_m) can be calculated as

$$\Delta S_m = \frac{\Delta H_m}{T_m} \tag{9.8}$$

which allows effective removal of the ΔS parameter from the equation, and with some algebraic rearrangement, one obtains:

$$\Delta G(T) = \Delta H_m\left(\frac{T_m - T}{T_m}\right) - \Delta Cp\left[T_m - T\left(1 - \ln\frac{T_m}{T}\right)\right] \tag{9.9}$$

Here again, the data are ΔG versus temperature, but the fitting parameters are now ΔH_m, ΔCp, and T_m. Some reasons for the popularity of this particular "modified, integrated" form of the Gibbs–Helmholtz equation in protein folding include the fact that it fits directly for parameters in which one is often interested (especially T_m and ΔCp), and that it allows one to fix the values for T_m or ΔH_m or both if this information has been determined independently by calorimetry or by spectroscopically monitored thermal denaturation.

Equations (9.7) and (9.9) are certainly not the only useful versions of the ΔCp-modified, integrated Gibbs–Helmholtz relationship, and other slightly different, but equivalent rearrangements appear in the literature. Furthermore, as noted above, exactly equivalent "ΔCp-modified and integrated" van't Hoff equations would allow one to do equivalent analyses of the same data. The Gibbs–Helmholtz forms in Eqs. (9.7) and (9.9), however, are by far the more frequently employed, likely due to several factors, including their relative ease of use, the fact that they often converge on realistic determinations of the fitted parameters even with relative messy data, and possibly due to a little bit of sociological cooperativity.

Equation (9.9) is most commonly used to fit ΔG of *unfolding* versus temperature data to produce what is now commonly called a stability curve (Becktel and Schellman, 1987). Use of the unfolding direction rather than the folding direction is simply convention. Figure 9.2 shows an example of fitting Eq. (9.9) to unfolding data for a hypothetical protein. Numerous

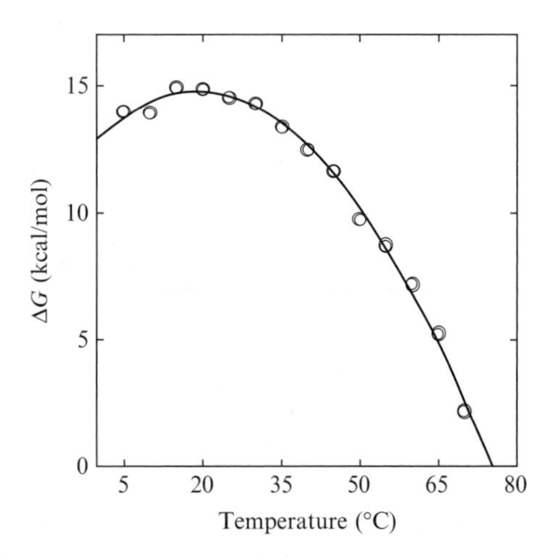

Figure 9.2 Example of a fitted stability plot for unfolding of a protein. ΔG versus temperature data (open circles) were simulated (with $\pm 10\%$ error) for unfolding of a protein with a T_m of 75 °C, with a ΔH_m of 180 kcal/mol, and a ΔCp of 3 kcal/mol K. The line shows the fit of the simulated data to Eq. (9.9).

examples of such analyses exist in the literature (e.g., Becktel and Schellman, 1987; Nojima *et al.*, 1977; reviews by Kumar *et al.*, 2001; Razvi and Scholtz, 2006).

Although there are certainly many highly precise data sets in the literature, free energy versus temperature data for protein folding often exhibits significant scatter, yet still are almost always easily fit with Eq. (9.9). ΔG unfolding values are typically determined using chemical denaturation with urea or guanidine HCl and monitored by circular dichroism or some other spectroscopic technique. This process itself is a multiparameter fit of the chemical denaturation curves at different temperatures. Numerous excellent methodological reviews of denaturation curve analysis exist (e.g., Pace, 1986; Santoro and Bolen, 1988), and the reader is directed to these as a starting place. This chapter focuses on what to do with the ΔG versus temperature data after one obtains it.

There are several features of the stability curve that warrant mention. Where the curve crosses zero at high temperature is the protein's melting temperature T_m, as noted above in the discussion of Eq. (9.9). As originally discussed by Privalov, the curve also crosses zero at low temperature, and this is denoted as the "cold denaturation temperature" for the protein (see Privalov, 1990 for a review). Direct study of cold denaturation is difficult because the protein has to remain in solution for this curve to continue to accurately describe it at low temperature. The highest point on the curve is the "optimal free energy" or ΔG_{max}. As is discussed further below, the temperature of ΔG_{max} is also the temperature where the entropy of unfolding is zero.

3. Using Stability Curves to Compare Mesophilic and Thermophilic Protein Pairs

Stability curves representing the most commonly discussed thermodynamic models that can explain how a protein can achieve a higher melting temperature (T_m) are illustrated schematically in Fig. 9.3. Again, the higher temperature where each curve crosses $\Delta G = 0$ is the melting temperature (T_m), and the lower temperature where each curve crosses zero is the cold denaturation temperature. A simulated mesophilic protein (bold, solid line with a maximal ΔG at 25 °C) can increase T_m in three main ways: (1) by pulling the curve straight up, that is, increasing ΔG at all temperatures but not otherwise altering the curve (shown as the lighter solid line), (2) by shifting the curve to the right, that is, increasing the temperature where ΔG is maximal, but not otherwise altering the curve (shown as the dotted line)—this model is also described as "increasing T_S" since the temperature where ΔG is maximal will also be the temperature where the enthalpy of unfolding crosses zero (the T_S), and (3) by broadening the curve, that is,

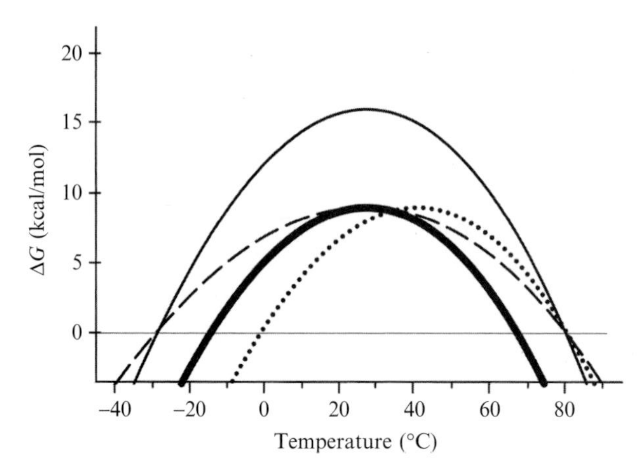

Figure 9.3 Stability curves representing different thermodynamic models that can explain how a protein can achieve a higher melting temperature (T_m). Each curve shows the free energy of unfolding versus temperature. The temperature where each curve crosses zero at high temperature is the T_m. A simulated mesophilic protein (bold solid line) can increase T_m in three main ways: (1) by increasing ΔG at all temperatures (lighter solid line), (2) by shifting the curve, that is, shifting all ΔG values to higher temperatures (dotted line), or (3) by reducing the ΔCp of unfolding (dashed line).

reducing the ΔCp of unfolding, but not otherwise altering the curve (shown as the dashed line). These thermodynamic models were originally proposed by Nojima *et al.* (1977). The strategy is to collect ΔG versus temperature data for both the mesophilic and thermophilic proteins, to compare curve shapes with those in Fig. 9.3, and to fit each curve with Eq. (9.9) to obtain the precise thermodynamic parameters. Some proteins follow a single model, others follow a mixture of models. For example, the thermophilic member of a homologous protein pair might exhibit both an increase in ΔG at all temperatures (model #1) and a decrease in ΔCp (model #3). To date, only two to three dozen thermophilic–mesophilic protein pairs have been compared in this way. This is largely due to the need for reliable, reversible denaturation data for both proteins over as wide a temperature range as possible. Reviews by Nussinov and associates (Kumar *et al.*, 2001) and Razvi and Scholtz (2006) have documented which thermophiles follow which thermodynamic model or combinations of models.

4. Temperature Dependence of Folding Enthalpies and Entropies

A slight extension of the stability curve analysis described above opens up further information about the thermodynamic origins of a particular protein's thermal stabilization. For example, model #1 in Fig. 9.3, enhancement of the ΔG of unfolding at all temperatures, has been shown in two different surveys to be the most common thermodynamic pattern found for mesophilic–thermophilic protein pairs (Kumar *et al.*, 2001; Razvi and Scholtz, 2006). By examining the enthalpy and entropy plots for protein pairs in this category, as described below, one immediately finds that two very different thermodynamic relationships can produce this same change in the Gibbs–Helmholtz curve.

Figure 9.4 shows example enthalpy and entropy plots for simulated mesophilic–thermophilic protein pairs, illustrating different thermodynamic models that these types of plots reveal. Once the ΔCp, ΔH_m, and T_m have been obtained by fitting the ΔG versus temperature data with Eq. (9.9), the ΔH and

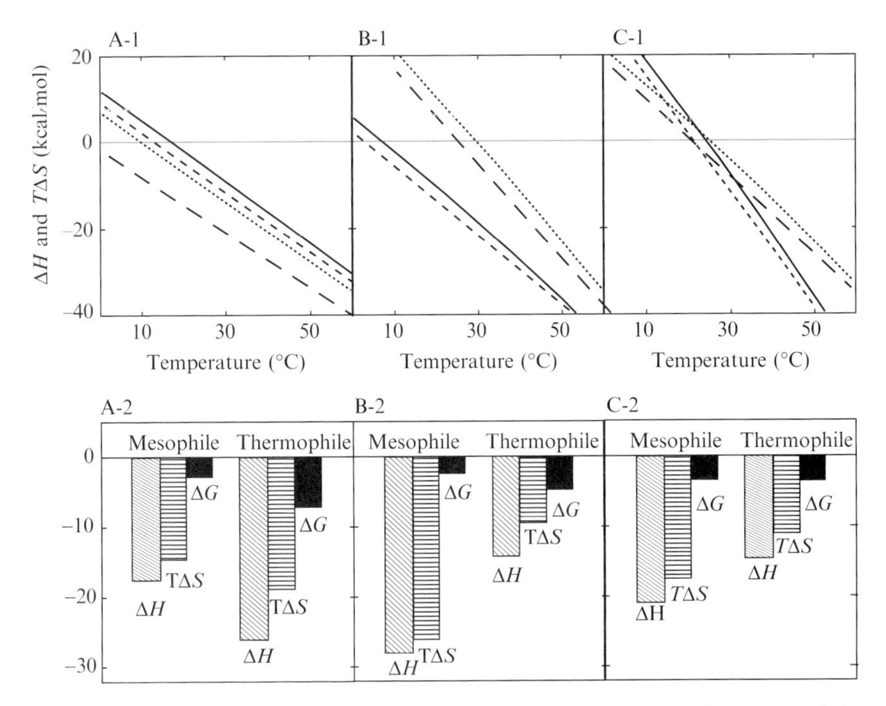

Figure 9.4 Example enthalpy and entropy plots for simulated mesophilic–thermophilic protein pairs, illustrating different thermodynamic models. In contrast to Figs. 9.2 and 9.3, here all thermodynamic parameters are shown for the folding direction. In all panels: (1) for the mesophile: the line with small dashes is the ΔH of folding, and the solid line is the ΔS of folding; and (2) for the thermophile: the line with large dashes is the ΔH of folding and the dotted line is the ΔS of folding. Panels A-1 and B-1 show two possible ways of achieving a larger ΔG for the thermophile (which is model #1 in Fig. 9.3), and which are discussed in the text. Panel C-1 shows the enthalpy and entropy plots obtained when the thermophile has a reduced ΔCp of folding, and thus the slope for ΔH and ΔS for the thermophile are reduced. The panels below each plot (A-2, B-2, and C-2) illustrate the relationships between ΔH, ΔS, and ΔG at a single example temperature (37 °C) for each simulated protein pair.

$T\Delta S$ values at each temperature can be straightforwardly calculated using Eqs. (9.5) and (9.6) from above, where the T_m is now used as the T_r value:

$$\Delta H(T) = \Delta H_m + \Delta Cp(T - T_m) \tag{9.10}$$

$$\Delta S(T) = \Delta S_m + \Delta Cp \ln\left(\frac{T}{T_m}\right) \tag{9.11}$$

Then, ΔH and ΔS are simply plotted at each temperature. In contrast to Fig. 9.3, here in Fig. 9.4, all thermodynamic parameters are shown in the folding direction. Panels A and B show two possible ways of achieving a larger ΔG for the thermophile. In both Panels A and B, the distance between the ΔH and the ΔS lines for the thermophile is much larger than the distance between the ΔH and ΔS lines for the mesophile, thus yielding a larger ΔG for the thermophile.

In Panel A of Fig. 9.4, the ΔH for the thermophile (large dashes) is the lowest line, indicating that at all temperatures, the thermophile has a much more favorable ΔH of folding. Enhancement of noncovalent interactions in the native state of the thermophile relative to the mesophile would, in general, be expected to enhance the ΔH of folding for the thermophile. Structural studies of many (but not all) thermophilic proteins have revealed a number of different noncovalent "enhancements" of the native state of the thermophile that help it achieve thermal stability (e.g., more electrostatic bonding, more hydrophobic bonding, better packing, etc.). There does not appear to be any universal structural mechanism for enhancing the stability of the native state (e.g., see Petsko, 2001; Sterner and Liebl, 2001; Szilagyi and Zavodsky, 2000). Although enhanced electrostatic bonding was believed, for some time, to be a universal mechanism, eventual structural examination of more thermophiles revealed that it was simply one of many. Any or all of these identifiable native state structural enhancements, however, are congruent with the significantly enhanced enthalpy of folding of the thermophile illustrated by the enthalpy and entropy plots in Panel A of Fig. 9.4.

In contrast, in Panel B, the ΔH of folding for the thermophile is much less favorable than that for the mesophile (compare the position of the large dashed line for the thermophile with that of the small dashed line for the mesophile). Panel B shows that the larger ΔG for the thermophile derives from a smaller entropic barrier to folding for the thermophile (the ΔS of folding for the thermophile, the dotted line, is the highest line at all temperatures). To help illustrate these ΔH, $T\Delta S$, and ΔG relationships, the bar graphs below each enthalpy and entropy plot show these values at a single representative temperature for the mesophile and the thermophile. The thermodynamic pattern in Panel B is found for a number of real mesophilic–thermophilic protein pairs and is largely incongruous with the concept that the native state of the thermophile contains more or enhanced noncovalent interactions. In summary, Panels A and B of Fig. 9.4 show two very different enthalpy and entropy thermodynamic patterns which look identical as stability plots (e.g., in Fig. 9.3).

Panel C shows one of the most popular thermodynamic models for explaining thermophily, although it is not the most experimentally prevalent model (Kumar *et al.*, 2001; Razvi and Scholtz, 2006). Here, a reduced ΔCp can easily be seen as a reduced slope for ΔH and ΔS of the thermophile. There is also a slightly larger ΔG for the thermophile in Panel C, but again, an assumption of predominantly native state structural stability enhancement is not particularly congruous with this thermodynamic pattern. Instead, the longstanding relationship between ΔCp and burial of surface area upon folding (discussed in further detail below in the sections on macromolecular interactions) has been used to structurally explain this thermodynamic pattern. In this model, it is hypothesized that there is partial structure in the denatured state of the thermophile which will reduce the amount of surface area it must bury when it folds, and that this results in a lower ΔCp of folding. A lower ΔCp of folding then produces the thermodynamic patterns seen in Panel C of Fig. 9.4 and in the broadened stability curve in Fig. 9.3. It should be noted that while these data are simulated for illustrative purposes, multiple mesophilic–thermophilic protein pairs exist that follow each of these patterns (Liu and LiCata, submitted).

Expanding Gibbs–Helmholtz data into enthalpy and entropy plots has been used previously in protein–DNA binding studies (e.g., see Record *et al.*, 1991), and a related variant of these plots have been used to study the entropy and enthalpy convergence temperature phenomenon in protein folding (e.g., see Fu and Freire, 1992), beyond this, however, their use has been quite rare. Figure 9.4 also highlights another interesting aspect of the folding reactions that is not evident in Fig. 9.3: the entropy and enthalpy for nearly all proteins cross zero at some point, often in the normal ambient temperature region. These temperatures are denoted as T_H and T_S, and T_S also corresponds to the temperature where the ΔG is maximized. What is interesting, however, is that whatever driving forces one postulates for explaining the thermal stabilization of a particular protein, they are thermodynamically reversed, from favorable to unfavorable (or vice versa) at T_H and T_S, and a full understanding of this at the molecular level is a wide-open question.

5. Analysis of ΔG Versus Temperature Data in Macromolecular Interactions

As with protein folding, one of the primary uses of modified Gibbs–Helmholtz analysis in binding studies is to determine the ΔCp for the equilibrium. Understanding heat capacity effects and their molecular origins has been of particular interest in the fields of protein folding and macromolecular interaction for some time. Heat capacity changes for binding of small ligands (e.g., enzyme substrates) are often very small, so most applications for these analyses are for macromolecular interactions, such as protein–DNA binding or protein–protein interaction. Here, we will focus more on protein–DNA interactions, since that is the area where our laboratory has applied these relationships; however, all of the analytical approaches here are applicable to any biochemical-binding reaction.

For binding studies, heat capacities are commonly measured both calorimetrically, by directly measuring ΔH versus temperature and obtaining the slope, and by Gibbs–Helmholtz curve analysis—by measuring ΔG versus temperature and obtaining ΔCp from a fit to a modified Gibbs–Helmholtz equation such as Eq. (9.7). Both of these approaches assume that the ΔCp itself does not change with temperature. In most of the studies of heat capacity effects in the past three decades, the fact that ΔCp probably does change with temperature is almost universally acknowledged, followed more or less immediately by assumption of a temperature-invariant ΔCp in subsequent analyses. There are two primary reasons typically offered for this duality: (1) it is very difficult to statistically resolve a temperature-variant heat capacity (a "$\Delta\Delta$Cp") because it is so small and (2) because $\Delta\Delta$Cp is so small, it will have little or no effect on the results. Item #1 is true, and resolving small values of $\Delta\Delta$Cp is the primary subject of this section. However, depending on one's application for the obtained $\Delta\Delta$Cp, item #2 can be true or false.

6. Fitting ΔH and ΔG Versus Temperature for a $\Delta\Delta$Cp

Figure 9.5 illustrates the extremely small effects adding a temperature variable heat capacity term ($\Delta\Delta$Cp) has on fits to the Gibbs–Helmholtz equation. The upper two panels show data that was simulated by including a $\Delta\Delta$Cp of -13 cal/mol K^2 and then error perturbing each point with up to 10% random error. The error perturbed, simulated data were then fit to Gibbs–Helmholtz equations with and without a $\Delta\Delta$Cp term. The bottom two panels show fits to empirical data for the binding of Klentaq polymerase to DNA (Liu *et al.*, 2008). In all of the plots, the solid lines show the fit with a $\Delta\Delta$Cp parameter, while the dotted lines show the fit without a $\Delta\Delta$Cp parameter. Several things are immediately clear from this figure: (1) It is easier to see the effect of a $\Delta\Delta$Cp in calorimetric data (ΔH versus temperature data), than from Gibbs–Helmholtz data (ΔG versus temperature), because without a $\Delta\Delta$Cp, the ΔH versus temperature data will be perfectly linear (the slope is the ΔCp, by definition). (2) It is very difficult, if not

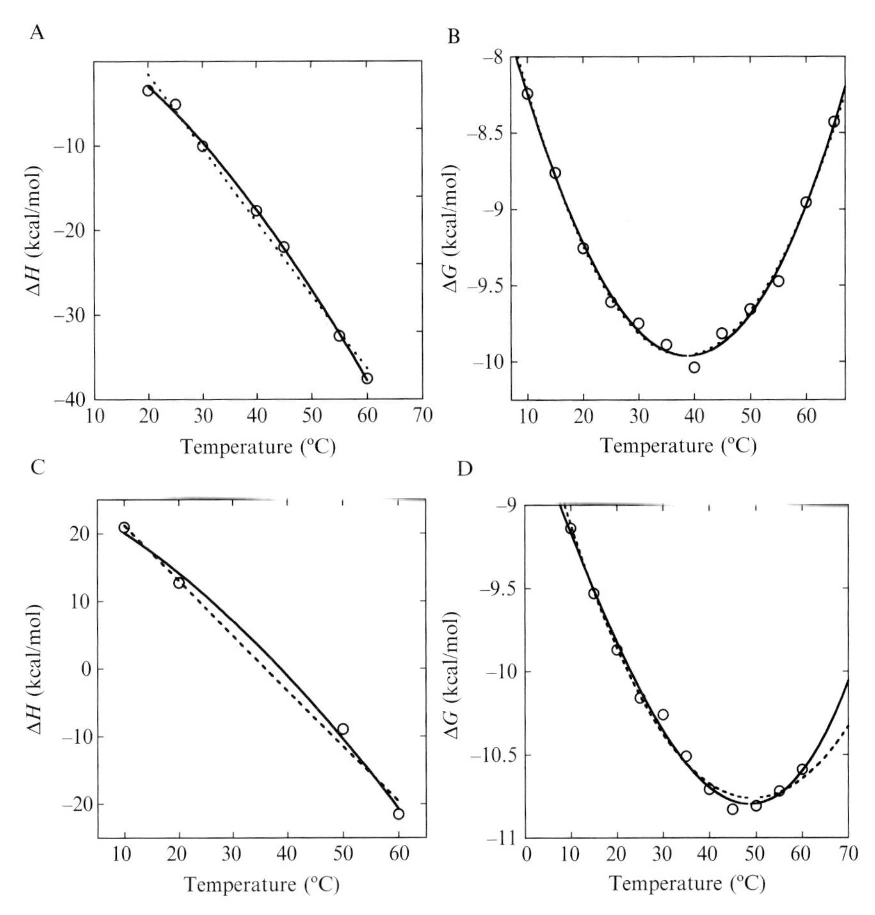

Figure 9.5 Illustrations of the effects of adding a temperature variable heat capacity term ($\Delta\Delta$Cp) on fits to the Gibbs–Helmholtz equation. Panels A and C show ΔH versus temperature data, while panels B and D show ΔG versus temperature data. The upper two panels show simulated data points with $\pm 10\%$ random error on each point. The data for these two upper panels were simulated by including a $\Delta\Delta$Cp of -13 cal/mol K^2. The bottom two panels show data for binding of Klentaq polymerase to DNA (data from Liu *et al.*, 2008), which return fitted $\Delta\Delta$Cp values of -11 cal/mol K^2 (Panel C) and -19 cal/mol K^2 (Panel D). In all plots, the solid lines are fits with a $\Delta\Delta$Cp parameter, while dotted lines are fits to the same data without a $\Delta\Delta$Cp parameter.

impossible, to detect visually a difference between fitting with a $\Delta\Delta$Cp and without a $\Delta\Delta$Cp for Gibbs–Helmholtz data, but it is statistically possible to do so.

The fits without a $\Delta\Delta$Cp parameter (all dotted lines) were performed using Eq. (9.5) for the ΔH versus temperature data and Eq. (9.7) for the ΔG

versus temperature data. For fits with a $\Delta\Delta$Cp parameter, the ΔCp was assumed to vary linearly with temperature:

$$\Delta \mathrm{Cp}(T) = \Delta \mathrm{Cp}_r + \Delta\Delta \mathrm{Cp}(T - T_r) \tag{9.12}$$

Again, T_r is an arbitrarily chosen reference temperature, in Kelvin. One can certainly use other nonlinear functional forms for the dependence of ΔCp on temperature, but this quickly turns a difficult multiparameter fitting problem into a more difficult one. One then basically substitutes the expression of ΔCp in Eq. (9.12) for the ΔCp parameters in Eqs. (9.5) and (9.7), as shown in Eqs. (9.13)–(9.21). So, Eq. (9.5) is the definite integral of

$$\Delta H(T) = \Delta H_r + \int_{T_r}^{T} \Delta \mathrm{Cp}(T)\mathrm{d}T \tag{9.13}$$

Inserting Eq. (9.12) into Eq. (9.13) yields

$$\Delta H(T) = \Delta H_r + \int_{T_r}^{T} [\Delta \mathrm{Cp}_r + \Delta\Delta \mathrm{Cp}(T - T_r)]\mathrm{d}T \tag{9.14}$$

Then, expressing Eq. (9.14) as a definite integral, one obtains

$$\Delta H(T) = \Delta H_r + \Delta \mathrm{Cp}_r(T - T_r) + \Delta\Delta \mathrm{Cp}\left[\left(\frac{T^2 - T_r^2}{2}\right) - T_r(T - T_\mathrm{r})\right] \tag{9.15}$$

With Eq. (9.15), the data are ΔH versus temperature data, and the fitting parameters are ΔH_r, $\Delta \mathrm{Cp}_r$, and $\Delta\Delta$Cp (e.g., Panels A and C in Fig. 9.5). T_r is chosen by the investigator and fixed in the equation (often at 298 K or some other temperature within the data range). If one is particularly interested in the ΔCp or ΔH at a specific temperature, that temperature can be chosen as T_r.

To get to a useful extension of Eq. (9.7) that fits for a $\Delta\Delta$Cp value in ΔG versus temperature data, we start one step back from the rearranged form of Eq. (9.7):

$$\Delta G(T) = \Delta H_r + \Delta \mathrm{Cp}(T - T_r) - T\left(\Delta S_r + \Delta \mathrm{Cp} \ln \frac{T}{T_r}\right) \tag{9.16}$$

Equation (9.16) is in the form $\Delta G = \Delta H - T\Delta S$, where ΔH and ΔS have been expanded using Eqs. (9.5) and (9.6). Equation (9.16) can be used similarly to Eq. (9.7) in fitting ΔG versus temperature data for ΔH_r, ΔS_r, and ΔCp. Equation (9.16) is the definite integral of

$$\Delta G(T) = \Delta H_r + \int_{T_r}^{T} \Delta Cp(T)\mathrm{d}T - T\left[\Delta S_r + \int_{T_r}^{T} \frac{\Delta Cp(T)}{T}\mathrm{d}T\right] \quad (9.17)$$

One then substitutes Eq. (9.12) for ΔCp into Eq. (9.17) to get

$$\begin{aligned}\Delta G(T) &= \Delta H_r + \int_{T_r}^{T} [\Delta Cp_r + \Delta\Delta Cp(T - T_r)]\mathrm{d}T \\ &-T\left[\Delta S_r + \int_{T_r}^{T}\left[\frac{\Delta Cp_r + \Delta\Delta Cp(T - T_r)}{T}\right]\mathrm{d}T\right]\end{aligned} \quad (9.18)$$

Then, to "simplify" this unwieldy indefinite integral into a slightly more unwieldy equation, but one that can be used in nonlinear regression, we take the definite integral

$$\begin{aligned}\Delta G(T) &= \Delta H_r + \Delta Cp_r(T - T_r) + \Delta\Delta Cp \\ &\left[\left(\frac{T^2 - T_r^2}{2}\right) - T_r(T - T_r)\right] \\ &-T\Delta S_r - \Delta Cp_r T \ln\frac{T}{T_r} - T\Delta\Delta Cp\left[(T - T_r) - T_r \ln\frac{T}{T_r}\right]\end{aligned} \quad (9.19)$$

Equation (9.19) can then be fit to ΔG versus temperature data, with fitting parameters ΔH_r, ΔS_r, ΔCp_r, and $\Delta\Delta Cp$. T_r in this equation is any arbitrarily chosen temperature in Kelvin and is fixed in the regression analysis. In practice, we have found better behavior in nonlinear convergence using a specific variation (reparameterization) of Eq. (9.19) that eliminates ΔS by setting T_r equal to the temperature where $\Delta G = 0$ (similar to the melting temperature in protein folding, but the protein is not melting here, it is simply the temperature where the free energy of binding is zero). Since this temperature is not within the range of the data, it will be a fitted parameter rather than a fixed one in this analysis. We denote this temperature as T_{r0} in the subsequent equations, so

$$\Delta S_{r0} = \frac{\Delta H_{r0}}{T_{r0}} \quad (9.20)$$

Like the Gibbs–Helmholtz curves for unfolding, there will be a high temperature and a low temperature where $\Delta G = 0$, but it does not matter which one is used. The initial guess one provides for the T_{r0} parameter will effectively determine whether the fit converges on the high or low T_{r0} value. With this substitution, the definite integral becomes

$$\Delta G(T) = \Delta H_{r0} + \Delta Cp_{r0}(T - T_{r0}) \\ + \Delta\Delta Cp\left[\left(\frac{T^2 - T_{r0}^2}{2}\right) - T_{r0}(T - T_{r0})\right] - T\frac{\Delta H_{r0}}{T_{r0}} - \Delta Cp_{r0} \\ T\ln\frac{T}{T_{r0}} - T\Delta\Delta Cp\left[(T - T_{r0}) - T_{r0}\ln\frac{T}{T_{r0}}\right] \tag{9.21}$$

Equation (9.21) then is fit to ΔG versus temperature data, with ΔH_{r0}, ΔCp_{r0}, T_{r0}, and $\Delta\Delta Cp$ as the fitted parameters. Because all of the other parameters are at T_{r0}, the primary utility of Eq. (9.21) is to determine the potential magnitude for $\Delta\Delta Cp$ for the reaction, but it converges more easily for some data sets. The $\Delta\Delta Cp$ values obtained from Eqs. (9.19) and (9.21) are the same.

As noted above, with ΔH versus temperature data, sometimes one can visually detect curvature in the data, which will reinforce the presence of a $\Delta\Delta Cp$ if Eq. (9.15) fits the data significantly better than Eq. (9.5) (i.e., if the χ^2 for the fit improves dramatically). For ΔG versus temperature data, however, one is often comparing the fits of two very similar curves. If the χ^2 does not improve upon adding a $\Delta\Delta Cp$ parameter, then one can almost immediately discount the presence of a $\Delta\Delta Cp$ (at least of a resolvable one). However, the χ^2 for the fit with a $\Delta\Delta Cp$ parameter will generally be improved (i.e., have a lower χ^2) relative to the fit without a $\Delta\Delta Cp$, since adding an extra parameter to the fitting equation usually improves the fit to any data. The *F*-test is one common method for determining whether the fit (χ^2) has been improved beyond the statistical improvement expected from the reduction in degrees of freedom of adding an extra parameter (Bevington and Robinson, 2003; Motulsky and Christopoulos, 2004). In a recent survey of 48 different protein–DNA systems, our laboratory found that about half of them could be fit with a $\Delta\Delta Cp$ parameter with statistical significance. The mean $\Delta\Delta Cp$ for the systems in this survey that do exhibit a $\Delta\Delta Cp$ is -13 cal/mol K^2, which is the value chosen for the simulations in Panels A and B of Fig. 9.5. If one only has a small number of data points or can only obtain data over a restricted temperature range, finding a lack of better fit with a $\Delta\Delta Cp$ may simply be a sampling problem. However, if one has good data density over a several decades of temperature, these fits can give one confidence in the absence of a $\Delta\Delta Cp$ as well.

A straightforward "trick" our laboratory has found useful in detecting the presence or absence of a $\Delta\Delta Cp$ and getting a preliminary estimate for its magnitude is to fit the lower temperature data and the higher temperature data separately to Eq. (9.7) or (9.16) (the equations without $\Delta\Delta Cp$ parameters). For most data sets that exhibit a $\Delta\Delta Cp$, these two fits will yield different ΔCp values. One should use two or three subsets of "lower" and "higher" temperature data points to feel confident in the absence or presence of a $\Delta\Delta Cp$ using this method. If this approach reveals different ΔCp values dependent upon which range of the data one examines, then a full fit to Eq. (9.21) is in order to determine the value of the $\Delta\Delta Cp$.

7. Examples of Potential Consequences of a Small $\Delta\Delta Cp$

A fair amount of effort is expended in the previous section to show how to determine the presence or absence of a rather tiny $\Delta\Delta Cp$ (note that they are so small that they are often expressed in calories rather than in kcal). One area where a small $\Delta\Delta Cp$ can have a large effect on the obtained results is in the application of the correlation between ΔCp and changes in accessible surface area (ΔASA), as illustrated in Fig. 9.6. First, Panel A shows how the value of ΔCp will change with temperature resulting from a $\Delta\Delta Cp$ of -13 cal/mol K^2 (-0.013 kcal/mol K^2), using an arbitrarily chosen starting ΔCp of -0.5 kcal/mol K at 25 °C. Panels B–D show the subsequent consequences of such a change in ΔCp on the predicted changes in surface area using the most widely utilized ΔCp–ΔASA relationship for protein–DNA interactions. This value of -13 cal/mol K^2 was, again, chosen for $\Delta\Delta Cp$ because it is the mean $\Delta\Delta Cp$ found from the analysis of approximately two-dozen different protein–DNA interaction data sets (Liu *et al.*, 2008), and so it is used here to represent the "typical" $\Delta\Delta Cp$ one might find for a protein–DNA interaction, if one's system actually has a $\Delta\Delta Cp$.

At least five different quantitative relationships between ΔCp and the sum of buried nonpolar + polar surface areas have been proposed (Makhatadze and Privalov, 1995; Murphy and Freire, 1992; Myers *et al.*, 1995; Roberston and Murphy, 1997; Spolar *et al.*, 1992). All such relationships have the form $\Delta Cp = -(x\Delta ASA_{nonpolar} - y\Delta ASA_{polar})$, where $\Delta ASA_{nonpolar}$ and ΔASA_{polar} are the amounts of nonpolar and polar surface area buried in the interface (in Å^2), x and y are empirically determined constants, and ΔCp is assumed to be temperature invariant. While these quantitative relationships continue to work reasonably well for protein folding, many protein–DNA systems deviate from these relationships (e.g., see Datta *et al.*, 2006; Jen-Jacobson *et al*, 2000; Jin *et al.*, 1993; Merabet and Ackers, 1995; Morton and Ladbury, 1996; Petri *et al.*, 1995). To simulate the effects of a $\Delta\Delta Cp$ in Fig. 9.6, the most frequently applied ΔCp–ΔASA relationship for protein–DNA interactions was used, where $x = -0.32$ and $y = 0.14$ (Spolar and Record, 1994). Panels B–D show the effect of propagating the temperature dependence of ΔCp (from Panel A)

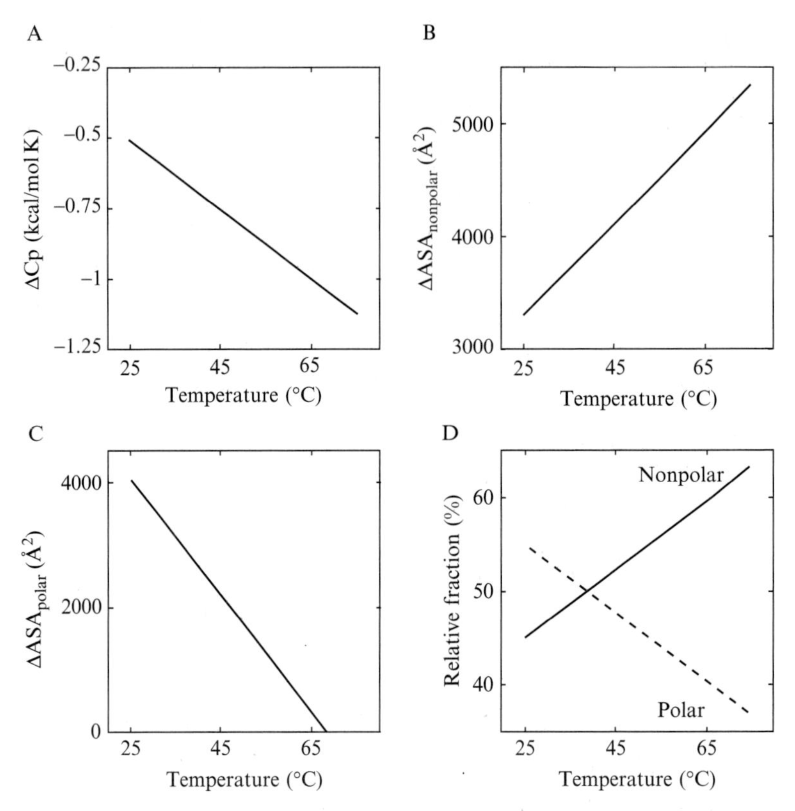

Figure 9.6 Illustrations of the consequences of a small ΔΔCp. Panel A illustrates a starting ΔCp of -0.5 kcal/mol K at 25 °C and shows how a -0.013 kcal/mol K^2 ΔΔCp will change the ΔCp over a 50° range. Panels B–D show the effect of propagating the ΔCp change in Panel A into the predicted changes in accessible surface area (ΔASA) calculated using a popular ΔCp–ΔASA relationship for protein–DNA interactions. Panel B illustrates the consequences of a -13 cal/mol K^2 ΔΔCp if all of the change is reflected in the nonpolar term. Panel C illustrates the effect if all of the change occurs in the polar term. Panel D shows how the relative ratios of polar and nonpolar surface could change to account for the larger magnitude of ΔCp. In Panel D, the ratio of polar to nonpolar surface would have to change from 55:45 to 36:64 over the 50° temperature range.

into the predicted change in accessible surface area (ΔASA) that one would calculate using this relationship. For this ΔCp–ΔASA relationship to continue to hold as the ΔCp changes as shown in Panel A would require one of the following to be true: (1) (Panel B) the nonpolar ΔASA for the reaction would have to increase by over 2000 $Å^2$ over the 50° range; or (2) (Panel C) the polar ΔASA for the reaction would have to disappear, and yet still would not account for all of the ΔCp change with temperature; or (3) (Panel D) the relative fractions (ratios) of polar and nonpolar surface areas would have to simultaneously change as shown with temperature. Again, the behaviors shown in Fig. 9.6 would be the consequences of a -0.013 kcal/mol K^2 $\Delta\Delta Cp$.

It is certainly possible that for some protein–DNA binding complexes, the ΔASA of interaction may undergo extremely large changes of this type solely by changing the temperature, but the most important point of Fig. 9.6 is that a seemingly tiny, but empirically resolvable $\Delta\Delta Cp$ value can have large consequences on the ΔCp value one obtains at any particular temperature, and very likely negates or seriously confounds the possible use of a ΔCp–ΔASA correlation that relies on a temperature-invariant ΔCp. However, it should also be noted that of the 48 protein–DNA systems examined in a prior study, about half of them did not have a detectable $\Delta\Delta Cp$ (Liu *et al.*, 2008). The point of Fig. 9.6 is then twofold: for systems without a $\Delta\Delta Cp$, the problems illustrated by Fig. 9.6 are irrelevant (although there may be other reasons why a ΔCp–ΔASA correlation does not hold for such systems). If one is applying a ΔCp–ΔASA correlation to a protein–DNA system that does have a $\Delta\Delta Cp$, however, the calculated ΔASA has a high probability of being erroneous. The summary of Figs. 9.5 and 9.6, and the use of Eqs. (9.12)–(9.21), are that (1) it is possible to statistically resolve a small $\Delta\Delta Cp$ from ΔG or ΔH versus temperature data (or conversely, to show that $\Delta\Delta Cp \approx 0$); and (2) even a very small $\Delta\Delta Cp$ can have serious consequences for some interpretive applications.

REFERENCES

Becktel, W. J., and Schellman, J. A. (1987). Protein stability curves. *Biopolymers* **26,** 1859–1877.

Bevington, P., and Robinson, D. K. (2003). Data Reduction and Error Analysis for the Physical Sciences. 3rd edn. McGraw Hill, Bostonpp. 194–217.

Datta, K., Wowor, A. J., Richard, A. J., and LiCata, V. J. (2006). Temperature dependence and thermodynamics of Klenow polymerase binding to primed-template DNA. *Biophys. J.* **90,** 1739–1751.

Fu, L., and Freire, E. (1992). On the origin of the enthalpy an entropy convergence temperatures in protein folding. *Proc. Natl. Acad. Sci. USA* **89,** 9335–9338.

Jen-Jacobson, L., Engler, L. E., Amers, J. T., Kurpiewski, M. R., and Gigorescu, A. (2000). Thermodynamic parameters of specific and nonspecific protein-DNA binding. *Supramol. Chem.* **12,** 143–160.

Jin, L., Yang, J., and Carey, J. (1993). Thermodynamics of ligand binding to trp repressor. *Biochemistry* **32,** 7302–7309.

Kumar, S., Tsai, C.-J., and Nussinov, R. (2001). Thermodynamic differences among homologous thermophilic and mesophilic proteins. *Biochemistry* **40,** 14152–14165.

Liu, C. -C., and LiCata, V. J. (submitted) The thermodynamic basis for the extreme stability of Taq polymerase is a reduced entropic folding penalty.

Liu, C.-C., Richard, A. J., Datta, K., and LiCata, V. J. (2008). Prevalence of temperature-dependent heat capacity changes in protein–DNA interactions. *Biophys. J.* **94,** 3258–3265.

Makhatadze, G. I., and Privalov, P. L. (1995). Energetics of protein structure. *Adv. Protein Chem.* **47,** 307–425.

Merabet, E., and Ackers, G. K. (1995). Calorimetric analysis of lambda cI repressor binding to DNA operator sites. *Biochemistry* **34,** 8554–8563.

Morton, C. J., and Ladbury, J. E. (1996). Water-mediated protein–DNA interactions: The relationship of thermodynamics to structural detail. *Protein Sci.* **5,** 2115–2118.

Motulsky, H., and Christopoulos, A. (2004). Fitting Models to Biological Data using Linear and Nonlinear Regression. Oxford University Press, New York, NY, pp. 138–142.

Murphy, K. P., and Freire, E. (1992). Thermodynamics of structural stability and cooperative folding behavior in proteins. *Adv. Protein Chem.* **43,** 313–361.

Myers, J. K., Pace, C. N., and Scholtz, J. M. (1995). Denaturant *m* values and heat capacity changes: Relation to changes in accessible surface area of protein unfolding. *Protein Sci.* **4,** 2138–2148.

Nojima, H., Ikai, A., Oshima, T., and Noda, H. (1977). Reversible thermal unfolding of thermostable phosphoglycerate kinase. Thermostability associated with mean zero enthalpy change. *J. Mol. Biol.* **116,** 429–442.

Pace, C. N. (1986). Determination and analysis of urea and guanidine hydrochloride denaturation curves. *Methods Enzymol.* **14,** 266–280.

Petri, V., Hsieh, M., and Brenowitz, M. (1995). Thermodynamic and kinetic characterization of the binding of the TATA binding protein to the adenovirus E4 promoter. *Biochemistry* **34,** 9977–9984.

Petsko, G. A. (2001). Stuctural basis of thermostability in hyperthermophilic proteins, or "there's more than one way to skin a cat. *Methods Enzymol.* **334,** 469–478.

Privalov, P. L. (1990). Cold denaturation of proteins. *Crit. Rev. Biochem. Mol. Biol.* **25,** 281–305.

Razvi, A., and Scholtz, J. M. (2006). Lessons in stability from thermophilic proteins. *Protein Sci.* **15,** 1569–1578.

Record, M. T., Ha, J.-H., and Fisher, M. A. (1991). Analysis of equilibrium and kinetic measurements to determine thermodynamic origins of stability and specificity and mechanism of formation of site-specific complexes between proteins and helical DNA. *Methods Enzymol.* **208,** 291–343.

Roberston, A. D., and Murphy, K. P. (1997). Protein structure and the energetics of protein stability. *Chem. Rev.* **97,** 1251–1267.

Santoro, M. M., and Bolen, D. W. (1988). Unfolding free energy changes determined by the linear extrapolation method: 1. Unfolding of phenylmethanesulfonyl alpha-chymotrypsin using different denaturants. *Biochemistry* **27,** 8063–8068.

Spolar, R. S., and Record, M. T. (1994). Coupling of local folding to site-specific binding of proteins to DNA. *Science* **263,** 777–784.

Spolar, R. S., Livingstone, J. R., and Record, M. T. (1992). Use of liquid hydrocarbon and amide transfer data to estimate contributions to thermodynamic functions of protein folding from the removal of nonpolar and polar surface from water. *Biochemistry* **31,** 3947–3955.

Sterner, R., and Liebl, W. (2001). Thermophilic adaptation of proteins. *Crit. Rev. Biochem. Mol. Biol.* **36,** 39–106.

Szilagyi, A., and Zavodsky, P. (2000). Structural differences between mesophilic, moderately thermophilic, and extremely thermophilic protein subunits: Results of a comprehensive survey. *Structure* **8,** 493–504.

CHAPTER TEN

Application of the Sequential n-Step Kinetic Mechanism to Polypeptide Translocases

Aaron L. Lucius, Justin M. Miller, *and* Burki Rajendar

Contents

Abstract

Clp/Hsp100 proteins are essential motor proteins in protein quality control pathways in all organisms. Such enzymes couple the energy derived from ATP binding and hydrolysis to translocate and unfold polypeptide substrates. Often they perform this role in collaboration with proteases for protein removal or with other chaperones for protein disaggregation. Unlike other well-characterized motor proteins, fundamental parameters such as the microscopic rate constants and overall rate of translocation, step-size (amino acids translocated per step), processivity, and directionality are not available for many of these enzymes. We have recently developed a fluorescence stopped-flow method

Department of Chemistry, The University of Alabama at Birmingham, Birmingham, Alabama, USA

Methods in Enzymology, Volume 488
© 2011 Elsevier Inc.
ISSN 0076-6879, DOI: 10.1016/B978-0-12-381268-1.00010-0
All rights reserved.

to elucidate these fundamental mechanistic details. In addition, we have developed a quantitative method to examine the single-turnover time courses that result from the rapid mixing experiments. With these two advances in hand, we have recently reported the first determination of the microscopic rate constants, overall rate of translocation, kinetic step-size, and processivity for the *E. coli* ClpA polypeptide translocase. Here, we report a description of both the fluorescence stopped-flow method to examine the mechanism of enzyme catalyzed polypeptide translocation and the mathematics required to quantitatively examine the resulting time courses.

1. Introduction

Motor proteins that translocate directionally on a linear lattice have been studied extensively. Such motor proteins couple the energy derived from NTP binding and hydrolysis to mechanical movement. Enzymes such as kinesin translocate directionally on microtubules (Block *et al.*, 1990; Howard *et al.*, 1989; Vale and Fletterick, 1997). Nucleic acid polymerases (Kornberg and Baker, 1992) and helicases (Lohman and Bjornson, 1996; Lohman *et al.*, 2008; Matson and Kaiser-Rogers, 1990; Pyle, 2008) translocate directionally on linear nucleic acid filaments. Clp/Hsp100 proteins translocate directionally on polypeptide chains while disrupting protein structure (Hoskins *et al.*, 2000; Kim *et al.*, 2000; Rajendar and Lucius, 2010; Reid *et al.*, 2001).

A complete characterization of the molecular mechanism of translocation on a linear lattice requires the determination of several fundamental physical parameters. These parameters include the microscopic rate constants and overall rate of translocation, the distance traveled per translocation cycle (step-size), processivity, and the amount of NTP used per translocation step (coupling efficiency).

Elucidation of such parameters for nucleic acid motors like helicases and polymerases, and other enzymes like myosin and kinesin has long been an active area of research. As such, these fundamental kinetic parameters have been elucidated for many of these enzymes (Ali and Lohman, 1997; Jankowsky *et al.*, 2000; Svoboda *et al.*, 1993; Uyeda *et al.*, 1990). In contrast, until recently, quantitative estimates of many of these fundamental physical parameters had not been reported for Clp/Hsp100 enzymes (Rajendar and Lucius, 2010). Here, we report a detailed description of both the fluorescence stopped-flow method to examine the mechanism of enzyme catalyzed polypeptide translocation and the mathematics required to quantitatively examine the resulting time courses.

2. Single-Turnover Fluorescence Stopped-Flow Method to Monitor Polypeptide Translocation

A major difficulty in examining the mechanism of polypeptide translocation by Clp/Hsp100 enzymes like *Escherichia coli* ClpA, ClpB, or ClpX is that the substrate enters and leaves the reaction without being covalently modified. ClpA, for example, binds a polypeptide substrate, directionally translocates the substrate and then releases the substrate with no covalent change in the substrate. If the polypeptide being translocated is a folded protein upon entering the reaction, then the only change in the substrate that is induced by the protein translocase is the disruption of protein structure. However, upon release, the protein substrate, most likely, promptly refolds making it difficult to monitor how the enzyme has transiently affected the protein secondary structure.

To overcome these difficulties we have developed a single-turnover fluorescence stopped-flow method to examine polypeptide translocation catalyzed by protein unfoldases that do not covalently modify the substrate they translocate. Although this has been developed and applied to *E. coli* ClpA catalyzed polypeptide translocation, in principle, this approach is applicable to a variety of molecular chaperones that directionally translocate on a linear lattice.

The strength of the single-turnover method is that it allows for the examination of the first turnover of translocation in the absence of any binding or rebinding of the enzyme. Therefore, single-turnover experiments are sensitive to the microscopic rate constants that govern the reaction. This is in stark contrast to multiple-turnover or steady-state kinetic measurements, where multiple rounds of dissociation, binding, and catalysis occur. Such steady-state measurements are a reflection of the slowest step in the repeating cycles of binding, catalysis, and dissociation. Often, the slowest step is binding or even macromolecular assembly. Thus, interpreting the steady-state kinetic parameters in terms of microscopic rate constants is difficult. More importantly, elucidating parameters such as the kinetic step-size (average number of amino acids translocated per step) and processivity (the probability that the enzyme will translocate vs. dissociate) that are essential for our understanding of any motor protein is impossible.

Two strategies are employed to insure that the observed kinetic time courses are not sensitive to binding or rebinding of the enzyme. First, the enzyme and a fluorescently modified substrate are premixed to allow for binding equilibrium to be achieved before the reaction is initiated with ATP. This insures that the acquired kinetic time course will not reflect any bimolecular steps and will only be sensitive to the events in the active site of the enzyme, thus simplifying the kinetic model. Second, after the enzyme and substrate have achieved binding equilibrium the translocation reaction is initiated by rapidly mixing with ATP and a large excess of enzyme trap, see Fig. 10.1. This enzyme trap is anything that can serve to bind specifically to the enzyme in the substrate binding site that will inhibit rebinding of the fluorescently modified substrate. For example, one choice would be identical substrate without fluorescent modification. In summary, by pre-binding the enzyme to the substrate and including an enzyme trap, the observed signal is only a reflection of translocation catalyzed by enzyme that was bound to the substrate at time zero in the absence of binding or rebinding. Thus, any kinetic mechanism that will describe the time course that results upon rapid mixing with ATP will not contain any bimolecular steps thereby intensely simplifying the system of differential equations that describe the reaction (see Section 3).

Traditionally, single-turnover experiments are defined as maintaining the enzyme in large excess over the substrate. Although enzyme is often in large excess over the substrate in our experimental design, because a trap for free enzyme is included, no rebinding of the protein occurs (see Section 2.1). Thus, the single-turnover method presented here does not, necessarily, require the enzyme be in large excess over the substrate.

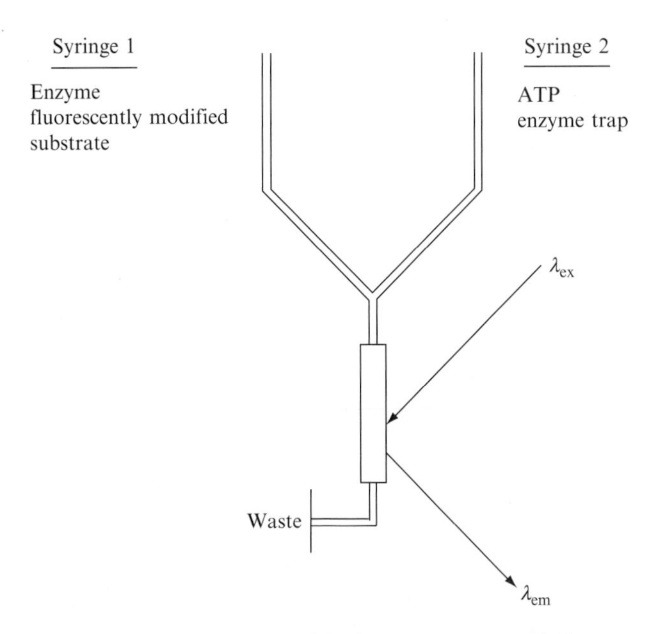

Figure 10.1 Schematic representation of single-turnover stopped-flow experiment. In Syringe 1, the enzyme and fluorescently modified substrate are preincubated. Syringe 2 contains ATP and a trap for free enzyme. The two are rapidly mixed together in the chamber, where the reactants are irradiated by light at a wavelength, λ_{ex} and emission is observed by a photomultiplier tube at λ_{em}.

2.1. Substrate design

The first task at hand in the development of a single-turnover stopped-flow experiment to monitor translocation is the development of a suitable substrate. Some knowledge about where the enzyme binds and initiates translocation on the substrate is necessary. The question is; does the enzyme bind and initiate at a specific site or randomly along the entire substrate? The case of random binding has been described for single stranded DNA translocation catalyzed by the UvrD helicase and will not be discussed here (Fischer and Lohman, 2004; Fischer *et al.*, 2004). This is because, in the case of ClpA, the enzyme is known to bind specifically to the 11 amino acid (AA) SsrA sequence (AANDENYALAA) placed at the carboxy terminus of protein substrates (Piszczek *et al.*, 2005). Thus, we synthesized a series of polypeptide substrates of various lengths each containing the SsrA sequence at the carboxy terminus, see Table 10.1. For the purposes of fluorescent modification we placed a single cysteine residue at the carboxy terminus. The single cysteine reside can then be fluorescently modified with the maleimide functional group that is commercially available on many fluorophores such as fluorescein-maleimide or Cy3-maleimide.

There are a plethora of fluorophores that the experimenter can choose from. Fluorescence is intensely sensitive to the environment. Thus, the predominant criterion in selecting a fluorophore is finding one that is sensitive to the presence of the protein when bound to the substrate. We have investigated two different fluorophores that react in completely opposite ways. First, we have used Fluorescein-maleimide to fluorescently label our peptides. Polypeptide substrates that contain fluorescein exhibit a fluorescence quenching of the fluorescein when ClpA is bound to the polypeptide substrate. Thus, a distinct fluorescence enhancement is observed when prebound ClpA either dissociates from the substrate or translocates the substrate and subsequently dissociates. In contrast, the identical substrates labeled with Cy3-maleimide exhibit a fluorescence enhancement when ClpA is bound and a loss of signal upon ClpA dissociation.

The dependence of the observed kinetics on substrate length is essential for any experimental design that is aimed at determining the molecular mechanism for a translocating enzyme (see Section 3). With this in mind a series of substrates that ranged in length from the 11 AA SsrA sequence up to 50 AA were synthesized. As can be seen in Table 10.1, the carboxy terminus always contains the 11 AA SsrA binding sequence. From the SsrA sequence the substrate is extended at the amino-terminus. The rational for this substrate design, for ClpA, is that we anticipate that it will translocate from the carboxy terminal binding site to the amino-terminus based on previous work performed with ClpAP (Reid *et al.*, 2001). However, we will also discuss a method for examining directionality (see Section 3.6).

Table 10.1 Polypeptide translocation substrates

Substrate	Name	Length (AA)	Sequence
I	N-Cys-30	30	CTKSAANLKVKELRSKKKL**AANDENYALAA**
II	N-Cys-40	40	CTGEVSFQAANTKSAANLKVKELRSKKKL**AANDENYALAA**
III	N-Cys-50	50	CLILHNKQLGMTGEVSFQAANTKSAANLKVKELRSKKKL**AANDENYALAA**
IV	C-Cys-30	30	TKSAANLKVKELRSKKKL**AANDENYALAA**C
V	C-Cys-40	40	TGEVSFQAANTKSAANLKVKELRSKKKL**AANDENYALAA**C

2.2. Enzyme trap

As discussed above and shown in Fig. 10.1, a trap for free enzyme is included to maintain single-turnover conditions. As mentioned, the trap serves to insure that rebinding of the enzyme to the fluorescently modified substrate cannot occur. The 11 AA SsrA peptide is a logical choice for a trap for ClpA since the enzyme binds specifically to this substrate. However, in principle, any substrate that would efficiently compete for binding with the fluorescently modified substrates presented in Table 10.1 could serve as a trap. For example, one could always use a 100-fold excess of the sequence contained within the fluorescently modified substrate, but without the fluorophore. In this scenario, it could be assumed that, at most, only 1% rebinding could occur. In our case, the polypeptide substrates contain a cysteine that would be reactive and this would not be a practical choice. Moreover, synthetic polypeptide substrates become inordinately expensive as the length is increased. Thus, to use any polypeptide substrate above 20 AA as a trap would not be a practical choice.

Single-turnover experiments performed to examine single-stranded DNA translocation and double stranded DNA unwinding catalyzed by DNA helicases have employed heparin as a trap (Fischer and Lohman, 2004; Lucius *et al.*, 2004). Heparin is a long polyanion and is, therefore, a good mimic of single stranded DNA. Because ClpA binds to a heparin column during protein purification, we attempted to use heparin as a trap. However, at the highest concentrations achievable, heparin does not efficiently compete for ClpA binding.

The obligatory rapid mixing kinetic experiment that is performed to test for adequate trapping is shown in Fig. 10.2. The schematic representation shown in Fig. 10.2 is essentially identical to the standard translocation experiment shown in Fig. 10.1. In the standard translocation experiment shown in Fig. 10.1, the enzyme at concentration "x" is prebound to the substrate at concentration "y." This complex is rapidly mixed with ATP and trap. The question to be answered is; what concentration of trap is required to inhibit the "x" concentration of enzyme from binding to the "y" concentration of substrate when the enzyme encounters the substrate in an excess concentration of enzyme trap? To answer this question one only need move the "y" concentration of fluorescently modified substrate into Syringe 2, where ATP and "z" concentration of trap is present. The two reactants in Fig. 10.2 can then be rapidly mixed at various concentrations of trap until no signal change is observed. At the concentration of trap where no signal change is observed, there can be no rebinding of the substrate by the enzyme at these concentrations of trap, substrate, and enzyme. Once established the standard translocation experiment illustrated in Fig. 10.1 can be performed at the determined concentration of trap. In our examination of ClpA we found that a final mixing concentration of 100 μM SsrA

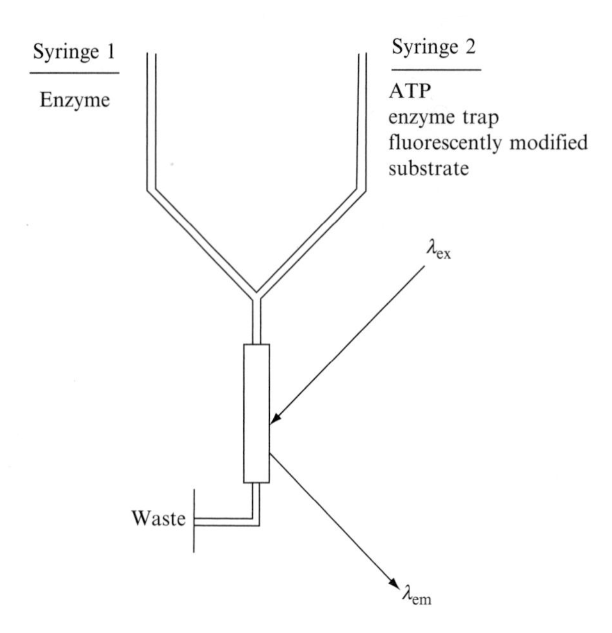

Figure 10.2 Schematic representation of "Trap-Test." Schematic shows the standard method for determining if the enzyme trap is effective at inhibiting binding to the substrate. The experimental design is the same as Fig. 10.1, with the exception that the fluorescently modified substrate has been moved to Syringe 2. The two reactants are rapidly mixed together at increasing concentrations of enzyme trap until no signal is observed.

substrate is sufficient for inhibiting 500 n*M* ClpA monomer from binding 50 n*M* fluorescently modified substrate.

3. Application of the Sequential n-Step Mechanism

The ClpA polypeptide translocase binds specifically to the carboxy terminus of the model substrates presented in Table 10.1. Since ClpA requires nucleoside triphosphate binding to assemble into a hexamer with polypeptide binding activity, ATPγS is included in Syringe 1 to preassemble and bind ClpA to the model substrate. The single-turnover polypeptide translocation experiment illustrated by Fig. 10.1 is performed by preincubating ClpA with each of the model substrates in Table 10.1 and subsequently rapidly mixing with ATP and enzyme trap. When this experiment is performed in this manner, a distinct lag in the fluorescence signal is observed.

The presence of the observed lag phase followed by a fluorescence increase indicates that the fluorescence remains constant for a period of time followed by enzyme dissociation, see Fig. 10.3. In order to observe a lag phase under single-turnover conditions the enzyme must cycle through at least two steps with similar rate constants before dissociation. Although counter intuitive, a lag phase will not be observed due to a single slow step. Rather, a single slow step will be described by a single exponential function that intersects the origin.

3.1. n-step mechanism with equivalent rate-constants

The simplest model that will give rise to a lag phase that incorporates the experimental design described above and illustrated in Fig. 10.1 is given by Scheme 10.1. In Scheme 10.1, E denotes the enzyme and P denotes the peptide substrate. Thus, $(E \cdot P)_L$ denotes the prebound enzyme–peptide complex with substrate of length, L, which would be present in Syringe 1, see Fig. 10.1. Upon rapid mixing with ATP, the enzyme will proceed through a translocation step with rate constant k_T to form the first intermediate, $I_{(L-m)}$, where L is the substrate length and m is the kinetic step-size (average number of amino acids translocated per step). The enzyme will cycle through n steps with rate constant k_T until reaching the end and releasing the unchanged peptide substrate.

Many events must occur for every step that the enzyme translocates forward. At a minimum this would include ATP binding, hydrolysis, mechanical movement, various conformational changes and ADP + P_i release. The mechanism presented in Scheme 10.1 assumes that a single step within each repeating cycle is rate-limiting. Thus, each step in Scheme 10.1 is considered to be the same repeating step. In the upcoming sections we will present a method for testing this assumption.

To either model experimental time courses or simulate the behavior of the various models, the first task is to determine an equation that describes the fraction of peptide released as a function of time, $f_P(t)$, defined by Eq. (10.1).

$$f_P(t) = \frac{P(t)}{(E{\cdot}P)_0} \tag{10.1}$$

where $(E \cdot P)_0$ is the concentration of enzyme peptide complex at $t = 0$, and $P(t)$ is the concentration of peptide released as a function of time. The system of coupled differential equations that results from Scheme 10.1 are solved using the Laplace transform method as previously described (Lucius *et al.*, 2003). The strength of using this method is that it reduces the system of coupled differential equations to a system of coupled algebraic equations that can be solved using matrix methods. Moreover, the resulting Laplace

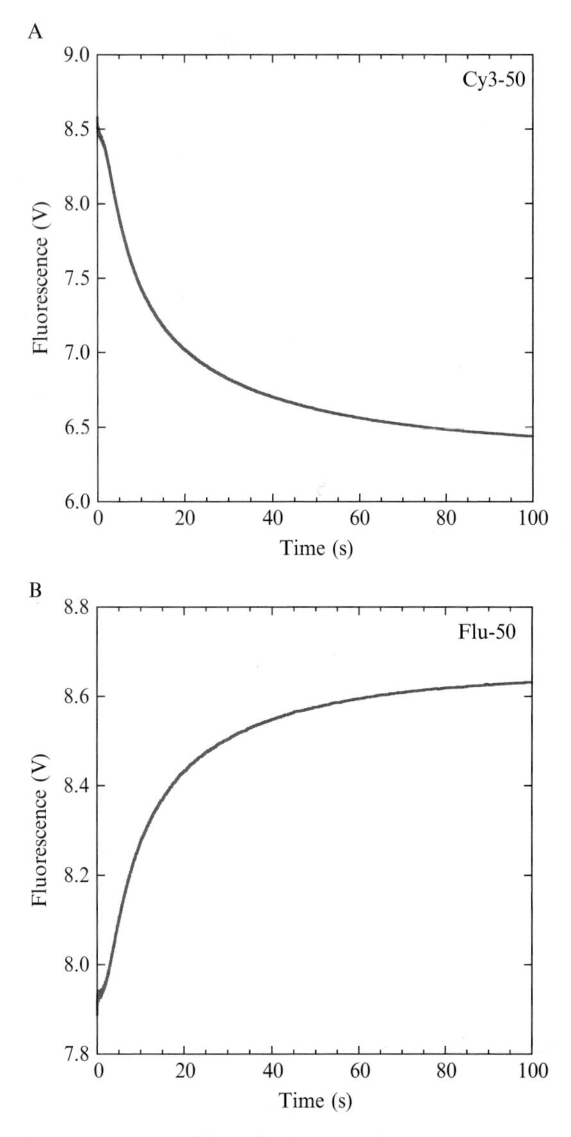

Figure 10.3 Representative time courses from single-turnover stopped-flow fluorescence experiment. Shows a representative time course for a stopped-flow fluorescence experiment performed as shown schematically in Fig. 10.1 with 1 m*M* ClpA, 100 n*M* (A) Cy3-50 or (B) Fluorescein-50 in Syringe 1 and rapidly mixing with 10 m*M* ATP and 300 μ*M* SsrA peptide. All concentrations are syringe concentrations and the final mixing concentration is twofold lower.

$$(E \bullet P)_L \xrightarrow{k_T} I_{(L-m)} \xrightarrow{k_T} \dots I_{(L-im)} \dots \xrightarrow{k_T} I_{(L-(n-1)m)} \xrightarrow{k_T} E+P$$

Scheme 10.1 Sequential n-step mechanism with all steps having the same rate constant.

transform of $f_P(t)$ is a continuous function of the number of steps, n, and is given by Eq. (10.2).

$$\mathcal{L}f_P(t) = F_P(s) \tag{10.2}$$

where $\mathcal{L}$ is the Laplace transform operator, and $F_P(s)$ is the Laplace transform of $f_P(t)$. For Scheme 10.1, the resulting Laplace transform of the fraction of peptide released as a function of time, $f_P(t)$, is given by Eq. (10.3).

$$F_P(s) = \frac{k_T^n}{s(k_T + s)^n} \tag{10.3}$$

where s is the Laplace variable. In order to analyze experimental time courses one must determine $f_P(t)$, which is accomplished by finding the inverse Laplace transform of $F_P(s)$ as described by Eq. (10.4).

$$\mathcal{L}^{-1}F_P(s) = f_P(t) \tag{10.4}$$

where $\mathcal{L}^{-1}$ is the inverse Laplace transform operator. Traditionally, the inverse Laplace transform can be found by consulting tables of Laplace transforms. However, these tables are also present in most modern symbolic mathematics software packages such as Mathematica, Maple, MathCad, etc.

Although for Scheme 10.1 a closed form expression of $f_P(t)$ is easily found using Eqs. (10.3) and (10.4) and has been reported previously (Lucius *et al.*, 2003), a solution is not possible for most schemes. Thus, for the analysis of experimental time courses or to simulate time courses, we numerically solve Eq. (10.4). For example, experimental time courses were simulated by combining Eqs. (10.3) and (10.4) for Scheme 10.1 with $k_T = 1.67\ s^{-1}$ and $n = 1 - 5$, see Fig. 10.4. As seen in Fig. 10.4, for one step, $n = 1$, a time course that can be described by a single exponential function is observed. However, as the number of steps, n, increases the extent of the lag increases. Thus, Scheme 10.1 adequately describes a lag phase that increases with increasing numbers of steps.

Although Scheme 10.1 fulfills the requirement of describing an increasing lag phase with increasing numbers of steps, it does not adequately describe the shape of the curves that are experimentally observed. As seen in Fig. 10.3 for ClpA catalyzed polypeptide translocation, the time courses clearly exhibit biphasic kinetics. The time course in Fig. 10.3 exhibits a

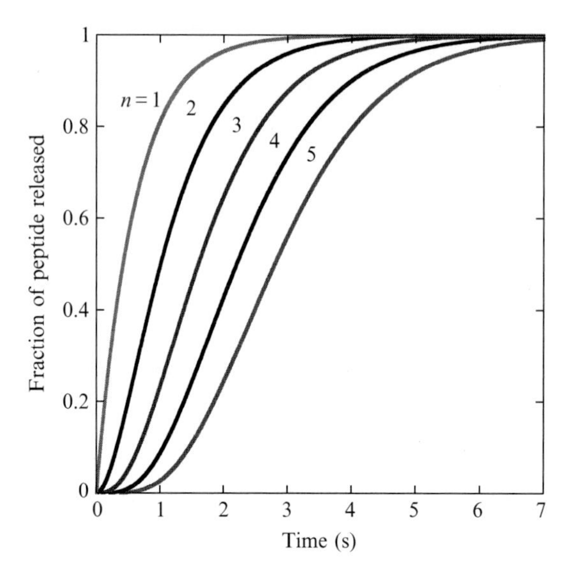

Figure 10.4 Simulated time courses from Scheme 10.1. Simulated time courses were generated by numerically solving the inverse Laplace transform for fraction of peptide released as a function of time, $f_P(t)$, for Scheme 10.1. This was accomplished using Eqs. (10.4) and (10.3) with $k_T = 1.67\ s^{-1}$ and $n = 1 - 5$. Plot illustrates an increase in the extent of the lag with increasing numbers of steps, n.

$$
\begin{array}{l}
(E \bullet P)_{NP} \\
\downarrow k_{NP} \\
(E \bullet P)_L \xrightarrow{k_T} I_{(L-m)} \xrightarrow{k_T} \ldots.\ I_{(L-im)} \ldots. \xrightarrow{k_T} I_{(L-(n-1)m)} \xrightarrow{k_T} E + P
\end{array}
$$

Scheme 10.2 Sequential n-step mechanism with all steps having the same rate constant with inclusion of slow isomerization step, k_{NP}.

rapid phase that is complete in ~5 s followed by a slow second phase that is complete in greater than 100 s. Since the signal is only sensitive to enzyme that was bound at time zero, the simplest explanation for biphasic kinetics is that there are two conformations of enzyme bound at time zero. Scheme 10.2 describes a scenario where the enzyme peptide complex can exist in a nonproductive conformation, $(E \cdot P)_{NP}$, that must proceed through a slow isomerization step with rate constant k_{NP} to form the productive complex, $(E \cdot P)_L$, that can immediately initiate translocation.

3.2. Biphasic kinetics

Using the same Laplace transform approach described above, we derived an equation for $F_P(s)$ for Scheme 10.2, Eq. (10.5).

$$F_P(s) = \frac{k_T^n(k_{NP} + sx)}{s(k_T + s)^n(k_{NP} + s)} \tag{10.5}$$

where x is defined as the fraction of productively bound complexes, which is given by Eq. (10.6).

$$x = \frac{[(E{\cdot}P)_L]}{[(E{\cdot}P)_L] + [(E{\cdot}P)_{NP}]} \tag{10.6}$$

Figure 10.5A shows simulations using Scheme 10.2 (Eqs. (10.4) and (10.5)) with $k_T = 1.67\ s^{-1}$, $k_{NP} = 0.167\ s^{-1}$, $x = 0.5$, and $n = 1 - 5$. By comparing Figs. 10.4–10.5A, it can be seen that the time courses simulated using Scheme 10.2 (Fig. 10.5A) clearly exhibit biphasic kinetics in contrast to the time courses simulated from Scheme 10.1 (Fig. 10.4), which exhibit only a single phase. Thus, Scheme 10.2 adequately describes the macroscopic features of the experimental time courses shown in Fig. 10.3, which is a lag phase exhibiting biphasic kinetics.

Figure 10.5B is a series of simulations that were performed using Scheme 10.2 (Eqs. (10.4) and (10.5)) with $k_T = 1.67\ s^{-1}$, $k_{NP} = 0.167\ s^{-1}$, and $n = 5$, with varying values of the fraction of productively bound complexes. The simulations show the effect that the various values of x have on the shape of the kinetic time courses. When the fraction of productively bound complexes, x, is equal to unity, the time course is identical to the time course in Fig. 10.4 for $n = 5$. This is the expected result since when $x = 1$ all of the complexes initiate from the productively bound state and Scheme 10.2 collapses to Scheme 10.1. This can also be observed by recognizing that once x is set to 1 in Eq. (10.5), which is the Laplace transform of the time dependent equation describing peptide release for Scheme 10.2, Eq. (10.5) simplifies to Eq. (10.3), which is the Laplace transform of the time dependent equation describing peptide release for Scheme 10.1.

In contrast to Fig. 10.5B, Fig. 10.5C shows a series of simulations for peptide release from Scheme 10.2 (Eqs. (10.4) and (10.5)) with $k_T = 1.67\ s^{-1}$, $x = 0.5$, and $n = 5$, with varying values of k_{NP}. What is observed is that when k_{NP} is faster than k_T, then the time course is identical to the time course simulated from the simplest model, Scheme 10.1, Fig. 10.4 $n = 5$. This makes sense because the limit as k_{NP} goes to infinity of Eq. (10.5) is Eq. (10.3), which describes Scheme 10.1, see Eq. (10.7).

$$\lim_{k_{NP}\to\infty} \frac{k_T^n(k_{NP} + sx)}{s(k_T + s)^n(k_{NP} + s)} = \frac{k_T^n}{s(k_T + s)^n} \tag{10.7}$$

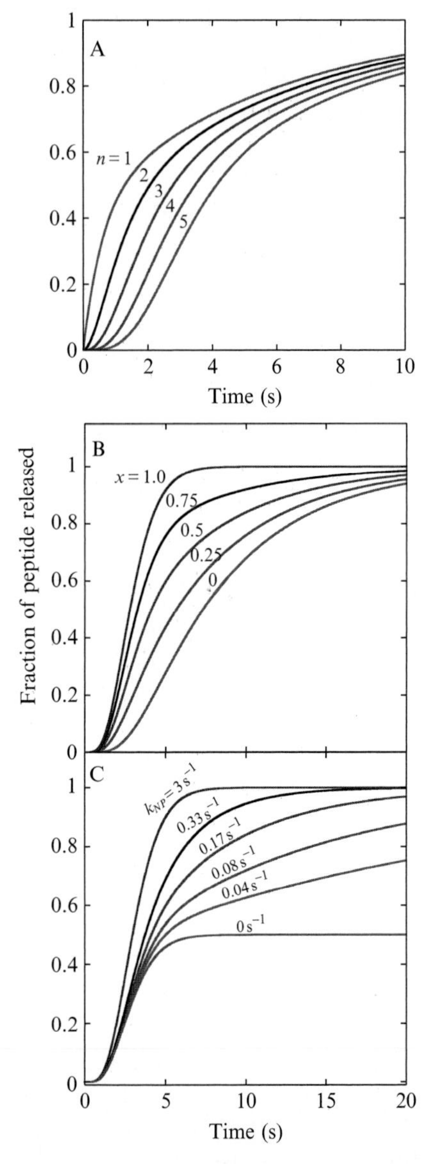

Figure 10.5 Simulated time courses from Scheme 10.2. Simulated time courses were generated by numerically solving the inverse Laplace transform for fraction of peptide released as a function of time, $f_P(t)$, for Scheme 10.2 using Eqs. (10.4) and (10.5). (A) $k_T = 1.67\ s^{-1}$, $k_{NP} = 0.167\ s^{-1}$, $x = 0.5$, and $n = 1 - 5$. (B) $k_T = 1.67\ s^{-1}$, $k_{NP} = 0.167\ s^{-1}$, $n = 5$, and $x = 0 - 1$. (C) $k_T = 1.67\ s^{-1}$, $x = 0.5$, $n = 5$, and $k_{NP} = 0 - 3\ s^{-1}$.

However, when k_{NP} is slower than k_T, a second slower phase is clearly apparent. Finally, when $k_{NP} = 0$, the fraction of productively bound complexes becomes little more than an amplitude term. This can be seen mathematically by taking the limit of Eq. (10.5) as k_{NP} goes to zero.

$$\lim_{k_{NP}\to 0} \frac{k_T^n(k_{NP} + sx)}{s(k_T + s)^n(k_{NP} + s)} = \frac{k_T^n x}{s(k_T + s)^n} = x\frac{k_T^n}{s(k_T + s)^n} \tag{10.8}$$

where x in Eq. (10.8) is a constant and thus is treated as a scalar multiplier of Eq. (10.3). That is to say, when the Inverse Laplace transform of the result in Eq. (10.8) is determined the constant, x, is moved out in front of the operator as shown in Eq. (10.9).

$$\mathcal{L}^{-1}F_P(s) = \mathcal{L}^{-1}x\frac{k_T^n}{s(k_T + s)^n} = x\mathcal{L}^{-1}\frac{k_T^n}{s(k_T + s)^n} = xf_P(t) \tag{10.9}$$

This can be seen, graphically, in Fig. 10.5C. In Fig. 10.5C, when $x = 0.5$ and $k_{NP} = 0$ the time course is scaled down from 1 to 0.5. Thus, at these limits, Scheme 10.2 describes a scenario where some fraction of enzyme binds but never initiates translocation.

In summary, both Schemes 10.1 and 10.2 have been introduced and discussed previously (Lucius *et al.*, 2003). In fact, Scheme 10.2 has been used extensively to model helicase catalyzed DNA unwinding since biphasic kinetics, to our knowledge, has always been observed in single-turnover DNA unwinding experiments (Ali and Lohman, 1997; Jankowsky *et al.*, 2000; Lucius *et al.*, 2002). Although the molecular explanation for why a second phase is observed in these experiments is not clear, it is interesting that polypeptide translocases and helicases both exhibit such kinetic behavior.

3.3. Finite processivity

The fluorescence stopped-flow method that we have developed and applied to polypeptide translocases is sensitive to the dissociation of the motor from the linear lattice. For example, when ClpA is bound to the fluorescently modified peptide substrate containing fluorescein, there is a fluorescence quenching. Upon rapidly mixing with ATP and enzyme trap an increase in fluorescence is observed upon dissociation. To describe this, we introduce Scheme 10.3 that includes a dissociation step with rate constant k_d at each intermediate. Because signal is observed at each intermediate dissociation step, the equation that describes product formation is substantially more complicated. That is to say, the equation,$F_P(s)$, for Scheme 10.3 has a contribution from dissociation at each step as well as dissociation from

$$
\begin{array}{l}
(E\bullet P)_{NP} \\
\downarrow k_{NP} \\
(E\bullet P)_L \xrightarrow{k_T} I_{(L-m)} \xrightarrow{k_T} \ldots I_{(L-im)}\ldots \xrightarrow{k_T} I_{(L-(n-1)m)} \xrightarrow{k_T} E+P \\
\downarrow k_d \qquad \downarrow k_d \qquad \downarrow k_d \qquad \downarrow k_d \\
\mathrm{P} \qquad P \qquad P \qquad P
\end{array}
$$

Scheme 10.3 Incorporation of dissociation.

the end. The Laplcace transform of $f_\mathrm{P}(t)$ for Scheme 10.3 is given by Eq. (10.10).

$$
F_\mathrm{P}(s) = \frac{\left(k_T^n(k_\mathrm{d}+k_T)(k_\mathrm{d}+s) + k_\mathrm{d}(k_\mathrm{d}+k_T+s)^n\left(k_T - \left(\frac{k_T}{k_\mathrm{d}+k_T+s}\right)^n(k_\mathrm{d}+k_T+s)\right)\right)(k_{NP}+sx)}{sk_T(k_\mathrm{d}+s)(k_{NP}+s)(k_\mathrm{d}+k_T+s)^n} \tag{10.10}
$$

Equation (10.10) collapses to Eq. (10.5), which describes the Laplace transform of $f_\mathrm{P}(t)$ for Scheme 10.2, as the k_d approaches 0, see Eq. (10.11).

$$
\begin{aligned}
&\lim_{k_\mathrm{d}\to 0} \frac{\left(k_T^n(k_\mathrm{d}+k_T)(k_\mathrm{d}+s) + k_\mathrm{d}(k_\mathrm{d}+k_T+s)^n\left(k_T - \left(\frac{k_T}{k_\mathrm{d}+k_T+s}\right)^n(k_\mathrm{d}+k_T+s)\right)\right)(k_\mathrm{NP}+sx)}{sk_T(k_\mathrm{d}+s)(k_{NP}+s)(k_\mathrm{d}+k_T+s)^n} \\
&= \frac{k_T^n(k_\mathrm{NP}+sx)}{s(k_T+s)^n(k_\mathrm{NP}+s)}
\end{aligned} \tag{10.11}
$$

Thus, there is a continuum of models from Schemes 10.3 to 10.1. That is to say, when $k_\mathrm{d} = 0$ in Eq. (10.10) (Scheme 10.3) it collapses to Eq. (10.5) (Scheme 10.2). When $x = 1$, which says that no enzyme binds in a nonproductive fashion, Eq. (10.5) (Scheme 10.2) collapses to Eq. (10.3) (Scheme 10.1).

By numerically solving $F_\mathrm{P}(s)$ for Scheme 10.3 (Eqs. (10.4) and (10.10)), we simulated a series of time courses with $k_\mathrm{T} = 1.67\ \mathrm{s}^{-1}$, $n = 5$, and varying values of k_d. As can be seen in Fig. 10.6, since complete dissociation is being monitored all of the simulated time courses proceed to unity, that is eventually everything dissociates. As expected, when $k_\mathrm{d} = 0$, the time course is equivalent to the time course simulated from Scheme 10.1 with $k_\mathrm{T} = 1.67\ \mathrm{s}^{-1}$ and $n = 5$, compare Fig. 10.1 $n = 5$ to Fig. 10.6 $k_d = 0$ and $n = 5$. Strikingly, when $k_\mathrm{d} = 0.08\ \mathrm{s}^{-1}$ the lag no longer has zero slope as it does for $k_\mathrm{d} = 0$, but has positive slope. Likewise, if k_d is increased to $0.17\ \mathrm{s}^{-1}$ then the slope in the lag region is further increased. Finally, if k_d is increased to $0.25\ \mathrm{s}^{-1}$ there is little ability to determine if there is any change in slope from the lag region to the rapid increase in signal. In fact, with the introduction of experimental error one would not be able to discern the

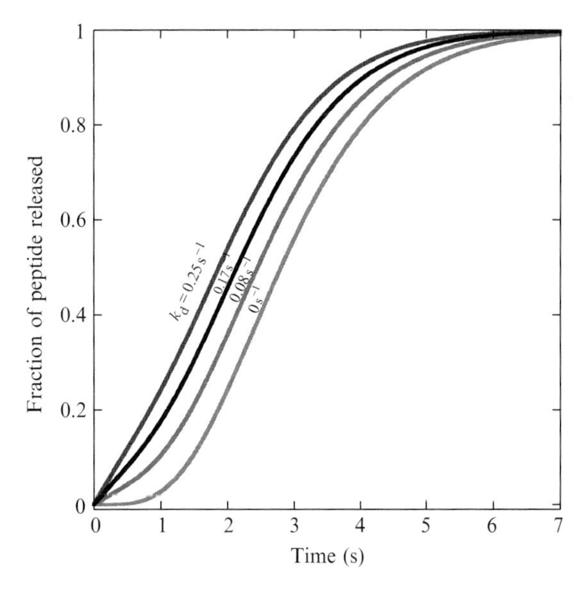

Figure 10.6 Simulated time courses from Scheme 10.3. Simulated time courses were generated by numerically solving the inverse Laplace transform for fraction of peptide released as a function of time, $f_P(t)$, for Scheme 10.3 using Eqs. (10.4) and (10.10). Time courses were simulated with $k_T = 1.67\ s^{-1}$, $n = 5$, and $k_d = 0 - 0.25\ s^{-1}$.

presence or absence of a lag. It should be noted that the disappearance of the lag at $k_d = 0.25\ s^{-1}$ is relative to $k_T = 1.67\ s^{-1}$. Which likely suggests that a discernable lag can only be observed if $k_d < \sim 0.15 k_T$.

With a measurement of the dissociation rate constant, k_d, a measure of the processivity of the enzyme is possible. The processivity, P_r, is defined as the probability that the enzyme will either proceed through the next step or dissociate and is defined by Eq. (10.12).

$$P_r = \frac{k_T}{k_T + k_d} \tag{10.12}$$

It can be easily seen from Eq. (10.12) that when $k_d = 0$ the processivity, $P_r = 1$, and this describes an infinitely processive enzyme, were every enzyme that binds fully translocates the substrate. In contrast, when k_d is finite P_r varies between 0 and 1 and describes finite processivity.

Interestingly, as stated above, in order to observe a lag phase for Scheme 10.3, k_d must be less than approximately $0.15 k_T$. By making this substitution into Eq. (10.12) a $P_r = 0.87$ results. This shows that the simple observation of a lag phase in the kinetic time course qualitatively suggests a minimum processivity of ~ 0.87, although the processivity may be much higher.

3.4. Determination of kinetic step-size

The kinetic step-size is defined as the average number of amino acids translocated per rate-limiting step. It is important to note that others have defined the step-size as the number of ATP molecules hydrolyzed per catalytic cycle. However, we define the number of ATP molecules hydrolyzed per catalytic cycle as the coupling efficiency and, here, the step-size will always be referred to as the number of amino acids translocated per step. To determine the kinetic step-size, m (average number of amino acids translocated between two rate-limiting steps), the dependence of the observed number of steps, n, on substrate length is examined. This is accomplished by subjecting the time courses to nonlinear-least-squares analysis using the models discussed.

Graphically, the information on the number of steps required to translocate a given substrate is contained within the extent of the lag. That is to say, as the enzyme proceeds through more steps a longer lag in the kinetic time course is expected. Single-turnover polypeptide translocation experiments are performed on multiple substrates that differ in length to determine if the extent of the lag increases with increasing length of substrate. In the case of ClpA, a distinct increase in the extent of the lag with increasing substrate length is observed. To determine the kinetic step-size, m, the relationship between the number of steps to describe each time course, n, and substrate length, L, must be determined. To do this, one must always collect time courses for multiple substrates since the number of steps, n, and the translocation rate constant, k_T, are highly correlated. Thus it would be impossible to have confidence in values of k_T and n or m if determined from a single time course collected for a single substrate of a given length, L. Therefore, the best way to determine a unique value for both of these parameters is through global nonlinear-least-squares analysis of a series of time courses collected for different lengths of substrates.

Upon collecting a series of time courses for various substrate lengths the first diagnostic examination of the data is done by subjecting the data to NLLS analysis using Scheme 10.1 if no second phase is present or Scheme 10.2 if a second phase is present. The inherent assumption that is made in doing this is that the observed steps all have the same rate constant. This analysis is done by first assuming that k_T is the same for all lengths, but each length is described by a different number of steps, n, with rate constant k_T. That is to say, k_T is a global parameter and n is a local parameter.

Figure 10.7 shows a series of time courses collected for the 30, 40, and 50 AA substrates with Fluorescein attached to the carboxy terminus, shown in Table 10.1. The solid lines are the result of a global NLLS analysis using Scheme 10.2, where the rate constants k_T and k_{NP} are constrained to be the same for all lengths and n is a local parameter for each length. Figure 10.7C shows a plot of the number of steps to describe each time course versus

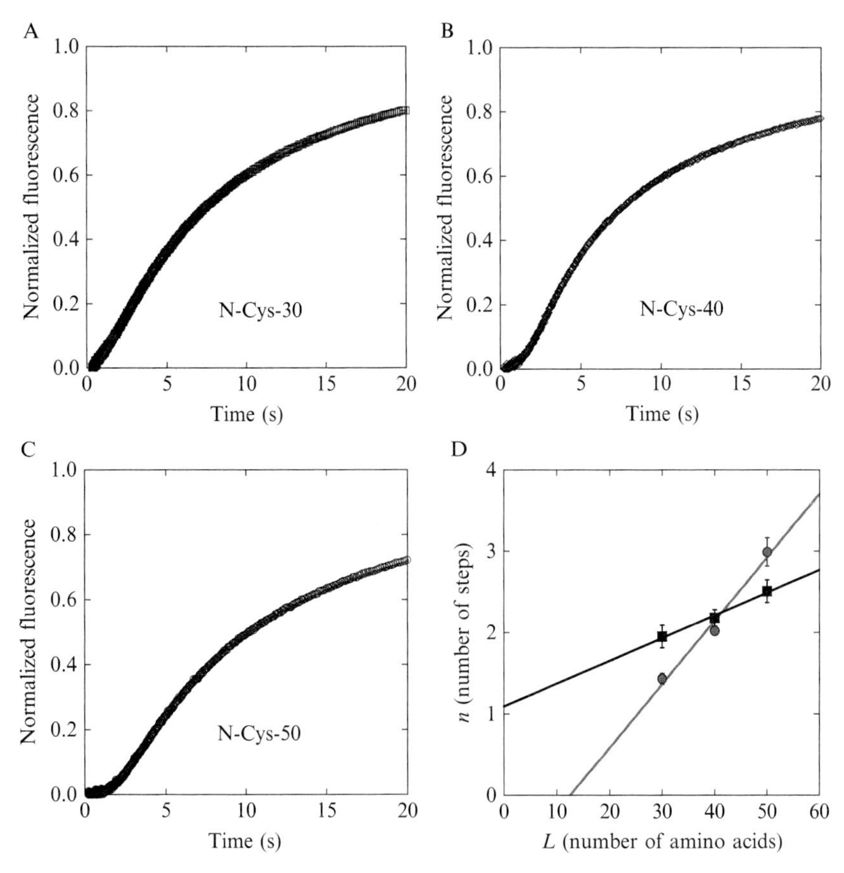

Figure 10.7 Fluorescence time courses for ClpA catalyzed polypeptide translocation. Time courses were collected as described in Fig. 10.1. (A) N-Cys-30, (B) N-Cys-40, (C) N-Cys-50. The solid lines in panels (A)–(C) represent a global NLLS fit using Scheme 10.4 (Eqs. (4) and (11)) for time courses collected with substrates I–III in Table 10.1. The resultant parameters are $k_T = (1.45 \pm 0.05)\ s^{-1}$, $k_C = (0.210 \pm 0.003)\ s^{-1}$, $k_{NP} = (0.0455 \pm 0.0005)\ s^{-1}$, $k_d = 0$, $m = (12.6 \pm 0.5)$ AA step^{-1}, $d = (11.7 \pm 0.4)$ AA. (E) Dependence on polypeptide length of the numbers of steps, n, determined from analysis of time courses presented in panels (B)–(D). Each time course was analyzed by constraining the parameters k_T, k_C, k_{NP}, and h to be global parameters, while A_t, x, and n were allowed to float for each time course. The solid squares represent a determination of the number of steps, n, required to describe each time course in panels (A)–(C) using Scheme 10.4 (Eqs. (10.4) and (10.11) with $h = 0$), that is to say no slow step with rate constant, k_C. The solid line through the solid squares represents a linear least squares fit with a slope = 0.028 and intercept = 1.09. The solid circles represent the analysis using Scheme 10.4 (Eqs. (10.4) and (10.11) with $h = 1$), that is to say one slow step with rate constant, k_C. The solid line through the solid circles represents a linear least squares fit with a slope = 0.078 and intercept = -0.973.

$$
\begin{array}{l}
(E \bullet P)_{NP} \\
\downarrow k_{NP} \\
(E \bullet P)_L \xrightarrow{k_c} (E \bullet P)_L^1 \xrightarrow{k_T} I_{(L-m)} \xrightarrow{k_T} \ldots\, I_{(L-im)} \ldots \xrightarrow{k_T} I_{(L-(n-1)m)} \xrightarrow{k_T} E + P \\
\downarrow k_d \qquad\qquad \downarrow k_d \qquad\qquad \downarrow k_d \qquad\qquad \downarrow k_d \\
\mathrm{P} \qquad\qquad\quad P \qquad\qquad\quad P \qquad\qquad\quad P
\end{array}
$$

Scheme 10.4 Incorporation of slow step with rate-constant k_C.

substrate length, L. If all of the observed rate-limiting steps are the same then a plot of n versus L should be linear with a zero intercept. That is to say, the observed number of steps should be related to substrate length by the equation $n = L/m$, where m is the kinetic step-size (average number of amino acids translocated between two rate-limiting steps). However, as can be seen in Fig. 10.7C, a positive y-intercept is observed. Thus, the simple linear equation, $n = L/m$, does not apply and a better description is $n = L/m + b$, where b describes an intercept term.

Qualitatively, the n versus L plot shown in Fig. 10.7C suggests that at a length of $L = 0$ there are still some number of steps taken by the enzyme. As we have shown previously, the observation of a positive y-intercept in an n versus L plot is a diagnostic observation indicating that there are additional steps not equal to k_T (Rajendar and Lucius, 2010). To describe this, we introduce Scheme 10.4, which includes an additional step with rate constant k_C. The equation that describes the Laplace transform of product formation as a function of time, $f_P(t)$, is given by Eq. (10.13).

$$
F_P(s) = \frac{\left(k_C^h k_T^n (k_d + k_T)(k_d + s) + k_d (k_d + k_T + s)^n (-k_T)(k_C + k_d + s)^h \left(\left(\frac{k_C}{k_C + k_d + s} \right)^h - 1 \right) + k_c^h \left(k_T - \left(\frac{k_T}{k_d + k_T + s} \right)^n (k_d + k_T + s) \right) \right)(k_{NP} + sx)}{s k_T (k_d + s)(k_{NP} + s)(k_C + k_d + s)^h (k_d + k_T + s)^n} \tag{10.13}
$$

where h is the number of steps with rate constant k_C. Strictly speaking, the Laplace transform of the equation that describes product formation as a function of time for Scheme 10.4 would have $h = 1$. However, we have derived Eq. (10.13) such that it can accommodate h number of steps with rate constant k_C. Thus, Eq. (10.13) represents a function that is continuous on both the number of steps with rate constant k_T, n, and the number of steps with rate constant k_C, h. Although Scheme 10.4 contains the step with rate constant k_C at the beginning of the reaction, because of the symmetry in Eq. (10.13), there is no information on whether the step is at the beginning, the end, or somewhere in the middle. To say it another way, if the step with rate constant k_C is placed at any position in the reaction scheme, given by Scheme 10.4, upon solving the system of couple differential equations the same solution given by Eq. (10.13) will result.

Conveniently, Eq. (10.13) exhibits the expected behavior at the extremes of the parameters. Namely, Eq. (10.13) collapses to Eq. (10.10), which describes Scheme 10.3, when $k_C \gg k_T$. Therefore, as discussed above, at the appropriate limits the Laplace transform of the equation that describes product formation for Scheme 10.3 can be used to describe Scheme 10.2 and subsequently Scheme 10.1. Thus, all four schemes presented here can be modeled by one equation, Eq. (10.13), which, in its current form, describes Scheme 10.4.

We simulated and analyzed a series of mock time courses to determine if time courses simulated using Scheme 10.4, which contains a step with rate constant k_C, and examined them using Scheme 10.3, which assumes all steps are the same, would yield a positive y-intercept in an n versus L plot, as observed experimentally (see Fig. 10.7D). These time courses were simulated by numerically solving Eqs. (10.4) and (10.13) with $n = 5$, $k_T = 1.67\ s^{-1}$, $k_d = 0$ and various values of k_C, see Fig. 10.8A. In this simulation, it was assumed that all bound enzyme started in the productive state by setting $x = 1$. Time courses were simulated and 1% error was introduced. The time courses were then subjected to NLLS analysis using Eqs. (10.4) and (10.10), which is the Laplace transform of $f_P(t)$ for Scheme 10.3. The global NLLS analysis was performed by constraining the rate constants to be the same for all time courses, that is global parameters, and the number of steps to be different for each time course, that is local parameters.

Figure 10.8B is a plot of the number of steps to describe each time course as a function of substrate length for various values of k_C. As expected, the number of steps increases linearly with increasing substrate length. Although the plot is linear, it clearly exhibits a positive y-intercept and therefore cannot be described by the relationship $n = L/m$. This result shows that when data comes from a model with an additional slow step, Scheme 10.4, upon analyzing this data with a simpler model, Scheme 10.3, the resulting n versus L plot will exhibit a positive y-intercept indicating the presence of the slow step.

Interestingly, the n versus L plots that were generated for different values of k_C all result in a y-intercept that is above one, but does not exceed two. Thus, the intercept represents a qualitative predictor of the number of k_C steps. As the value of the rate constant is increased the resulting n versus L plot becomes steeper. Thus, one cannot simply fit the line to an equation such as $n - L/m + b$ and determine the kinetic step-size, m as this would result in an underestimate. Clearly, there is some dependence of the slope on the value of the rate constant k_C. Finally, once the value of k_C exceeds that of k_T, the n versus L plot again intersects the origin. In summary, the n versus L plot should be used as a diagnostic plot to determine what model may better describe the experimental data.

Simulations were performed to determine the dependence of the y-intercept in the n versus L plot on the number of steps with rate constant

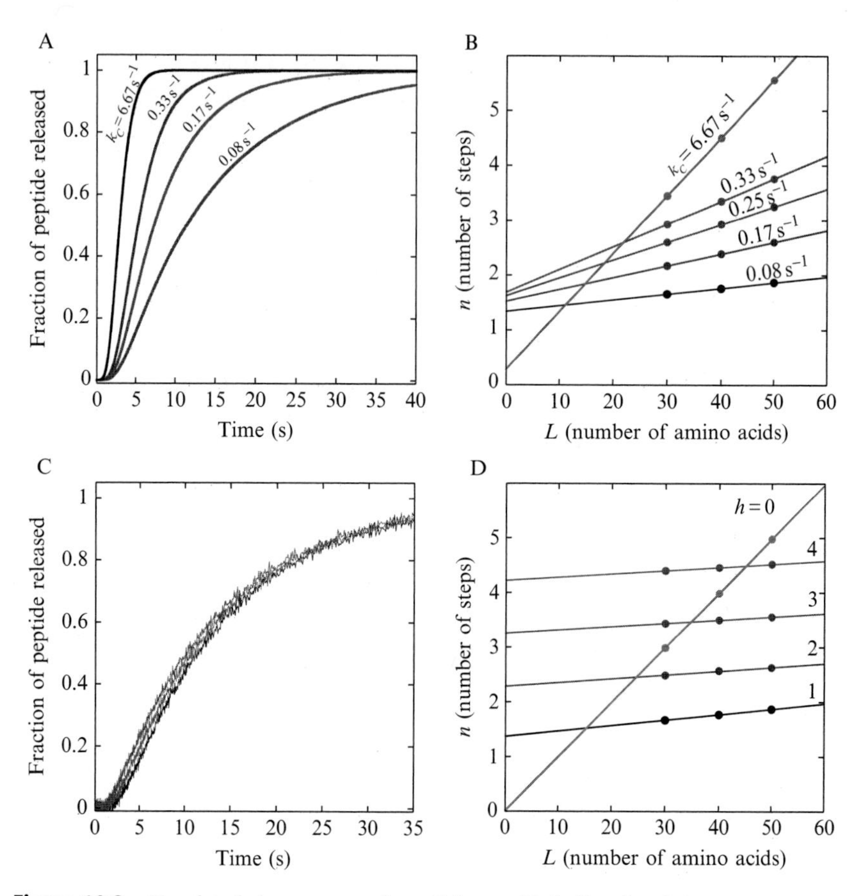

Figure 10.8 Simulated time courses from Scheme 10.4. Simulated time courses were generated by numerically solving the inverse Laplace transform for fraction of peptide released as a function of time, $f_P(t)$, for Scheme 10.4 using Eqs. (10.4) and (10.10). Time courses in panel (A) were simulated with $n = 5$, $k_T = 1.67\ s^{-1}$, $k_d = 0$ and various values of k_C. Simulations were performed for Scheme 10.4 using Eqs. (10.4) and (10.10) by replacing n with L/m, and $m = 10$ AA step^{-1}, $L = 30$, 40, and 50 AA, $k_T = 1.67\ s^{-1}$, $k_d = 0$ and various values of k_C. Time courses were subjected to global NLLS analysis using Scheme 10.3 and the number of steps, n, to describe each time course was determined. (B) A plot of the dependence of n on substrate length, L. (C) Representative set of time courses for varying values of $k_C = 0.08\ s^{-1}$. (D) Dependence of the number of steps to describe each time course on substrate length, L, for various values of h. In all cases, the y-interecept approximates the number of steps with rate constant k_C.

k_C, h. Time courses were simulated using Eqs. (10.4) and (10.13) by replacing n with L/m, and $m = 10$ AA step^{-1}, $L = 30$, 40, and 50 AA with $k_T = 1.67\ s^{-1}$, $k_d = 0$, $k_C = 0.08\ s^{-1}$, and h equal to integer values between 0 and 4. Representative time courses with 1% random error introduced are shown in Fig. 10.8C. In this simulation, for simplicity, it was again assumed that all bound enzyme started in the productive state by setting $x = 1$. As expected, when $h = 0$ the line describing the data points intersects the origin. In all other cases, with $h > 0$, the intercept always exhibits a value that exceeds the number of steps with rate constant k_C, h. However, the intercept never exceeds the true value of h by a whole integer value, reiterating the conclusion that the intercept may serve as a qualitative estimate of the number of steps with rate constant k_C.

3.5. ClpA catalyzed polypeptide translocation

The time courses shown in Fig. 10.7 were initially analyzed using Eq. (10.10) that describes the Laplace transform of $f_P(t)$ for Scheme 10.3. All attempts to float the value of k_d resulted in k_d floating to an infinitely small number. Thus, we concluded that, on such short substrates, no appreciable dissociation was occurring until the enzyme full translocates the substrate. This conclusion is not only drawn from the NLLS analysis but also from simple inspection of the curves. If one compares the experimental time courses shown in Fig. 10.7 to the simulated time courses shown in Fig. 10.6 that were generated from Eq. (10.10) that includes k_d, it is immediately obvious that the experimental time courses do not exhibit any measurable slope in the lag region.

The experimental time courses shown in Fig. 10.7 were analyzed using Eq. (10.11) that describes the Laplace transform of $f_P(t)$ for Scheme 10.4. Multiple observations led to applying Scheme 10.4 to the experimental time courses. First, we observe biphasic kinetics and thus we must incorporate k_{NP}. Second, the n versus L plot generated from the analysis of the experimental time courses exhibit a positive y-intercept slightly above one. This observation indicates that there is at least one additional step not involved in polypeptide translocation. Third, the data do not require a significant dissociation rate constant to be described adequately.

The first strategy in the analysis of the experimental time courses was to analyze the kinetic time courses using Scheme 10.4 but allowing the number steps, n, to float for each time course. Upon doing this, the n versus L plot could again be constructed by removing the contribution from the additional step with rate constant k_C. Strikingly, the positive y-intercept vanishes and an x-intercept is observed. Empirically, we take the observation that the line intersects the x-axis to mean that some number of amino acids in the substrate are in contact with the enzyme and are not translocated. That is to say, the enzyme does not bind at the extreme end of the carboxy terminus and translocate the full substrate. Rather, in the case of ClpA, it is in contact with the 11 AA SsrA sequence. Thus, the actual length of substrate being translocated is $L - d$, where d is the contact site size. Therefore, the number of steps, n, is replaced with $n = (L - d)/m$.

Upon incorporating the relationship of the number of steps to the substrate length, $n = (L - d)/m$, the data were globally analyzed using Scheme 10.4. The resultant parameters are $k_T = (1.45 \pm 0.05)\ s^{-1}$, $k_C = (0.210 \pm 0.003)\ s^{-1}$, $k_{NP} = (0.0455 \pm 0.0005)\ s^{-1}$, $k_d = 0$, $m = (12.6 \pm 0.5)$ AA step^{-1}, and $d = (11.7 \pm 0.4)$ AA. The product of the kinetic step-size, m, and the translocation rate constant, k_T, yields the macroscopic rate of polypeptide translocation and from this analysis is $mk_T = (18.2 \pm 1.3)$ AA s^{-1} at saturating ATP concentrations and 25 °C.

3.6. Determination of translocation directionality

To examine translocation directionality the dependence of the observed signal on the position of the fluorophore is tested. For ClpA catalyzed polypeptide translocation, we observed a distinct lag phase when the Fluorophore is placed at the amino-terminus and the SsrA binding sight is at the carboxy terminus. However, when the fluorophore is placed at the carboxy terminus, see C-Cys-30 and C-Cys-40 in Table 10.1, no lag is observed. In fact, not only is a lag not observed, but the time courses are identical (Rajendar and Lucius, 2010). This indicates that all of the enzymes are translocating away from the fluorophore and thus an immediate change in the signal is observed with no lag time.

4. Concluding Remarks

We have developed a novel method to examine the molecular mechanism of polypeptide translocation catalyzed by polypeptide translocases. Two major advances are summarized here and reported previously (Rajendar and Lucius, 2010). First, we have developed an experimental strategy to perform fluorescence stopped-flow experiments to examine a single turnover of polypeptide translocation catalyzed by ClpA. Second, we describe a method to quantitatively analyze the experimental data by applying sequential n-step kinetic models.

Polypeptide translocases or protein unfoldases, such as ClpA, are often associated with a proteolytic component, like ClpP. For example, the 26 S proteasome is composed of a motor component termed the 19 S cap and a proteolytic component termed the 20 S core (Eytan *et al.*, 1989; Hough *et al.*, 1987). These motor proteins, like ClpA, likely employ a similar mechanism to translocate a ubiquitinilated protein into the proteolytic cavity for degradation. Because these motor components are often associated with a protease the observation of proteolytic fragments has often been used to infer information on the activities of the motor component. With the development of the method presented here, we are able to acquire information on the translocation of a polypeptide substrate without the need for covalent modification of the substrate. This will allow us to answer a variety of fundamental questions regarding the molecular mechanism. For example, does the proteolytic component exert allosteric control over the motor component? That is to say, is the mechanism employed by the motor alone the same as the mechanism employed when associated with the protease. Preliminary observations have suggested that the rate for ClpA alone is different than that of ClpAP.

Equally important, ClpB is a protein unfoldase with high homology to ClpA. However, ClpB does not associate with any known protease. In contrast to ClpA, ClpB disaggregates protein aggregates in vivo and, in collaboration with DnaK, resolubilizes protein aggregates (Glover and Lindquist, 1998; Goloubinoff *et al.*, 1999). However, because the enzyme does not covalently modify the substrate it translocates, little detail on the molecular mechanism of polypeptide translocation catalyzed by ClpB is available. This is underscored in a recent report where multiple mutations were made in ClpB to force it to interact with the protease ClpP so that proteolytic degradation could be used to monitor the activities of ClpB (Weibezahn *et al.*, 2004). However, with the approach presented here, we may be able to shine new light on the mechanism of ClpB catalyzed polypeptide translocation.

In addition to the application of these new approaches to polypeptide translocases, we have opened up the potential to quantitatively examine a variety of enzymes involved in protein translocation across membranes that do not covalently modify their substrate, for review see (Rapoport, 2007).

ACKNOWLEDGMENTS

This work was supported by NSF grant MCB-0843746 to A. L. L.

REFERENCES

Ali, J. A., and Lohman, T. M. (1997). Kinetic measurement of the step size of DNA unwinding by *Escherichia coli* UvrD helicase. *Science* **275,** 377–380.

Block, S. M., *et al.* (1990). Bead movement by single kinesin molecules studied with optical tweezers. *Nature* **348,** 348–352.

Eytan, E., *et al.* (1989). ATP-dependent incorporation of 20S protease into the 26S complex that degrades proteins conjugated to ubiquitin. *Proc. Natl. Acad. Sci. USA* **86,** 7751–7755.

Fischer, C. J., and Lohman, T. M. (2004). ATP-dependent translocation of proteins along single-stranded DNA: Models and methods of analysis of pre-steady state kinetics. *J. Mol. Biol.* **344,** 1265–1286.

Fischer, C. J., *et al.* (2004). Mechanism of ATP-dependent translocation of *E. coli* UvrD monomers along single-stranded DNA. *J. Mol. Biol.* **344,** 1287–1309.

Glover, J. R., and Lindquist, S. (1998). Hsp104, Hsp70, and Hsp40: A novel chaperone system that rescues previously aggregated proteins. *Cell* **94,** 73–82.

Goloubinoff, P., *et al.* (1999). Sequential mechanism of solubilization and refolding of stable protein aggregates by a bichaperone network. *Proc. Natl. Acad. Sci. USA* **96,** 13732–13737.

Hoskins, J. R., *et al.* (2000). Protein binding and unfolding by the chaperone ClpA and degradation by the protease ClpAP. *Proc. Natl. Acad. Sci. USA* **97,** 8892–8897.

Hough, R., *et al.* (1987). Purification of two high molecular weight proteases from rabbit reticulocyte lysate. *J. Biol. Chem.* **262,** 8303–8313.

Howard, J., *et al.* (1989). Movement of microtubules by single kinesin molecules. *Nature* **342,** 154–158.

Jankowsky, E., *et al.* (2000). The DExH protein NPH-II is a processive and directional motor for unwinding RNA. *Nature* **403,** 447–451.

Kim, Y. I., *et al.* (2000). Dynamics of substrate denaturation and translocation by the ClpXP degradation machine. *Mol. Cell* **5,** 639–648.

Kornberg, A., and Baker, T. A. (1992). DNA Replication, W.H. Freeman, New York.

Lohman, T. M., and Bjornson, K. P. (1996). Mechanisms of helicase-catalyzed DNA unwinding. *Annu. Rev. Biochem.* **65,** 169–214.

Lohman, T. M., *et al.* (2008). Non-hexameric DNA helicases and translocases: Mechanisms and regulation. *Nat. Rev. Mol. Cell Biol.* **9,** 391–401.

Lucius, A. L., *et al.* (2002). DNA unwinding step-size of *E. coli* RecBCD helicase determined from single turnover chemical quenched-flow kinetic studies. *J. Mol. Biol.* **324,** 409–428.

Lucius, A. L., *et al.* (2003). General methods for analysis of sequential "n-step" kinetic mechanisms: Application to single turnover kinetics of helicase-catalyzed DNA unwinding. *Biophys. J.* **85,** 2224–2239.

Lucius, A. L., *et al.* (2004). Fluorescence stopped-flow studies of single turnover kinetics of *E. coli* RecBCD helicase-catalyzed DNA unwinding. *J. Mol. Biol.* **339,** 731–750.

Matson, S. W., and Kaiser-Rogers, K. A. (1990). DNA helicases. *Annu. Rev. Biochem.* **59,** 289–329.

Piszczek, G., Rozycki, J., Singh, S. K., Ginsburg, A., and Maurizi, M. R. (2005). The molecular chaperone, ClpA, has a single high affinity peptide binding site per hexamer. *J. Biol. Chem.* **280,** 12221–12230.

Pyle, A. M. (2008). Translocation and unwinding mechanisms of RNA and DNA helicases. *Annu. Rev. Biophys.* **37,** 317–336.

Rajendar, B., and Lucius, A. L. (2010). Molecular mechanism of polypeptide translocation catalyzed by the *Escherichia coli* ClpA protein translocase. *J. Mol. Biol.* **399,** 665–679.

Rapoport, T. A. (2007). Protein translocation across the eukaryotic endoplasmic reticulum and bacterial plasma membranes. *Nature* **450,** 663–669.

Reid, B. G., *et al.* (2001). ClpA mediates directional translocation of substrate proteins into the ClpP protease. *Proc. Natl. Acad. Sci. USA* **98,** 3768–3772.

Svoboda, K., *et al.* (1993). Direct observation of kinesin stepping by optical trapping interferometry. *Nature* **365,** 721–727.

Uyeda, T. Q., *et al.* (1990). Myosin step size. Estimation from slow sliding movement of actin over low densities of heavy meromyosin. *J. Mol. Biol.* **214,** 699–710.

Vale, R. D., and Fletterick, R. J. (1997). The design plan of kinesin motors. *Annu. Rev. Cell Dev. Biol.* **13,** 745–777.

Weibezahn, J., *et al.* (2004). Thermotolerance requires refolding of aggregated proteins by substrate translocation through the central pore of ClpB. *Cell* **119,** 653–665.

Zimmer, J., *et al.* (2008). Structure of a complex of the ATPase SecA and the protein-translocation channel. *Nature* **455,** 936–943.

CHAPTER ELEVEN

A Coupled Equilibrium Approach to Study Nucleosome Thermodynamics

Andrew J. Andrews*[,1] *and* Karolin Luger*[,†]

Contents

Abstract

The repeating structural unit of eukaryotic chromatin, the nucleosome, is composed of two copies each of the histone proteins H2A, H2B, H3, and H4. These proteins form an octamer around which 147 bp of DNA is wrapped in 1.65 superhelical turns (Luger *et al.*, 1997). The nucleosome represents a major obstacle for any protein seeking access to the DNA. Several strategies have evolved to regulate access to nucleosomal DNA, such as posttranslational modification of histones and histone variants, ATP-dependent chromatin remodeling machines, and histone chaperones. It is likely that most if not all of these mechanisms directly impact the thermodynamics of the nucleosome. The DNA sequence itself may also impact its own inherent accessibility through

* Department of Biochemistry and Molecular Biology, Colorado State University, Fort Collins, Colorado, USA
† Howard Hughes Medical Institute, USA
[1] Current address: FOx Chase Cancer Center, Philadelphia

Methods in Enzymology, Volume 488
ISSN 0076-6879, DOI: 10.1016/B978-0-12-381268-1.00011-2
© 2011 Elsevier Inc.
All rights reserved.

modulating nucleosome positioning and/or thermodynamics. However, these working hypotheses could not be tested directly because no assays to measure nucleosome stability under physiological conditions were available. Attempts to determine the stability of nucleosomes have been hampered by the fact that the nucleosomes do not assemble *in vitro* under physiological conditions, but will only form nucleosomes through titration from high (2 *M*) to low (>0.3 *M*) ionic strength. We have developed a coupled equilibrium approach using the histone chaperone Nap1 to measure nucleosome thermodynamics under physiological conditions. This method will be useful for examining the impact of DNA sequence, histone variants, and posttranslational modifications on nucleosome thermodynamics.

1. Introduction

Understanding the thermodynamic parameters that govern higher order DNA structure is critical for elucidating how proteins gain access to DNA in the eukaryotic nucleus. It has been hypothesized that the components of the nucleosome likely affect these parameters directly, through the addition of particular combinations of site-specific posttranslational modifications of histones, through the incorporation of histone variants, or through the DNA sequence itself. Despite the fact that the nucleosome was first characterized over 35 years ago (reviewed in Olins and Olins, 2003), this hypothesis could not be tested directly because no assay was available to measure the thermodynamics of nucleosome assembly and disassembly.

The nucleosome is a multicomponent complex that assembles in a sequential manner. One $(H3\text{–}H4)_2$ tetramer (or two H3/H4 dimers; Nakatani *et al.*, 2006) first binds ~80 bp of DNA, forming a tetrasome. This complex binds two H2A/H2B dimers in a cooperative manner (Mazurkiewicz *et al.*, 2006). Nucleosomes do not spontaneously self-assemble at physiological salt concentrations due to the preponderance of noncanonical histone–DNA interactions (Wilhelm *et al.*, 1978). This is the main reason why simple thermodynamic assays cannot be applied to this system (Thastrom *et al.*, 2004b). Here, we describe a method in which a histone chaperone is used as a "biosensor" for free histones to measure nucleosome thermodynamics.

The term "nucleosome stability" is multifaceted and is applied to different interactions within this multicomponent assembly (Fig. 11.1). It is used to refer to either the interaction of the DNA with the surface of the histone octamer (Fig. 11.1(II)), the interaction between the H2A–H2B dimer and $(H3\text{–}H4)_2$ tetramer in a nucleosome (Fig. 11.1(I to IV)), or the appearance of histone-free DNA (Fig. 11.1(I, VI)). All these events impact access to the DNA by nuclear factors, but are mechanistically and conceptually different.

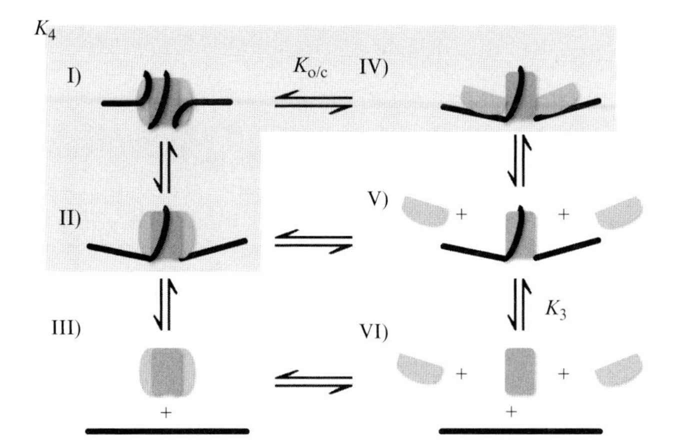

Figure 11.1 Nucleosome stability. The term stability can refer to several structural transitions of the nucleosome. It can refer to the interaction of the DNA ends with the histone octamer (I→II), the interaction of H2A–H2B with DNA–$(H3–H4)_2$ tetrasome (V→II, IV→I), or the appearance of free DNA (III). The coupled equilibrium approach focuses on the thermodynamics of interaction of H2A–H2B with the DNA–$(H3–H4)_2$ "tetrasome" and the $(H3–H4)_2$ tetramer interaction with DNA. *K*4 is the apparent equilibrium between V and some population of conformations I, II, and IV (shaded).

While transient DNA exposure (Fig. 11.1 (II)) allows the binding of transcription factors, transient opening of the nucleosome through disturbing the interface between the H2A–H2B dimer and the $(H3–H4)_2$ tetramer (Fig. 11.1 (IV)) may be the first step in its disassembly. Technical advances have revealed the rate(s) and equilibrium constants for DNA exposure from the nucleosome (Li *et al.*, 2005; Poirier *et al.*, 2008), but quantitative measurements of histone–histone interactions within the context of the nucleosome have not been published. The assay described here focuses on the interaction between the $(H3–H4)_2$ tetramer and the H2A–H2B dimer in the context of a nucleosome. The loss of this interaction (as shown in Fig. 11.1 (V)) is a biologically relevant intermediate during nucleosome (dis) assembly and during transcription (e.g., Kireeva *et al.*, 2002).

2. Salt-Mediated Nucleosome (Dis)Assembly

Any experiment that aims to provide thermodynamic data (either relative or absolute) on nucleosome assembly has to be shown to be at equilibrium. The dilution of preformed nucleosomes to monitor the fraction of free versus nucleosomal DNA is an obvious and straightforward approach. However, it can be shown that this reaction is not at equilibrium because the nucleosome cannot be assembled under these conditions (see Thastrom *et al.*, 2004a for a detailed discussion).

Thus, the most common approach for reversible *in vitro* nucleosome assembly is the use of salt gradients. This approach relies on the propensity of the $(H3–H4)_2$ tetramer to bind DNA at higher ionic strength than H2A–H2B dimer, allowing the ordered deposition of histone complexes to form a nucleosome. Support for the sequential addition of $(H3–H4)_2$ tetramer, followed by the binding of two H2A–H2B dimers in a salt-dependent manner is found throughout the literature (e.g., Oohara and Wada, 1987; Wilhelm *et al.*, 1978; Yager *et al.*, 1989. Since the histone octamer is not stable under physiological conditions, it is likely that nucleosomes are assembled by this mechanism *in vivo*.

It has been tacitly assumed that changes in salt dependence for nucleosome (dis)assembly directly reflect changes in nucleosome stability (e.g., Jin and Felsenfeld, 2007; Oohara and Wada, 1987; Park *et al.*, 2004; and many others). However, monitoring salt-mediated nucleosome (dis)assembly provides information about physical properties of the histones that may or may not reflect the thermodynamics of the nucleosome under physiological conditions. Furthermore, it is challenging to interpret the quantitative data from such approaches (e.g., Oohara and Wada, 1987; Park *et al.*, 2004). This can be observed by fitting data previously published by Park *et al.* (2004), where the salt-dependent dissociation of the nucleosome was monitored by FRET. Specifically, the concentration at which half of the nucleosome is disrupted (as monitored by loss of FRET between H2A–H2B dimer and $(H3–H4)_2$ tetramer) in the presence of DNA ($K_{1/2}{}^{NaCl}$) is 0.56 ± 0.01 *M* and is cooperative with a Hill coefficient (n_H) of 6.6 ± 0.2 while the octamer alone has a $K_{1/2}{}^{NaCl}$ of 0.44 ± 0.01 *M* and a $n_H = 2.6 \pm 0.1$ (Fig. 11.2). The differences in the Hill coefficients indicate that there are more groups of ions required to assemble (or disrupt) the nucleosome compared to the histone octamer. Together with the observed moderate differences in $K_{1/2}$, these data demonstrate the stabilization of the histone octamer by DNA. However, it is difficult to determine if the observed constants of nucleosome assembly by salt titration are the result of the octamer binding to DNA as a unit, or if assembly proceeds in an ordered reaction with tetrasome (H3/H4–DNA) formation preceding dimer binding.

Shrader and Crothers (1989) used the salt dilution method to determine the $\Delta\Delta G$ of nucleosomes assembled on different DNA sequences. This was done under conditions where a radiolabeled DNA fragment with defined sequence competed against a population of random sequence DNA fragments for limiting amounts of histone octamer while lowering the salt concentration from 1 to 0.1 *M*. The difference between the fraction of labeled DNA incorporation (relative to bulk DNA) and of a reference DNA (random genomic DNA) was used to calculate $\Delta\Delta G$ for the sequence of

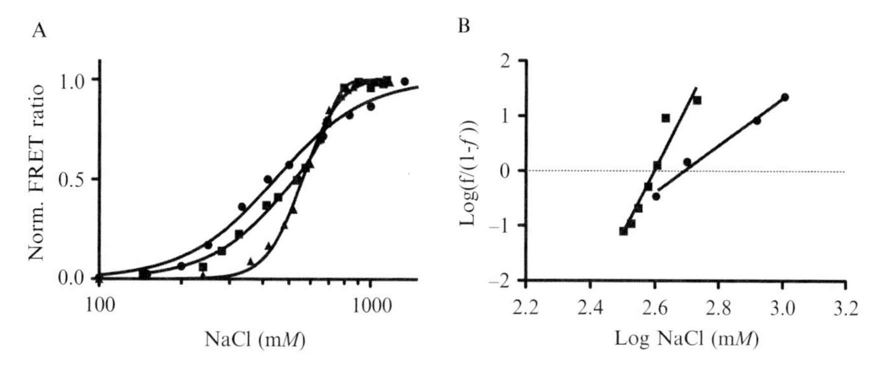

Figure 11.2 Histone–DNA interactions as a function of NaCl. Reanalysis of data previously published in Park *et al.* (2004) and our own unpublished data for the salt dependence of nucleosome (dis)assembly. Nucleosome disruption was observed by the loss of FRET between DNA ends or between H3–H4 and H2A–H2B. (A) Normalized FRET as a function of NaCl where squares are DNA–DNA FRET ($K_{1/2}^{NaCl} = 0.52 \pm 0.01$ *M*; $n_H = 2.7 \pm 0.2$; $K_{1/2}^{NaCl} = 0.70 \pm 0.02$ *M*; $n_H = 15.4 \pm 4.4$), circles are FRET between H3–H4 and H2A–H2B in the absence of DNA ($K_{1/2}^{NaCl} = 0.44 \pm 0.01$ *M*; $n_H = 2.6 \pm 0.1$), and triangles are FRET between H3–H4 and H2A–H2B in the presence of DNA ($K_{1/2}^{NaCl} = 0.56 \pm 0.01$ *M*; $n_H = 6.6 \pm 0.2$). (B) Hill plot of FRET between H3–H4 ($K_{1/2}^{NaCl} = 0.48 \pm 0.02$ *M*; $n_H = 7.7 \pm 1.7$ circles) or H2A–H2B ($K_{1/2}^{NaCl} = 0.41 \pm 0.01$ *M*; $n_H = 14.0 \pm 4.0$, squares) with DNA.

interest. Given that the titration of NaCl can be reversed resulting in free histones and DNA, this is a valid and powerful method for comparing the propensity of DNA sequences to form nucleosomes. Furthermore, this method proved to be instrumental in determining the rules for nucleosome positioning intrinsic in DNA sequences (for review, see Segal and Widom, 2009).

A limitation that is inherent to all of these approaches is that the observed effects may not be relevant under physiological conditions. Additionally, all of these experiments are ineffective under conditions where the histone octamer is in excess of DNA, due to aggregation, preventing the use of such an approach to test the effect of histone variants or posttranslational modifications on nucleosome stability. Finally, even under the best of circumstances, the obtained numbers reflect the thermodynamic sum of the effects of salt on histone–histone, histone–DNA, and DNA wrapping.

3. A Chaperone-Mediated Coupled Approach to Nucleosome Thermodynamics

Nucleosome assembly protein 1 (Nap1) is a histone chaperone that is widely employed in nucleosome assembly reactions *in vitro* (e.g., Ito *et al.*, 1996). Figure 11.3 shows a schematic of Nap1-mediated nucleosome

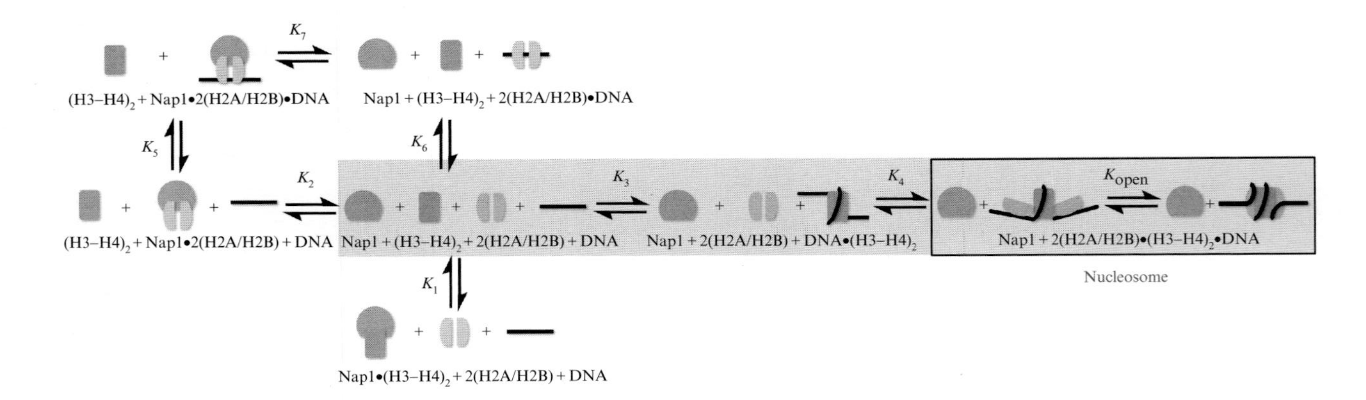

Figure 11.3 Nucleosome assembly in the presence and absence of Nap1. The shaded area represents the reaction scheme for nucleosome (dis)assembly that is expected to be the same in the presence and absence of Nap1. K_1, K_2, K_3, and K_6 can be measured directly (binary interactions), K_5 is measured by monitoring fluorescence under conditions where H2–H2B is labeled, Nap1 is fivefold above the K_2 and DNA is titrated, and K_4 is measured by the loss of FRET between Nap1–(H2A–H2B) as a function of tetrasome and/or the increase in FRET between $(H3–H4)_2$ and H2A–H2B.

assembly. Nap1 binds histones with low nanomolar affinities (K_1 and K_2; Andrews *et al.*, 2008). Nap1-mediated nucleosome formation *in vitro* is characterized by the transfer of a $(H3–H4)_2$ tetramer onto DNA (K_3), followed by the incorporation of H2A/H2B dimer (K_4) (Mazurkiewicz *et al.*, 2006).

Here, we employ Nap1 from yeast to monitor changes in free histone concentration during (dis)assembly under physiological conditions. This is done by measuring the FRET signal between Nap1 and histones under conditions where 50% of the histone is free, and 50% is bound. Any change in the free histone concentration upon addition of a competitor (which could be histone, histone–DNA complexes, or DNA; Fig. 11.4) can be monitored while maintaining the ionic strength, resulting in an accurate measurement of nucleosome (dis)assembly under physiological conditions. Yeast Nap1 can easily be prepared in large amounts, the binding constants for both H2A–H2B and H3–H4 are well described and the histone chaperone as well as the histones can be fluorescently labeled (Andrews *et al.*, 2008, Mazurkiewicz *et al.*, 2006). The experimental setup described below has been optimized for mononucleosomes reconstituted with recombinant histones onto defined sequence 207 bp DNA fragments. The use of recombinant histones allows for the separate preparation of H2A–H2B dimer and $(H3–H4)_2$ tetramer, and the preparation of fluorescently labeled histones;

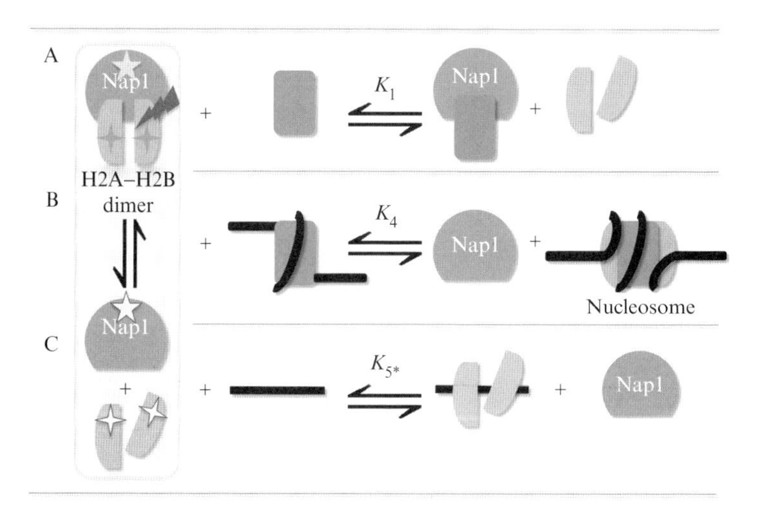

Figure 11.4 Coupled equilibrium approach to nucleosome thermodynamics. The fraction of Nap1-bound H2A–H2B (shaded area-vertical) is monitored as a function of either $(H3–H4)_2$ tetramer (A), tetrasome (B), or DNA (C). The equilibrium constant is shown in C is K_5* due to the fact that the FRET ratio as a function of DNA (Fig. 11.5C) is more consistent with K_5 measured as shown in Fig. 11.5B than with the value obtained for K_6.

the relatively short DNA sequence prevents the formation of more than one nucleosome on one DNA fragment to simplify the mathematical treatment of the results.

Understanding the thermodynamics/stoichiometry of Nap1–histone interactions is a necessary and essential starting point for the coupled equilibrium assay. One must keep in mind that Nap1 and histone complexes exist in the form of dimers or higher oligomerization states. Consider a case where K_d^{app} is obtained by titrating $(H3–H4)_2$ tetramer into Nap1, and the reverse case where Nap1 is titrated into $(H3–H4)_2$ tetramer. Since it is not known whether H3–H4 indeed binds as a tetramer or as two dimers, H3–H4 amounts can be calculated in two ways. This will impact the $K_d^{app.}$ by a factor of two (see Andrews *et al.*, 2008 for details).

The binding constant for H2A–H2B dimer and $(H3–H4)_2$ tetramer to DNA (K_6 and K_3) is determined by measuring the fluorescence change of labeled H2A–H2B dimer (K_6) or $(H3–H4)_2$ tetramer (K_3) as a function of DNA (see experimental setup and conditions). In our system, the interaction between $(H3–H4)_2$ tetramer and DNA to form the tetrasome is exceedingly tight (~1 n*M*, K_3 in Table 11.1), while H2A–H2B dimer bound with somewhat lower (44 n*M*) but still considerably high affinity (Fig. 11.3, K_6) (Andrews *et al.*, 2010).

Before using Nap1 to monitor free H2A–H2B in nucleosome assembly, proof of principle for this approach should be demonstrated by monitoring the change in FRET as a function of $(H3–H4)_2$ tetramer (Fig. 11.4A) (Andrews *et al.*, 2010) as it competes with H2A–H2B dimer for Nap1 (see experimental setup and conditions, Fig. 11.4A). When done correctly, the experiment should provide the same binding constant (K_1) for the Nap1–$(H3–H4)_2$ tetramer complex as measured under the exact same conditions by the direct approach described above.

The ability of $(H3–H4)_2$ tetramer–DNA complex to compete away H2A–H2B dimer from Nap1 is next determined. This measures the binding constant of H2A–H2B to the tetrasome, that is, the thermodynamic stability of the nucleosome (Figs. 11.4B and 11.5D, Table 11.1 K_4). To confirm that nucleosomes are indeed formed in this competition assay, it is advisable to repeat the experiment with FRET pairs attached to the $(H3–H4)_2$ tetramer and H2A–H2B dimer, respectively, in the presence and absence of Nap1. The ratio of the FRET signal with and without Nap1 as a function of tetrasome should provide the same value for K_4, confirming that this assay indeed monitors the formation of nucleosomes (Andrews *et al.*, 2010).

One interesting aspect of Nap1 (and probably of other histone chaperones) is observed in a competition experiment in which DNA is allowed to compete with Nap1 for the H2A–H2B dimer (Figs. 11.4C and 11.5C). For such an experiment, we expect values very similar to K_6. However, we find that under our conditions, DNA binds the Nap1–H2A–H2B complex instead of disrupting it, thus preventing the loss of FRET between Nap1

Table 11.1 Thermodynamic parameters for Nap1-mediated nucleosome assembly

DNA	K_1 [M]	K_2 [M]	K_3 [M]	K_4 [M]	K_5 [M]	K_6 [M]	K_7 [M]
601	$10.0 \pm 0.6 \times 10^{-9}$ $n_H = 1.4 \pm 0.1$	$7.8 \pm 0.4 \times 10^{-9}$	$0.9 \pm 0.1 \times 10^{-9}$ $n_H = 1.1 \pm 0.1$	$1.3 \pm 0.3 \times 10^{-8}$ $n_H = 1.6 \pm 0.3$	$4.7 \pm 1.9 \times 10^{-7}$ $n_H = 1.6 \pm 0.4$	$4.4 \pm 0.5 \times 10^{-8}$ $n_H = 1.9 \pm 0.3$	$\sim 0.8 \times 10^{-7}$
5S	$10.0 \pm 0.6 \times 10^{-9}$ $n_H = 1.4 \pm 0.1$	$7.8 \pm 0.4 \times 10^{-9}$	$1.1 \pm 0.1 \times 10^{-9}$ $n_H = 1.2 \pm 0.1$	$4.0 \pm 0.3 \times 10^{-8}$ $n_H = 2.0 \pm 0.3$	$2.4 \pm 0.2 \times 10^{-7}$ $n_H = 2.0 \pm 0.2$	$1.6 \pm 0.2 \times 10^{-8}$ $n_H = 1.6 \pm 0.2$	$\sim 1.2 \times 10^{-7}$
$\Delta\Delta G^\circ_{601/5S}$ (kcal/mol)	0	0	−0.1	−0.7	0.4	0.6	−0.4

Equilibrium constants K_1, K_3, K_4, K_5, and K_6 are from data shown in Fig. 11.1, K_1 and K_2 are from Andrews *et al.* and K_7 was calculated from the other constants(n_H). 601 and 5S are two DNA sequence with identical (207 bp) length but different nucleosome positioning potential (Lowary and Widom, 1998).

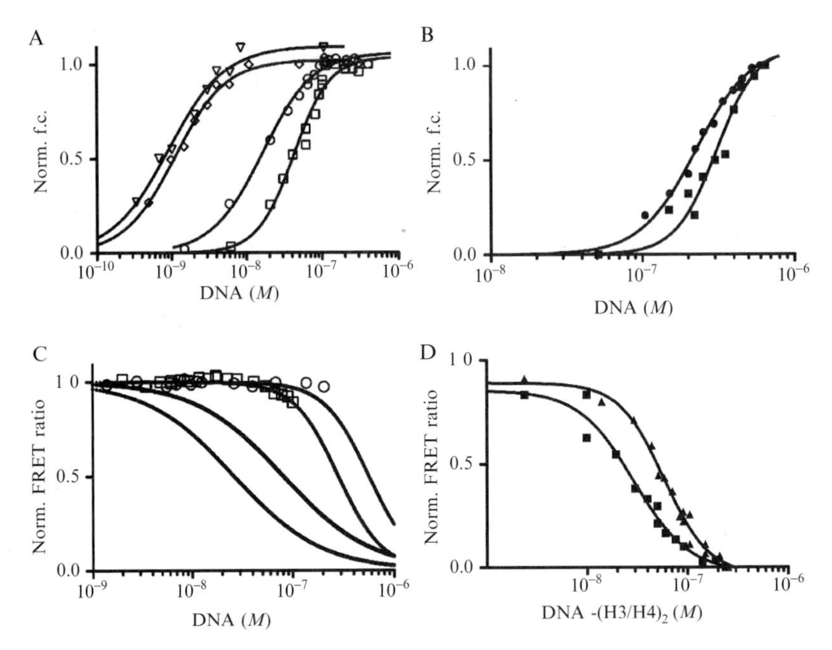

Figure 11.5 Measurements of thermodynamic constants required for nucleosome formation. (A) Fluorescence change as a function of histone–DNA interaction. Normalized fluorescence as a function of DNA binding to either H3/H4 (601 open triangles; 5S open diamonds) or H2A/H2B (5S open circles; 601 open squares). (B) Fluorescence change as a function of DNA interacting with Nap1–H2A/H2B (601 closed squares; 5S closed circles). (C) The change in FRET between Nap1 and H2A/H2B as a function of DNA. This demonstrates that DNA is unable to separate the H2A–H2B–Nap1 complex. Squares are 5S and circles are 601 DNA. The solid lines are modeled data using constants obtained for DNA–H2A/H2B interaction (A) and Nap1–H2A/H2B–DNA (B). (D) The change in FRET between Nap1–H2A/H2B as a function of tetrasome (601 triangles; 5S squares). Data are summarized in Table 11.1.

and H2B. To determine the affinity of DNA to the Nap1–H2A–H2B complex (K_5), DNA can be titrated under conditions where H2A–H2B is labeled and Nap1 is at concentrations 5- to 10-fold above K_2, so that all of the H2A–H2B is in complex with Nap1. Unlike the simple competition experiment, these data are more consistent with the Nap1–H2A–H2B FRET experiment (Fig. 11.5C). Thus, in the presence of Nap1, the interaction between H2A–H2B and DNA is much weaker than in the absence of the chaperone. In effect, the nonproductive H2A–H2B–DNA complex is strongly disfavored in the presence of Nap1. Thus, in addition to providing a thermodynamic assay to study nucleosome stability, our data suggest a mechanism for Nap1-mediated nucleosome assembly by disfavoring nonproductive binding of histones to DNA or nucleosomes by a $\Delta\Delta G°_{K5/K6}$ of $\sim$1.5 kcal/mol.

4. Experimental Setup and Considerations

4.1. General considerations for working with histones

The highly charged nature of histones and their ability to stick to surfaces make them difficult to work with (Allis *et al.*, 1979; Andrews *et al.*, 2008, and many others). This can cause significant problems for quantitative experiments. For this reason, it is highly recommended to take additional precautions to limit these effects, such as limiting the amount of required surface exposure, frequent confirmation of protein concentrations, and preexposing cuvettes and tubes to buffer containing 0.1 mg/ml BSA prior to diluting histones. When working with fluorescently labeled histones, it is recommended that various buffers and materials should be tested to reduce sticking. While materials such as polystyrene or methacrylate may not be as optically pure as quartz, signal loss is preferable to surface–histone interaction; the use of single-use cuvettes further reduces common problems due to ineffective cuvette cleaning. For our purposes, we use a mutant Nap1 protein in which all but one cysteine has been replaced by alanine (Andrews *et al.*, 2008). Purified Nap1 can be stored at −80 °C in 50% glycerol; fluorescently labeled protein (prepared as described in Andrews *et al.*, 2008) is stored at 4 °C and used within 2 weeks.

4.2. Fluorescence titrations

Starting with labeled protein in both the sample and reference cuvette, nonlabeled (binding partner) is added to the sample cuvette and buffer is added to the reference. Prior to adding the binding partner to the labeled protein, fluorescence should be monitored periodically (avoiding photobleaching) for at least the length of time (2–4 h) that it will take the experiment to run in order to confirm that your labeled protein is staying in solution and not either precipitating or sticking to the cuvette. Once the binding partner is added, it should be confirmed the fluorescence signal has reached equilibrium. The normalized fluorescence change can be determined by Eq. (11.1), where R_{obs} is the observed ratio of the fluorescence signal (sample cuvette signal divided by the reference cuvette signal),

$$\text{Norm.f.c.} = \frac{R_{obs} - R_i}{R_f - R_i} \quad (11.1)$$

R_i is ratio of the fluorescence initial and R_f is the ratio of the fluorescence final or where saturation is reached. While the magnitude of the signal change should be constant for each experiment, it can vary from 10% to 30% between different experiments (i.e., labeled yNap1 binding H2A/H2B vs. labeled H2A/H2B binding to yNap1) and with the label used. It is also recommended that the normalized fluorescence ratio of protein titrated into its corresponding binding partner in either buffer or 5 *M* guanidinium HCl be monitored as a confirmation for true interactions. The presence of guanidinium HCl neither typically alters the initial signal (preaddition of the binding partner) nor should the signal change with the addition of micromolar concentrations of the binding partner in guanidinium HCl. All the measurements schematically depicted in Fig. 11.4 are done identically except for the nature of competitor; the normalized FRET is calculated using Eq. (11.1) with *R* replaced by the FRET ratio.

4.3. Nap1-mediated nucleosome formation, as monitored by FRET between histones

This reaction serves the purpose to confirm that nucleosomes are indeed formed in the reaction shown in Fig. 11.4B. This is done by using limiting amounts of labeled dimer (H2B donor) and titrating in tetrasome containing labeled H3–H4 (H4 acceptor) in the presence and absence of Nap1. The difference between the FRET ratio with and without Nap1 is plotted as a function of tetrasome. The $K_{1/2}$ (or K_4) can be estimated by fitting a simple binding isotherm to the data or the more comprehensive fit derived from Fig. 11.3 (see data analysis). This experimental approach is not as robust as the one depicted in Fig. 11.4B, because either the concentration of acceptor is increasing, or, if kept constant, only a small fraction of the nucleosome formed produces a FRET signal ($\ll 1$ n*M*).

5. Data Analysis and Theory

5.1. Simple binary affinity measurements

The affinity for histone–DNA or histone chaperone complexes can be measured using concentrations of labeled protein (P) that are at least 5- to 10-fold less than the expected K_d^{app}. K_d^{app} is determined by fitting Eq. (11.2) derived from Scheme 11.1 to the normalized f.c. observed as a function of *L*,

$$\text{Norm.f.c.} = \text{f.c.}_{\max}\left(\frac{L_t^{n_H}}{L_t^{n_H} + K_d^{n_H}}\right) \tag{11.2}$$

where L_t is the total concentration of protein titrated, n_H is the Hill coefficient, and K_d is the apparent disassociation constant. The n_H is assumed to be 1 (and forced to 1) unless the data dictate otherwise. When $n_H > 1$, the linearized form of Eq. (11.2)

$$\mathrm{Log}\left[\frac{f}{(1-f)}\right] = n_{\mathrm{H}}\mathrm{Log}[L] + b \tag{11.3}$$

can be used (Eq. (11.3)), where f is equal to the normalized f.c. divided by the normalized f.c. max.

5.2. Chaperone-based assay to measure nucleosome thermodynamics (coupled equilibrium assay)

We can measure the equilibrium constants for the nucleosome by monitoring the fraction of Nap1–H2A–H2B complex in the presence of competitor (as shown in Fig. 11.4). As pointed out in the above section, it is critical to be mindful of how the concentrations of histones and Nap1 are calculated and that the associated binding constants are calculated the same way. Typically, we calculate Nap1 as a monomer (although is most likely a dimer under our conditions), H2A–H2B as a dimer, and H3–H4 as a tetramer.

In all three cases shown in Fig. 11.4, the displacement of H2A–H2B from Nap1 is observed by monitoring the FRET signal between Nap1 and H2A–H2B as a function of another protein, DNA, or protein–DNA complex (Schemes 11.1 and 11.2, Fig. 11.4). The experimental setup for all experiments is the same, Nap1 is kept close to K_2 ($\sim$8 nM), and H2A/H2B is $\sim$0.5 nM or less. Different equations have to be used depending on the interaction measured.

In the experiment depicted in Fig. 11.4A, $(\mathrm{H3{-}H4})_2$ is competing with H2A–H2B for Nap1. If Nap1 is kept close to K_2, we can assume that the FRET signal from H2B and Nap1 is proportional to $\sim$50% of H2A–H2B

$$\mathrm{P} + \mathrm{nL} \rightleftharpoons \mathrm{P(L)_n}$$

Scheme 11.1

$$\mathrm{Nap1} \bullet (\mathrm{H2A\text{-}H2B})_2 + (\mathrm{H3\text{-}H4})_2$$

$$\updownarrow \mathrm{K_2}$$

$$\mathrm{Nap1} + 2(\mathrm{H2A\text{-}H2B}) + (\mathrm{H3\text{-}H4})_2$$

$$\updownarrow \mathrm{K_1}$$

$$2(\mathrm{H2A\text{-}H2B}) + \mathrm{Nap1} \bullet (\mathrm{H3\text{-}H4})_2$$

Scheme 11.2

bound to Nap1. To fit the observed data at *any* concentration of Nap1, divide the fraction of H2A–H2B bound to Nap1 ($Fb^{Nap1-H2A-H2B}$; Eq. (11.4)) by the fraction of H2A–H2B bound to Nap1 in the presence of $(H3–H4)_2$ tetramer

$$Fb^{Nap1-H2A-H2B} = \frac{[Nap1]_{total}}{\left([Nap1]_{total} + K_d^{Nap1-H2A-H2B}\right)} \tag{11.4}$$

$$Fbi^{Nap1 - H2A - H2B} = \frac{[Nap1]_{free}}{\left([Nap1]_{free} + K_d^{Nap1 - H2A - H2B}\right)} \tag{11.5}$$

($Fbi^{Nap1-H2A-H2B}$) Eq. (11.5). In Eq. (11.4), the free concentration of Nap1 is equal to Nap1 total because the concentration of H2A–H2B is much less than the binding constant of Nap1 to H2A–H2B. This is not true for Eq. (11.5) because the concentration of Nap1 (8 n*M*) is close the binding affinity of Nap1 to $(H3–H4)_2$ ($\sim$10 n*M*). Under these conditions, the concentration of free Nap1 is equal to Eq. (11.6),

$$[Nap1]_{free} = [Nap1]_{total} - [Nap1 - H3 - H4] \tag{11.6}$$

$$[Nap1 - H3 - H4] = \frac{([Nap1]_{total} + [H3 - H4] + K_d^{Nap1-H3-H4}) - \sqrt{([Nap1]_{total} + [H3 - H4] + K_d^{Nap1-H3-H4})^2 + 4[H3 - H4]_{total}[Nap1]_{total}}}{2} \tag{11.7}$$

where the concentration of Nap1–(H3–H4) is equal to Eq. (11.7). The observed FRET ratio

$$S_{obs} = S_i + (S_f - S_i)(1 - (Fbi/Fb) \tag{11.8}$$

is equal to Eq. (11.8), where S_f is equal to signal final and S_i is equal to signal initial.

Equation (11.8) can also be used to determine the affinity of H2A–H2B to DNA or tetrasome (Scheme 11.3)

$$Fbi^{Nap1 - H2A - H2B} = \frac{[Nap1]_{total} K_2}{K_2 K_4^{n_{H^4}} + [Nap1]_{total} K_4^{n_{H^4}} + [tetrasome]^{n_{H^4}} K_2} \tag{11.9}$$

but the fraction of bound Nap1–H2A–H2B in the presence of tetrasome (or DNA, depending on what is measured) is equal to Eq. (11.9). In this case,

$$\mathrm{Nap1} \bullet (\mathrm{H2A\text{-}H2B})_2 + (\mathrm{H3\text{-}H4})_2 \bullet \mathrm{DNA}$$
$$\rightleftharpoons K_2$$
$$\mathrm{Nap1} + 2(\mathrm{H2A\text{-}H2B}) + (\mathrm{H3\text{-}H4})_2 \bullet \mathrm{DNA}$$
$$\rightleftharpoons K_4$$
$$\mathrm{Nap1} + 2(\mathrm{H2A\text{-}H2B}) \bullet (\mathrm{H3\text{-}H4})_2 \bullet \mathrm{DNA}$$

Scheme 11.3

Nap1 and $(H3–H4)_2$–DNA (tetrasome) are in competition for H2A–H2B and, therefore, the total Nap1 concentration is unchanged ([Nap1]total $\gg$ K_2). The concentration of tetrasome can be assumed to be equal to the concentration of $(H3–H4)_2$ tetramer when the concentration of DNA is greater than 10 nM or more than fivefold K_2. We have previously shown that DNA alone is not capable of disrupting the Nap1–H2A–H2B complex, and K_5 (for Nap1–H2A–H2B–DNA) is ~200–500 nM depending on the DNA sequence. Given the affinities typically observed for the nucleosome, saturation will be reached prior to DNA interacting with the Nap1–H2A–H2B complex. However, a more complete solution can be used that can take into account the formation of tetrasome, Nap1–histone interactions, and nonnucleosome and nucleosomal histone interactions (see next paragraph).

Modeling/calculating constants using equations derived from the thermodynamic scheme shown in Fig. 11.3, we can calculate the fraction of nucleosomes by

$$[\text{Nucleosome}] = \frac{([\text{H2A} - \text{H2B}] \times [\text{tetrasome}])^{n_{\text{H}^4}}}{K_4^{n_{\text{H}^4}}} \tag{11.10}$$

solving the scheme shown in Fig. 11.3. The concentration of nucleosome in this scheme is equal to Eq. (11.10), where the concentration of tetrasome is equal to Eq. (11.11).

$$[\text{tetrasome}] = \frac{([\text{DNA}] + [(\text{H3} - \text{H4})_2] + K_3) - \sqrt{([\text{DNA}] + [(\text{H3} - \text{H4})_2] + K_3)^2 - 4 \times [\text{DNA}] \times [(\text{H3} - \text{H4})_2]}}{2} \tag{11.11}$$

The next complex we need to solve for is the Nap1–H2A–H2B–DNA complex. The concentration of the Nap1–H2A–H2B–DNA complex is equal to Eq. (11.12),

$$[\mathrm{Nap1} \times \mathrm{H2A} - \mathrm{H2B} \times \mathrm{DNA}] = \left(\frac{([\mathrm{Nap1} \times \mathrm{H2A} - \mathrm{H2B}] \times [\mathrm{DNA}]_{\mathrm{free}})^{n_{\mathrm{H}^5}}}{K_5^{n_{\mathrm{H}5}}} \right) \tag{11.12}$$

where the concentration of Nap1–H2A–H2B is equal to Eq. (11.13) and the

$$[\mathrm{Nap1} \times \mathrm{H2A} - \mathrm{H2B}] = \left(\frac{[\mathrm{Nap1}] \times [\mathrm{H2A} - \mathrm{H2B}]}{K_2} \right) \tag{11.13}$$

$$[\mathrm{DNA}]_{\mathrm{free}} = [\mathrm{DNA}]_{\mathrm{total}} - [\mathrm{tetrasome}] \tag{11.14}$$

concentration of free DNA is equal to Eq. (11.14). Given that in our experimental conditions H2A–H2B is at low concentrations ($\ll$1 nM), we solved for the fraction of H2A–H2B that exists as nucleosomes. The total H2A–H2B concentration

$$\begin{aligned}[\mathrm{H2A} - \mathrm{H2B}]_{\mathrm{total}} = {} & [\mathrm{H2A} - \mathrm{H2B}]_{\mathrm{free}} + [\mathrm{Nap1} \times \mathrm{H2A} - \mathrm{H2B}] \\ & + [\mathrm{Nap1} \times \mathrm{H2A} - \mathrm{H2B} \times \mathrm{DNA}] + [\mathrm{nucleosome}]\end{aligned} \tag{11.15}$$

is equal to Eq. (11.15). The fraction of nucleosome can be calculated by using Eq. (11.16).

$$F_{\mathrm{nuc}} = \frac{[\mathrm{nucleosome}]}{[\mathrm{H2A} - \mathrm{H2B}]_{\mathrm{total}}} \tag{11.16}$$

The F_{nuc} (F_{X}) can be used to calculate the binding constant for H2A–H2B to tetrasome

$$S_{\mathrm{obs}} = S_{\mathrm{i}} + (S_{\mathrm{m}} - S_{\mathrm{i}}) \times F_{\mathrm{x}} \tag{11.17}$$

(in the experiment where the appearance of FRET between H2A–H2B dimer and $(\mathrm{H3–H4})_2$ tetramer is monitored) by using Eq. (11.17), where S_{obs} is the normalized FRET signal with Nap1 is divided by the normalized FRET without Nap1, S_{i} is the normalized FRET signal with no tetrasome added, and S_{max} is the maximum signal change. This method can also be used to calculate the loss of Nap1–H2A–H2B interaction as a function of

$$F_{\mathrm{nap1-H2A-H2B}} = \frac{[\mathrm{Nap1} \times \mathrm{H2A} - \mathrm{H2B}]}{[\mathrm{H2A} - \mathrm{H2B}]_{\mathrm{total}}} \tag{11.18}$$

tetrasome by Eq. (11.18) (coupled assay data), where the concentration Nap1 bound to H2A–H2B is equal to Eq. (11.13) and total H2A–H2B is equal to Eq. (11.15).

5.3. Hill coefficients

In our system, most of the equilibrium constants listed in Table 11.1 have Hill coefficients larger than 1. We have previously described the cooperativity of histone–Nap1 interactions and the difficulties in interpreting cooperativity in systems where multiple dimers or tetramers are interacting with each other (Andrews *et al.*, 2008). The appearance of cooperativity can be due to conventional reasons where the initial binding event is weaker than subsequent events, or from dimerization. Consider Scheme 11.4 where P (protein) binds L_2 (ligand dimer) but not L (ligand). In this case, the normalized fraction change (f.c.) can still be described by Eq. (11.2)

$$(L)_2 = \frac{L_t - L}{2} \qquad (11.19)$$

but n_H is equal to one and L_2 is equal to Eq. (11.19). In Eq. (11.19), L_t is the total concentration of protein titrated and L is equal to Eq. (11.20), where K_1 is the dimerization

$$L = \frac{-K_1 + \sqrt{(K_1)^2 - 8L_tK_1}}{2} \qquad (11.20)$$

constant. Dimerization or tetramerization could explain the observed cooperativity in the formation of tetrasome. This type of apparent cooperativity is easily seen when data is simulated using Scheme 11.3 and fit to a normal binding isotherm (Fig. 11.6). The reported tetramerization constant for H3–H4 is approximately 3×10^{-8} M (Scarlata *et al.*, 1989) which would explain the apparent cooperativity (Andrews *et al.*, 2008).

The observed cooperativity in Nap1-mediated nucleosome formation (K_4) is expected from the fact that the nucleosome contains two H2A–H2B dimers and has been inferred from other studies of Nap1-mediated assembly (Mazurkiewicz *et al.*, 2006). The more surprising fact is that we observe cooperativity in H2A–H2B interacting with DNA. Current information

$$P + 2L \underset{}{\overset{K_1}{\rightleftharpoons}} P + (L)_2 \rightleftharpoons P(L)_2$$

Scheme 11.4

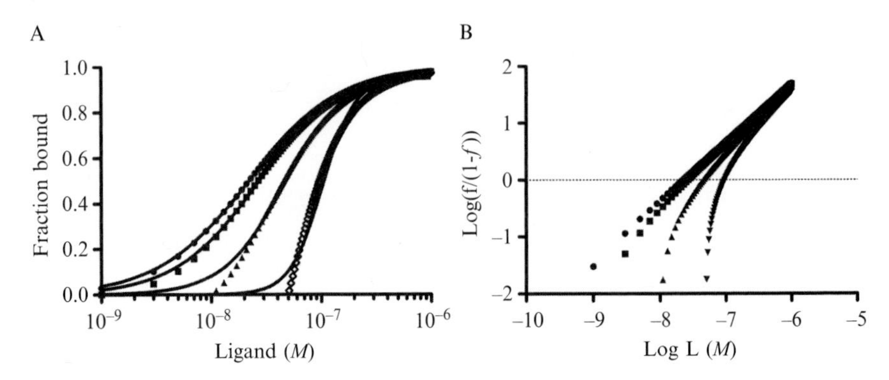

Figure 11.6 Model for dimerization-induced cooperativity. The fraction of bound protein as a function of ligand was modeled using Eq. (11.2) with the dimerized form of ligand modeled using Scheme 11.4, at different dimerization constants (0.1, 1, 10, and 500 n*M*). (A) Modeled data was plotted as a function of ligand and Eq. (11.2) was fit to the modeled data. As the dimerization constant was increased the apparent n_H increases from 1 to 3 (1.00 ± 0.01, 1.15 ± 0.01, 1.56 ± 0.01, 2.96 ± 0.01). (B) The Hill plot of the same data.

suggests that the interaction between H2A and H2B to form the H2A–H2B dimer is extremely stable and the H2A and H2B proteins are likely unfolded once separated from each other. One possibility is that the dimer–DNA interaction changes and possibly linearizes the DNA (as suggested by Samso and Daban, 1993), thereby aiding in the next binding event.

6. Summary and Implications

Nap1 both assembles and disassembles nucleosomes (Park *et al.*, 2005). We have shown that Nap1 assembles nucleosomes by disassembling non-nucleosomal H2A–H2B–DNA complexes (Andrews *et al.*, 2010). Our model predicts that disassembly occurs by depleting or sequestering histones from the nucleosome (Fig. 11.7A). It is currently unknown if Nap1 has any direct effect on other areas of nucleosome stability depicted in Fig. 11.1. If complexes such as Fig. 11.1(IV) exist even transiently, they could be a target for Nap1, resulting in the removal of H2A–H2B dimer from the DNA.

Our data suggest that nonnucleosomal H2A–H2B–DNA interaction is the main obstacle for nucleosome formation. However, nonnucleosomal H2A–H2B–DNA interaction may inhibit nucleosome formation kinetically rather than thermodynamically. In either case, the fact that Nap1 can

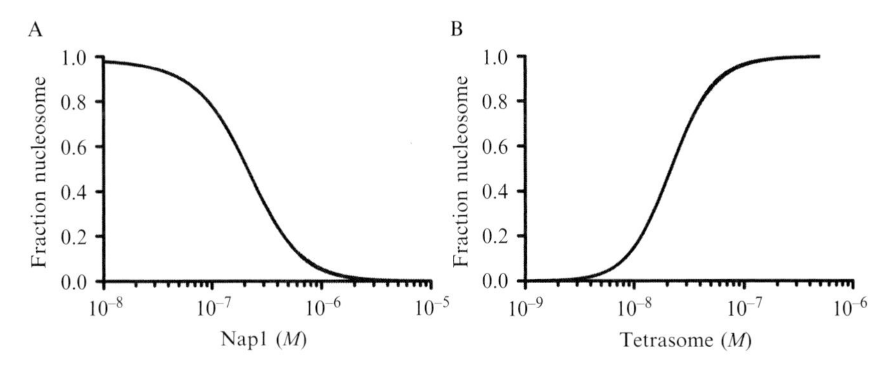

Figure 11.7 Simulations of Nap1-mediated nucleosome assembly. The fraction of nucleosome was calculated using Eq. (11.16) and parameters for the 601 sequence in Table 11.1. (A) The fraction of nucleosome at concentrations of tetrasome 10-fold above K_4 as a function of Nap1. (B) The fraction of nucleosome as a function of tetrasome at zero concentration of Nap1.

both assemble and disassemble nucleosomes implies that Nap1 can mediate an equilibrium between free histones and the nucleosome. This fact can be exploited to measure the free histone concentration in the presence of varying amounts of tetrasome to determine the binding affinity of H2A–H2B to tetrasome, thus forming the nucleosome.

The ability to measure the thermodynamic constants that govern nucleosome (dis)assembly will allow for a more complete understanding of how proteins gain access to DNA within the nucleosome. In particular, it will allow the testing of hypotheses that posttranslational modifications of histones, histone variants, and even DNA sequence alter the intrinsic thermodynamics of the nucleosome. The applicability and sensitivity of this assay is demonstrated in the comparison of two DNA sequences with identical lengths but different nucleosome positioning potential. The 601 207 bp DNA fragment (Lowary and Widom, 1998) was obtained through selection from a pool of DNA that could outcompete 5S DNA, a naturally occurring region from the promoter region of a ribosomal RNA gene with relatively high nucleosome positioning propensities (Simpson and Stafford, 1983). It was originally assumed that the positioning power of this sequence was due to the tight interaction of the $(H3–H4)_2$ tetramer with 601 relative to 5S, but we find this not to be the case. Interestingly, both the nonnucleosomal and nucleosomal H2A–H2B interactions are approximately twofold different, with the 601 sequence forming a more stable nucleosome and less favorable nonnucleosomal interactions. This could not have been determined by any other method and highlights the power and applicability of our approach.

REFERENCES

Allis, C. D., *et al.* (1979). Micronuclei of *Tetrahymena* contain two types of histone H3. *Proc. Natl. Acad. Sci. USA* **76,** 4857–4861.

Andrews, A. J., *et al.* (2008). A thermodynamic model for Nap1–histone interactions. *J. Biol. Chem.* **283,** 32412–32418.

Andrews, A. J., *et al.* (2010). The histone chaperone Nap1 promotes nucleosome assembly by eliminating nonnucleosomal histone DNA interactions. *Mol. Cell* **37,** 834–842.

Ito, T., *et al.* (1996). Drosophila NAP-1 is a core histone chaperone that functions in ATP-facilitated assembly of regularly spaced nucleosomal arrays. *Mol. Cell. Biol.* **16,** 3112–3124.

Jin, C., and Felsenfeld, G. (2007). Nucleosome stability mediated by histone variants H3.3 and H2A.Z. *Genes Dev.* **21,** 1519–1529.

Kireeva, M. L., *et al.* (2002). Nucleosome remodeling induced by RNA polymerase II: Loss of the H2A/H2B dimer during transcription. *Mol. Cell* **9,** 541–552.

Li, G., *et al.* (2005). Rapid spontaneous accessibility of nucleosomal DNA. *Nat. Struct. Mol. Biol.* **12,** 46–53.

Lowary, P. T., and Widom, J. (1998). New DNA sequence rules for high affinity binding to histone octamer and sequence-directed nucleosome positioning. *J. Mol. Biol.* **276,** 19–42.

Luger, K., *et al.* (1997). Crystal structure of the nucleosome core particle at 2.8 Å resolution. *Nature* **389,** 251–259.

Mazurkiewicz, J., *et al.* (2006). On the mechanism of nucleosome assembly by histone chaperone NAP1. *J. Biol. Chem.* **281,** 16462–16472.

Nakatani, Y., *et al.* (2006). How is epigenetic information on chromatin inherited after DNA replication? Ernst Schering Res Found Workshop, pp. 89–96.

Olins, D. E., and Olins, A. L. (2003). Chromatin history: Our view from the bridge. *Nat. Rev. Mol. Cell Biol.* **4,** 809–814.

Oohara, I., and Wada, A. (1987). Spectroscopic studies on histone–DNA interactions. II. Three transitions in nucleosomes resolved by salt-titration. *J. Mol. Biol.* **196,** 399–411.

Park, Y. J., *et al.* (2004). A new fluorescence resonance energy transfer approach demonstrates that the histone variant H2AZ stabilizes the histone octamer within the nucleosome. *J. Biol. Chem.* **279,** 24274–24282.

Park, Y. J., *et al.* (2005). Nucleosome assembly protein 1 exchanges histone H2A–H2B dimers and assists nucleosome sliding. *J. Biol. Chem.* **280,** 1817–1825.

Poirier, M. G., *et al.* (2008). Spontaneous access to DNA target sites in folded chromatin fibers. *J. Mol. Biol.* **379,** 772–786.

Samso, M., and Daban, J. R. (1993). Unfolded structure and reactivity of nucleosome core DNA–histone H2A, H2B complexes in solution as studied by synchrotron radiation X-ray scattering. *Biochemistry* **32,** 4609–4614.

Scarlata, S. F., *et al.* (1989). Histone subunit interactions as investigated by high pressure. *Biochemistry* **28,** 6637–6641.

Segal, E., and Widom, J. (2009). From DNA sequence to transcriptional behaviour: A quantitative approach. *Nat. Rev. Genet.* **10,** 443–456.

Shrader, T. E., and Crothers, D. M. (1989). Artificial nucleosome positioning sequences. *Proc. Natl. Acad. Sci. USA* **86,** 7418–7422.

Simpson, R. T., and Stafford, D. W. (1983). Structural features of a phased nucleosome core particle. *Proc. Natl. Acad. Sci. USA* **80,** 51–55.

Thastrom, A., *et al.* (2004a). Nucleosomal locations of dominant DNA sequence motifs for histone–DNA interactions and nucleosome positioning. *J. Mol. Biol.* **338,** 695–709.

Thastrom, A., *et al.* (2004b). Histone–DNA binding free energy cannot be measured in dilution-driven dissociation experiments. *Biochemistry* **43,** 736–741.

Wilhelm, F. X., *et al.* (1978). Reconstitution of chromatin: Assembly of the nucleosome. *Nucleic Acids Res.* **5,** 505–521.

Yager, T. D., *et al.* (1989). Salt-induced release of DNA from nucleosome core particles. *Biochemistry* **28,** 2271–2281.

CHAPTER TWELVE

Quantitative Methods for Measuring DNA Flexibility *In Vitro* and *In Vivo*

Justin P. Peters,* Nicole A. Becker,* Emily M. Rueter,* Zeljko Bajzer,* Jason D. Kahn,† *and* L. James Maher III*

Contents

* Department of Biochemistry and Molecular Biology, Mayo Clinic College of Medicine, Rochester, Minnesota, USA
† Department of Chemistry and Biochemistry, University of Maryland, College Park, Maryland, USA

Methods in Enzymology, Volume 488
ISSN 0076-6879, DOI: 10.1016/B978-0-12-381268-1.00012-4
© 2011 Elsevier Inc.
All rights reserved.

Abstract

The double-helical DNA biopolymer is particularly resistant to bending and twisting deformations. This property has important implications for DNA folding *in vitro* and for the packaging and function of DNA in living cells. Among the outstanding questions in the field of DNA biophysics are the underlying origin of DNA stiffness and the mechanisms by which DNA stiffness is overcome within cells. Exploring these questions requires experimental methods to quantitatively measure DNA bending and twisting stiffness both *in vitro* and *in vivo*. Here, we discuss two classical approaches: T4 DNA ligase-mediated DNA cyclization kinetics and lac repressor-mediated DNA looping in *Escherichia coli*. We review the theoretical basis for these techniques and how each can be applied to quantitate biophysical parameters that describe the DNA polymer. We then show how we have modified these methods and applied them to quantitate how apparent DNA physical properties are altered *in vitro* and *in vivo* by sequence-nonspecific architectural DNA-binding proteins such as the *E. coli* HU protein and eukaryotic HMGB proteins.

1. Introduction

The biophysics of DNA has been studied for more than half a century with the goals of understanding and predicting the behavior of the polymer chain. This work is applicable to the understanding of DNA packaging and gene expression *in vitro* and *in vivo*, and for design and implementation of DNA as a nanomaterial. Much progress has been made toward understanding the properties of purified DNA *in vitro*, but interpretation of even these relatively simple experiments can be controversial. What is the origin of the remarkable inflexibility of double-stranded DNA? To what extent do base pair stacking and electrostatics contribute? Over what length scales do simple polymer models apply? Beyond these questions, it remains of great interest to better understand the physical properties (e.g., bend and twist flexibility) of the double-stranded DNA polymer in the presence of proteins, in chromatin, and ultimately, in living cells.

Here, we focus on our recent implementations and revisions of two classical methods to quantitate DNA flexibility (Fig. 12.1). The first method involves ligase-catalyzed DNA cyclization kinetics *in vitro* (Fig. 12.1A). The second method involves analysis of DNA repression looping in living bacteria (Fig. 12.1B). We sketch the theory underlying each classic method and then show how we have updated the procedures and extended them to increasingly complex systems.

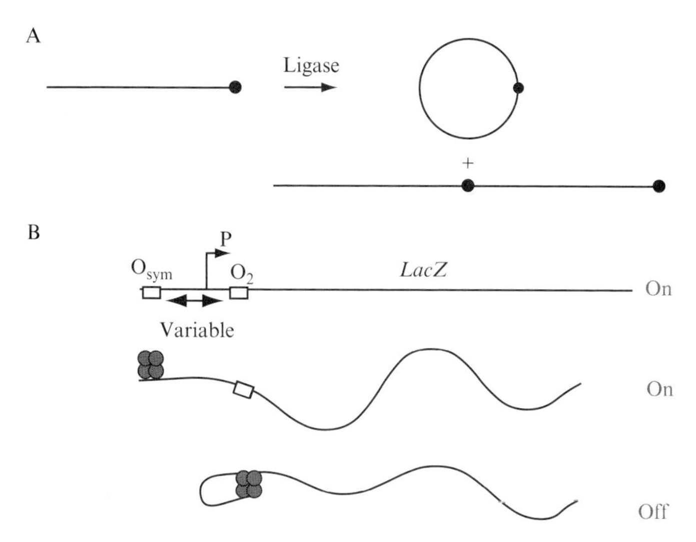

Figure 12.1 *Analysis methods described in this work.* (A) Classic *in vitro* assay of DNA *j*-factor by T4 DNA ligase-catalyzed cyclization kinetics. Linear radiolabeled DNA molecules of known concentration carry cohesive termini. Dot emphasizes monomer unit. Ligation gives rise to cyclic and linear products (two examples shown). Rates of accumulation of monomeric circles versus all other species are analyzed quantitatively by native polyacrylamide gel electrophoresis and fit to rate equations to estimate the *j*-factor. Determination of *j*-factor as a function of length allows estimation of persistence length (WLC model), helical repeat, and twist constant. (B) Classic *in vivo* assay of DNA looping mediated by lac repressor tetramer binding simultaneously to pairs of operators (O_{sym}, O_2) flanking a test promoter (P). Operator spacing is varied to monitor the length dependence of repression, allowing fitting to a thermodynamic model that links promoter repression to DNA looping.

2. DNA Polymer Theory

2.1. Worm-like chain (WLC) model

The mechanical properties of double-stranded DNA have been classically treated by the WLC model of DNA flexibility (Kratky and Porod, 1949; Rippe *et al.*, 1995; Shimada and Yamakawa, 1984). This model applies particularly well to polymers intermediate in behavior between a rigid rod and a random coil. The WLC model invokes a single variable, the *persistence length*, to account for both the local stiffness and the long-range bend flexibility of double-stranded DNA. The WLC model was derived by sequential imposition of constraints on chain geometry starting from a *random flight chain*. Limiting such a chain to segments of equal nonzero length yields a *freely jointed chain*. For such a chain with *n* segments of length

a, the contour length of the chain is na. Constraining the valence (included) angle between segments to the value θ (but not constraining the torsion angles between segments) in this chain results in a *freely rotating chain*, with free rotation permitted for every segment. The first segment r_1 may be considered to be along the z-axis. If $\mathbf{e}_z$ is a unit vector along the z-axis, the average z component of the end-to-end vector $\mathbf{R}$ for the chain is

$$\langle \mathbf{R} \cdot \mathbf{e}_z \rangle = \left(\sum_{i=1}^{n} \langle \mathbf{r}_i \rangle \right) \cdot \frac{\mathbf{r}_1}{a} = \frac{1}{a} \sum_{i=1}^{n} \langle \mathbf{r}_i \cdot \mathbf{r}_i \rangle \tag{12.1}$$

The terms in the sum can be evaluated analytically. $\langle \mathbf{r}_1 \cdot \mathbf{r}_2 \rangle$, the scalar projection of $\mathbf{r}_2$ onto $\mathbf{r}_1$, is given by

$$\langle \mathbf{r}_1 \cdot \mathbf{r}_2 \rangle = a^2(-\cos\theta) \tag{12.2}$$

because the angle between $\mathbf{r}_1$ and $\mathbf{r}_2$ is $\pi - \theta$ and $\cos(\pi - \theta) = -\cos\theta$. The value of $\langle r_1 \cdot r_3 \rangle$ is found by projecting $\mathbf{r}_3$ onto $\mathbf{r}_2$ and then projecting the result onto $\mathbf{r}_1$. Upon repeating scalar projections, the scalar products $\mathbf{r}_1 \cdot \mathbf{r}_i$ are given by

$$\langle \mathbf{r}_1 \cdot \mathbf{r}_i \rangle = a^2(-\cos\theta)^{i-1} \tag{12.3}$$

and the sum is a geometric series $a + a(-\cos\theta) + a(-\cos\theta)2 + \cdots + a(-\cos\theta)^{n-1}$. The sum of the first n terms in this series is

$$\langle \mathbf{R} \cdot \mathbf{e}_z \rangle = a \frac{1 - (-\cos\theta)^n}{1 - (-\cos\theta)} \tag{12.4}$$

In the limit $n \to \infty$

$$\lim_{n \to \infty} \langle \mathbf{R} \cdot \mathbf{e}_z \rangle = \frac{a}{1 + \cos\theta} \equiv P \tag{12.5}$$

where the length P is defined as the persistence length of the chain. P conveys the length over which the chain's "memory" of its initial direction persists.

The Kratky–Porod *WLC* is derived from the freely rotating chain by letting $n \to \infty$, $a \to 0$, and $\theta \to \pi$ under the constraint that both $na \equiv l$ (the chain contour length) and $a/(1 + \cos\theta)$ (the chain persistence length) remain constant. The WLC model can be used to express the mean-squared distance, R, between the ends of (or points along) a DNA chain in terms of the persistence length, P, and the contour length, l, separating the points:

$$\langle R^2 \rangle = 2Pl\left[1 - \frac{P}{l}\left(1 - e^{-1/P}\right)\right] \qquad (12.6)$$

Under physiological conditions, conventional values for P are near 50 nm (~150 bp). For contour lengths $l \leq P$, DNA behaves as a rod with elastic resilience. For contour lengths $l >> P$, DNA behaves as a flexible polymer for which chain entropy dominates over bending energy, and $\langle R^2 \rangle = 2Pl$, so that the RMS end-to-end distance scales with the square root of the number of segments as for any random walk.

2.2. The j-factor

The effective concentration of one site along the DNA contour in the neighborhood of another is an extremely useful measure of DNA stiffness. This parameter can be expressed by the experimental *j*-factor, most often obtained from the ring-closure probability for DNA fragments with cohesive termini in the presence of DNA ligase (Fig. 12.2A). DNA flexibility described by the WLC model predicts two global regimes that appear in plots of the *j*-factor as a function of the length of the intervening DNA tether (Fig. 12.2B). For DNA tether lengths less than ~400 bp, the *j*-factor increases dramatically with length, reflecting rod-like behavior: the energetic cost of bending a rod into a circle decreases rapidly as length increases. Beyond ~400 bp, the plot drops gradually with DNA length, reflecting dilution of DNA sites and more random coil-like behavior.

For DNA ligation, the requirement for 5′–3′ alignment of the helical ends demands consideration of DNA torsional flexibility. Typical values for the twist constant of DNA *in vitro* yield predictions shown in Fig. 12.2B, with the *j*-factor oscillating through maxima and minima for every ~10.5 bp (one helical turn) of increased length. Experimental estimates of the *j*-factor *in vitro* have been most commonly made using ligase-dependent DNA cyclization assays (Fig. 12.2B, inset) as a function of DNA length. Related processes, such as the formation of protein–DNA loops, can depend on the same physical properties that determine the *j*-factor, but the results of such experiments will depend upon the nature of the site–site interactions in question. For example, protein–DNA looping can be modeled using the same methods used for cyclization by simply changing the boundary conditions for end-joining (Levene and Crothers, 1986; Lillian *et al.*, 2008; Swigon *et al.*, 2006). In practice, contact through a bound protein has usually been found to be a much less stringent constraint than covalent ligation of the termini (Allemand *et al.*, 2006; Peters and Maher, 2010; Rippe, 2001; Rippe *et al.*, 1995).

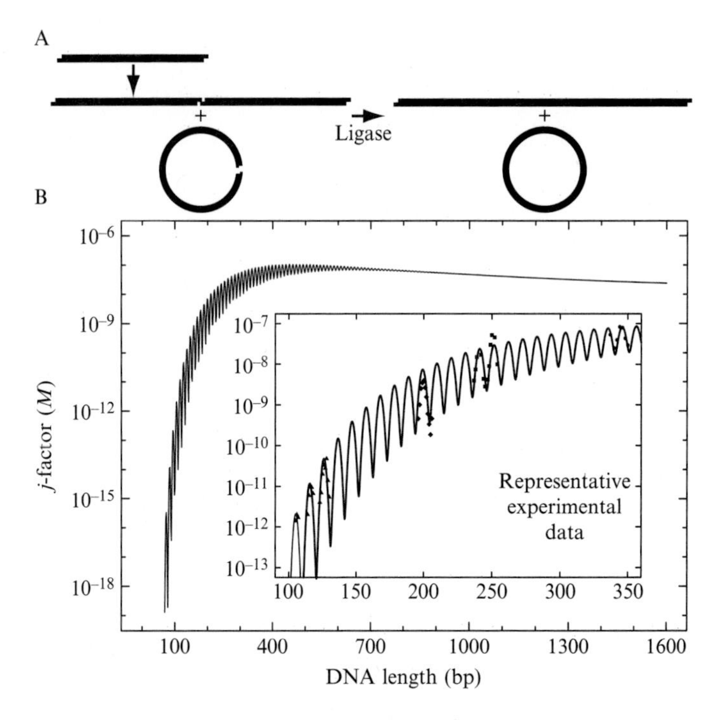

Figure 12.2 *Theoretical and experimental j-factor determination by DNA cyclization kinetics in vitro.* (A) Schematic depiction of the formation of transient circles and multimers by DNA terminal interactions, which are then captured by DNA ligation. (B) Predicted dependence of *j*-factor on DNA length for a persistence length of 46.5 nm, torsion constant of 2.4×10^{-19} erg-cm, and helical repeat of 10.48 bp. Inset: expanded trace of predicted *j*-factor with representative experimental data from multiple laboratories for purified DNA *in vitro*.

3. Ligase-Catalyzed DNA Cyclization Kinetics *In Vitro*

3.1. DNA cyclization kinetics theory

Ligase-catalyzed DNA cyclization is a powerful classic approach that has received widespread theoretical and experimental consideration (Crothers *et al.*, 1992; Du *et al.*, 2005; Hagerman and Ramadevi, 1990; Kahn and Crothers, 1992; Peters and Maher, 2010; Podtelezhnikov *et al.*, 2000; Shore *et al.*, 1981; Taylor and Hagerman, 1990; Vologodskaia and Vologodskii, 2002; Zhang and Crothers, 2003). Details of experimental conditions and key assumptions can be crucial to the validity of the results (Davis *et al.*, 1999; Du *et al.*, 2005). Recently, we have reviewed the theoretical derivation underlying the analysis of ligase-catalyzed DNA cyclization kinetics,

(Peters and Maher, 2010) as originally described by Shore and Baldwin (1983) (Shore *et al.*, 1981). In brief, cyclization converts a linear DNA molecule (L) with cohesive termini into a circular substrate (S_c) for DNA ligase (E). The ligation reaction produces a covalently closed DNA circle as product (P_c).

Cyclization is proposed to proceed through the following kinetic mechanism:

$$L \underset{k_{21}}{\overset{k_{12}}{\rightleftarrows}} S_c + E \underset{k_{32}}{\overset{k_{23}}{\rightleftarrows}} ES_c \overset{k_{34}}{\rightarrow} E + P_c \tag{12.7}$$

The subscripts on the rate constants indicate which sets of species are being converted.

Ligation kinetics experiments should be performed under conditions where the dissociation of cohesive ends is fast compared to ligase-catalyzed ring closure, so that there exists a rapid preequilibrium between L and S_c. In this case, the formation of ES_c, which involves the bimolecular formation of an enzyme–substrate complex (ES_c), is slow relative to end dissociation, and the rate of ligase-catalyzed covalent ring closure (k_c) is directly proportional to the equilibrium fraction of cyclized molecules (f_{S_c}) given by

$$f_{S_c} = \frac{[S_c]}{[L] + [S_c]} = \frac{k_{12}}{k_{12} + k_{21}} \tag{12.8}$$

This condition can be assured by using a sufficiently small concentration of enzyme E. If in addition the concentrations of S_c is sufficiently small relative to L, then the rate constants k_C and k_D for the ligation of both circular and biomolecular substrates can be expressed in terms of the equilibrium constants for transient end-joining by applying the steady-state approximation to the kinetic mechanism given in Eq. (12.7), along the lines of Michaelis–Menten kinetics. The full derivation (Peters and Maher, 2010) yields

$$j \equiv \frac{K_{a,\text{circular}}}{K_{a,\text{bimolecular}}} \approx \frac{k_C}{k_D} \tag{12.9}$$

where K_a values are equilibrium constants and k_C and k_D are the rate constants for ligase-catalyzed intramolecular cyclization and intermolecular dimerization, respectively. Equation (12.9) shows that, subject to the rapid preequilibrium conditions, the *j*-factor (defined as a ratio of equilibrium constants) can be obtained from the ratio of two experimentally measurable rate constants. These constants (K_C and K_D) are both dependent on DNA length and on temperature. However, for DNA lengths 126–4361 bp, K_D

has been shown to be essentially independent of DNA length (Shore *et al.*, 1981). The length dependence of *j* therefore reflects primarily changes in K_C, which Fig. 12.2B shows to be much more dramatic. As above, the dominant determinants of K_C for small DNA molecules are the energy required for bending DNA into a small circle and the energy required for twisting the DNA for alignment of termini.

The conditions that are required for the validity of Eq. (12.9) have been a matter of concern in the field. Rapid preequilibria among species require that the K_m of ligase for substrate is in excess of the substrate concentration, and the connection between kinetics and thermodynamics requires that the fraction of substrate must be small for each substrate type. Thus, in practice, the rate of product formation must be directly proportional to the ligase concentration. The requirement that K_m exceeds substrate concentration must be met for both circular and bimolecular substrates (which are likely to be at very different concentrations) so that the relative equilibrium populations of each are faithfully reported. Fundamentally, the *j*-factor estimate should not be dependent on experimental DNA or ligase concentrations, and this must be tested over a range of concentrations of each. This can be challenging, because if the *j*-factor is small or the DNA concentration is large, the amount of cyclized product may be small relative to bimolecular products. If the *j*-factor is large, then the bimolecular product can be completely suppressed.

Careful cyclization experiments require substrates of higher purity than typical molecular biology applications such as cloning, mobility shift experiments, or even many types of footprinting. This is because cyclization requires that both DNA termini be fully active, with the correct sequence overhangs and phosphorylation states. DNA with only one active terminus can still undergo bimolecular ligation, and the appearance of low levels of bimolecular products from DNA that is incapable of cyclization can lead to underestimation of the *j*-factor, especially for molecules with large *j*-factors. On the other hand, it can be difficult to measure the true cyclization-competent concentration (the endpoint of the cyclization reaction), because partially melted molecules or molecules with defective termini can still be ligated, albeit slowly, and ligation of a small population of aberrant molecules can lead to overestimation of the *j*-factor, especially for molecules with low *j*-factors. These issues have been discussed (Davis *et al.*, 1999), but because of the inhomogeneity of the problematic DNA molecules, to date there is no satisfactory way to consider them quantitatively in the analysis of ligation data.

3.2. Representative protocols for DNA probe design, preparation, labeling, and quantitation

Several methods have been described for preparing DNA probes for *j*-factor determination and for carrying out the ligation experiments (Shore *et al.*, 1981; Vologodskaia and Vologodskii, 2002; Zhang and Crothers, 2003).

We have found the approach described below to be relatively simple and to give consistent experimental results. The approach provides stronger signals than analyses requiring limited reaction progress with little depletion of starting material, and thus any problems with nonuniform reactivity of the probe are more readily apparent. We routinely use this procedure to measure the effect of high mobility group B (HMGB) proteins or DNA base modifications on the *j*-factor. Comparisons are facilitated because the effects of nonspecific DNA-binding proteins or the differential effects of base modifications can be examined using a uniform set of starting DNA constructs. These are provided by plasmid pUC19-based constructs pJ823, pJ825, pJ827, pJ829, pJ831, and pJ833, which contain intrinsically straight ~200 bp sequences (Vologodskaia and Vologodskii, 2002). The sequences were created using unique 5-bp direct repeats to eliminate long-range sequence-directed curvature.

Polymerase chain reaction (PCR) products containing these intrinsically straight subsequences, flanked by *Hin*dIII sites, are amplified using primers LJM-3222 (5′-$G_3TA_2CGC_2AG_3T_4$) and LJM-3223 (5′-TGTGAG-T_2AGCTCACTCAT_2AG_2) to give ~400-bp products. PCR reactions (300 μL) include 24-ng plasmid template, 0.4 μM each forward and reverse primers, 100-μg/mL bovine serum albumin (BSA), *Taq* DNA polymerase buffer (Invitrogen), 4-m*M* $MgCl_2$, 0.2 m*M* each dNTP (exact concentrations confirmed spectrophotometrically at pH 7.0: dATP, $\lambda_{max\ 259\ nm}$ 15.2×10^3 M^{-1} cm^{-1}; dCTP, $\lambda_{max\ 271\ nm}$ 9.1×10^3 M^{-1} cm^{-1}; dGTP, $\lambda_{max\ 253\ nm}$ 13.7×10^3 M^{-1} cm^{-1}; dTTP, $\lambda_{max\ 267\ nm}$ 9.6×10^3 M^{-1} cm^{-1}), 60-μCi [α-^{32}P]dATP (typically 3000 Ci/mmol stock specific activity, but any specific activity can be used as long as the chemical concentration of the label is much less than 0.2 m*M*), and 12 U *Taq* DNA polymerase (Invitrogen). Cycle conditions are initial denaturation at 98 °C for 3 min, 30 cycles of 94 °C (30 s), 55 °C (30 s), 72 °C (45 s), and a final extension at 72 °C for 5 min. The reaction is then adjusted to 5-m*M* EDTA, 10-m*M* Tris–HCl (pH 8.0), and 0.5% SDS and treated with 50-ng/μL proteinase K at 37 °C for 1 h.

Radioactivity in a defined fraction of the total PCR mixture is measured by scintillation counting. Reactions are extracted with an equal volume of phenol:chloroform (1:1) and the DNA is precipitated from ethanol before overnight digestion with *Hin*dIII under conditions recommended by the supplier (New England Biolabs). We recommend using the smallest amount of restriction enzyme that is sufficient for a nearly complete digest, as this minimizes the fraction of damaged termini arising from star activity or from contaminating nucleases or phosphatases. The digest is loaded onto a 5% native polyacrylamide gel (29:1 acrylamide:bisacrylamide) and visualized by exposure to BioMAX XR film. The ~400-bp PCR product is cleaved by *Hin*dIII digestion into three easily distinguishable products, allowing facile purification of the ~200 bp probe species without any uncut DNA, and with the added advantage that no synthetic DNA remains in the final product. If an inactive DNA problem is encountered, purification on a higher-percentage 40-cm gel and allowing the desired probe to migrate at least 30 cm may help separate the desired product from possible exonuclease/phosphatase products. The gel slice including the probe is cut out, crushed, and eluted overnight at 22 °C in 50-m*M* NaOAc, pH 7.0. The eluted DNA is precipitated with ethanol. The final concentration of probe DNA is calculated by determining the fraction of original radioactivity incorporated, multiplying this result by the unlabeled dNTP concentration in the PCR reaction, and dividing by the number of dA residues per duplex DNA probe. For example, typically 1 μL of the original 300-μL PCR reaction is diluted 1:100 and the radioactivity in 2 μL of this diluted sample is compared with the radioactivity in 2 μL of purified DNA probe. That fraction (corrected for dilution) determines the fraction of the 60-nmol dATP (the amount in the original PCR reaction) that is present in the purified DNA probe, giving the number of moles of A residues in the purified product.

3.3. Purification of HMGB proteins

The *Saccharomyces cerevisiae* Nhp6A coding region is cloned between the *Nco*I and *Bam*HI sites of bacterial expression vector pET-15b (Novagen). The resulting plasmid (pJ1400) encodes untagged Nhp6A protein (molecular weight 10,801) with the following amino acid sequence:

$MVTPREPK_2RT_2RK_3DPNAPKRALSAYMF_2ANENRDIVRSENP$
$DITFGQVGK_2LGE\text{-}KWKALTPE_2KQPYEAKAQADK_2RYESEKELYN$
$ATLA$

A 6-mL overnight LB culture of *Escherichia coli* strain BL21(DE3) containing carbenicillin (Cb) (50 μg/mL) is grown from a single colony, pelleted by centrifugation, washed with 1-mL fresh LB medium to remove secreted β-lactamase, and resuspended in 1 mL of medium. A 250-μL aliquot is used to inoculate a 250-mL LB–Cb culture. Cells are grown with shaking (250 rpm) at 37 °C until an OD_{600} of 0.600 is achieved. IPTG is added to a final concentration of 1 m*M* and cells are grown at 37 °C for a further 3 h with continued shaking. The culture is then subjected to centrifugation at 6000×*g* for 15 min at 4° C to pellet cells. The pellet is resuspended in a 6-mL lysis buffer (50-m*M* Na_2HPO_4/NaH_2PO_4, 100-m*M* NaCl, 1-m*M* EDTA, pH 7.0), sonicated three times with 15 s pulses separated by 1-min intervals on ice. Lyzate is clarified by centrifugation at 20,000×*g* for 45 min at 4 °C. The soluble fraction is retained and heated at 70 °C in a water bath for 10 min to irreversibly denature host proteins. The resulting opaque suspension is cooled at room temperature for 30 min, clarified by centrifugation at 20,000×*g* for 45 min (4 °C), and the supernatant retained. A 1:40 dilution of 3% polyethyleneimine (PEI) is added to the supernatant with gentle agitation for 20 min at 4 °C. Nucleic acids are then removed by centrifugation at 5000×*g* for 10 min at 4 °C, and the clear supernatant, containing the Nhp6a protein, is recovered. The protein is concentrated using Vivaspin spin columns (Vivascience; 5000 molecular weight cutoff) and purified by reverse phase-HPLC on a Jupiter C_{18} column (250 × 21.2 mm, 15 μm; Phenomenex, Torrance, CA) in 0.1% trifluoroacetic acid (TFA)/water with a 50-min gradient from 10 to 70% B, where B is 80% acetonitrile/0.1% TFA. Protein concentrations are estimated by Bradford assay (BioRad, Hercules, CA).

Plasmid pJ1192 encodes rat HMGB1 (boxes A + B, with an internal *Nde*I site removed by silent site-directed mutagenesis) subcloned between the *Nde*I and *Xho*I sites of expression vector pET-15b. The encoded HMGB1(A + B) protein (molecular weight 23,321) lacks the acidic C-terminal tail and has the amino acid sequence shown below (affinity tag in lower case), HMG boxes A and B underlined:

mgs$_2$h$_6$s$_2$glvprgshMGKGDPK$_2$PRGKMS$_2$YAF$_2$VQTCRE$_2$HK$_3$HPD ASVNFSEFSK$_2$CSERWKTMSAKEKGKFEDMAKADKARYEREMK TYIP$_2$KGETK$_3$FKDPNA PKRP$_2$SAF$_2$LFCSEYRPKIKGE HPGLSIGD VAK$_2$LGEMWN$_2$TA$_2$D$_2$KQPYEK$_2$A$_2$KLKEKYEKDIA$_2$YRAKGKPD A$_2$K$_2$GV$_2$KAEKSK$_4$

This plasmid is transformed into bacterial strain BL21 Gold (Stratagene), and a single bacterial colony is used to inoculate a 6-mL overnight LB–Cb culture. Cells are then pelleted, washed with 1-mL fresh LB medium to remove excess β-lactamase, and 250 μL is used to inoculate a 250-mL culture that is then grown as described above. Cells are pelleted by centrifugation at 6000×*g* and stored at −80 °C. The cell pellet is resuspended in 10 mL 1× Nickel binding buffer (Novagen) and sonicated as described above. The lyzate is clarified by centrifugation at 15,000×*g* for 20 min (4 °C). Protein is purified from the lyzate using Nickel chelate chromatography on a Novagen "Quick 900" cartridge with buffers provided by the manufacturer. Eluted fractions are concentrated to ∼2 mL using Vivaspin spin columns (Vivascience). Purified protein is dialyzed overnight against a 1-L storage buffer (20-m*M* HEPES pH 7.5, 100-m*M* KCl, 1-m*M* EDTA) in 3500 MWCO Slide-a-Lyzer dialysis cassettes (Pierce) at 4 °C, and then dialyzed a second time overnight against storage buffer containing 5% glycerol. Protein is further purified by size exclusion chromatography with a Superdex 75 10/30 column and dialyzed into a storage buffer containing glycerol.

3.4. DNA ligation kinetics

T4 DNA ligase-catalyzed cyclization reactions (50 or 100 μL) are performed at 22 °C with 30-n*M* DNA restriction fragment (with or without various concentrations of added HMGB protein), ligation buffer (20-m*M*

Tris–HCl, pH 8.0, 30-m*M* KCl, 100-μg/mL BSA, 1.8-m*M* ATP, 10-m*M* $MgCl_2$), and T4 DNA ligase [final concentrations of 50 U/mL (DNA alone) or 20 U/mL (DNA and HMGB protein), with ligase units as defined by New England Biolabs]. Ligase concentrations must be varied empirically depending on the *j*-factor and the stability of cohesive end hybridization. If a wide range of ligase concentrations are to be used, it is necessary to use siliconized tubes and to add NP40 detergent (Igepal) to the ligase buffer to prevent loss of ligase upon serial dilution. As described below, we test ligase concentrations to demonstrate that they do not influence *j*-factor estimates for ~200-bp probes, and a similar conclusion was previously reached for DNA probes as short as 100 bp (Du *et al.*, 2005). Aliquots (10 μL) are removed at various times and quenched by addition of EDTA to 20 m*M*. In cases where high concentrations of HMGB proteins were present, protein is purified away from DNA using Qiagen QIAQuick PCR Clean Up columns and DNA recovered in 30-μL elution buffer.

Ligation reaction aliquots are analyzed (Fig. 12.3) by electrophoresis through 5% native polyacrylamide gels (29:1 acrylamide:bisacrylamide) in 0.5× TBE buffer, drying, and storage phosphor imaging (Fig. 12.3A). Gel imaging is done with STORM Scan software, and quantitation is performed with NIH ImageJ software for Macintosh. After correction for background, signal intensities for individual species are measured and normalized to the total signal for each gel lane. The resulting values are multiplied by the original DNA concentration to estimate the concentration of each species (Fig. 12.3C).

Experiments done at varying DNA concentrations can determine which products are linear versus circular: circular products are relatively more abundant at lower DNA concentrations. In addition, results may be verified using treatment with Bal31 exonuclease as follows: Aliquots (~10 μL) of the final time point (prior to EDTA addition) are diluted in Bal31 reaction buffer (New England BioLabs) and brought to a final volume of 20 μL with addition of 1 U of Bal31 exonuclease. Reactions are incubated at 30 °C for 30 min. The Bal31 reactions can then be analyzed by electrophoresis with the other reactions. The circular products will be the only remaining labeled species (Fig. 12.3A, c.f. lanes 9 and 10). Very small DNA circles (<80 bp) or supercoiled minicircle DNA may also be digested.

3.5. Data fitting and ODE approach

Ligation kinetics results in which only one circular product is formed can be fit to analytical expressions described in our early work (Davis *et al.*, 1999; Kahn and Crothers, 1992). Systems in which multimer cyclization is considered require numerical simulation. Our current method of *j*-factor analysis is based on modeling cyclization kinetics with a reasonable reaction

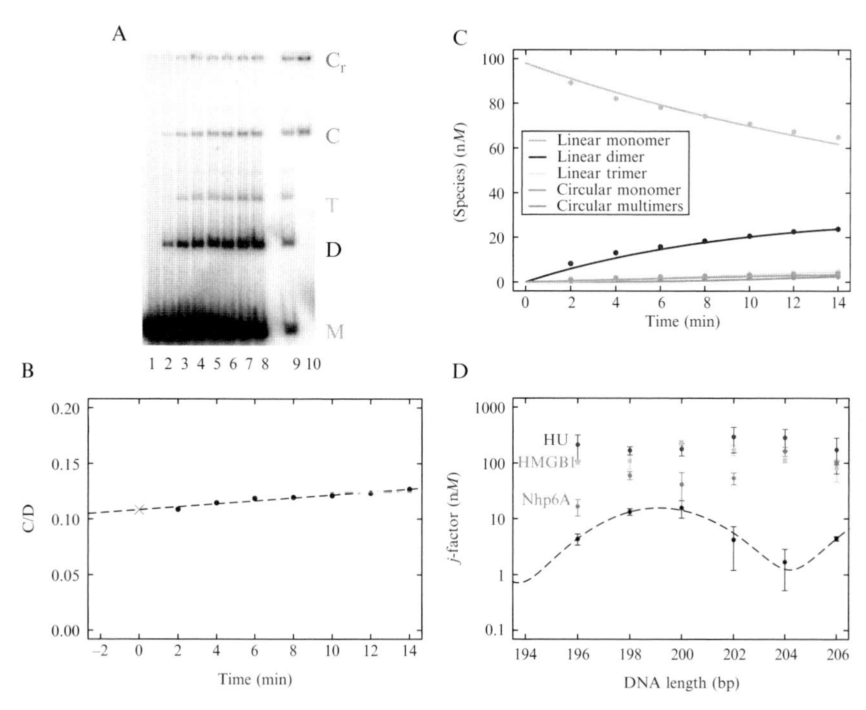

Figure 12.3 *Cyclization kinetics data analysis to estimate j-factor in vitro.* (A) Imaged native polyacrylamide gel. Lane 1 contains monomer (*M*) alone and lanes 2–8 are increasing 2 min time points showing the evolution of linear monomer (*M*), dimer (*D*), and trimer (*T*) as well as circular monomer (*C*) and circular dimer and trimer (collectively C_r). Lanes 9 and 10 are control treatment and Bal31 exonuclease treatment, respectively, of the final time point. Bal31 treatment (lane 10) provides verification of circular species. (B) Extrapolation method based on the ratio (**C/D**) of monomer circle products (**C** = *C*) to the sum of all other ligation products (**D** = $D + T + C_r$) under conditions that limit the extent of ligation reaction. The *j*-factor estimate is given by twice the initial probe DNA concentration multiplied by the *y*-intercept value (red cross). (C) Kinetic analysis applicable to full-time course of ligation reactions with multiple products and depletion of starting material. (D) Example of cyclization data published for intrinsically straight DNA (Vologodskaia and Vologodskii, 2002) and measured in our laboratory for either free DNA (fitted with WLC model) or in the presence of 40 n*M* Nhp6A (red), 40 n*M* HMGB1 (green), or 10 n*M* HU (blue). (See Color Insert.)

model derived from the principles of mass action (Fig. 12.4A) and a corresponding system of ordinary differential equations (ODEs; Fig. 12.4B). Rate constants are determined by least-squares fitting to the experimental data. This approach can be applied throughout the course of the reaction. In contrast, analysis based on linear extrapolation methods requires limited reaction progress. When applied to DNA concentrations where the production of pentameric (and larger) products is negligible, all

A

$$
\begin{array}{lllll}
M+M \xrightarrow{k_D} D & D+D \xrightarrow{k_D} Q & T+T \xrightarrow{k_D} \ldots & Q+Q \xrightarrow{k_D} \ldots & P+P \xrightarrow{k_D} \ldots \\
M+D \xrightarrow{k_D} T & D+T \xrightarrow{k_D} P & T+Q \xrightarrow{k_D} \ldots & Q+P \xrightarrow{k_D} \ldots & \\
M+T \xrightarrow{k_D} Q & D+Q \xrightarrow{k_D} \ldots & T+P \xrightarrow{k_D} \ldots & & \\
M+Q \xrightarrow{k_D} P & D+P \xrightarrow{k_D} \ldots & & & \\
M+P \xrightarrow{k_D} \ldots & & & &
\end{array}
\qquad
\begin{array}{l}
M \xrightarrow{k_{C1}} C_M \\
D \xrightarrow{k_{C2}} C_D \\
T \xrightarrow{k_{C3}} C_T \\
Q \xrightarrow{k_{C4}} C_Q \\
P \xrightarrow{k_{C5}} C_P
\end{array}
$$

B

$$
\begin{aligned}
\frac{d[M]}{dt} &= -4k_D[M]^2 - 4k_D[M][D] - 4k_D[M][T] - 4k_D[M][Q] - 4k_D[M][P] - k_{C1}[M] \\
\frac{d[D]}{dt} &= 2k_D[M]^2 - 4k_D[D][M] - 4k_D[D]^2 - 4k_D[D][T] - 4k_D[D][Q] - 4k_D[D][P] - k_{C2}[D] \\
\frac{d[T]}{dt} &= 4k_D[M][D] - 4k_D[T][M] - 4k_D[T][D] - 4k_D[T]^2 - 4k_D[T][Q] - 4k_D[T][P] - k_{C3}[T] \\
\frac{d[Q]}{dt} &= 2k_D[D]^2 + 4k_D[M][T] - 4k_D[Q][M] - 4k_D[Q][D] - 4k_D[Q][T] - 4k_D[Q]^2 - 4k_D[Q][P] - k_{C4}[Q] \\
\frac{d[P]}{dt} &= 4k_D[M][Q] + 4k_D[D][T] - 4k_D[P][M] - 4k_D[P][D] - 4k_D[P][T] - 4k_D[P][Q] - 4k_D[P]^2 - k_{C5}[P] \\
\frac{d[C_M]}{dt} &= k_{C1}[M] \\
\frac{d[C_D]}{dt} &= k_{C2}[D] \\
\frac{d[C_T]}{dt} &= k_{C3}[T] \\
\frac{d[C_Q]}{dt} &= k_{C4}[Q] \\
\frac{d[C_P]}{dt} &= k_{C5}[P]
\end{aligned}
$$

Figure 12.4 *Kinetic scheme for analysis of DNA cyclization kinetics.* (A) Left: Intermolecular ligation reactions giving rise to linear products. Right, intramolecular cyclization reactions giving rise to cyclic (*C*) products. Monomer, dimer, trimer, tetramer, and pentamer are indicated by *M*, *D*, *T*, *Q*, and *P*, respectively. Rate constants are indicated, assumed equal for all intermolecular ligations. (B) System of ordinary differential equations describing accumulation of species up to and including pentamers. When it occurs, the statistical factor of four accounts for the inability to distinguish between identical DNA termini in intermolecular ligation of linear species.

possible interactions of linear species up to pentamer are modeled, that is, linear monomer (*M*), dimer (*D*), trimer (*T*), tetramer (*Q*), and pentamer (*P*). Each of these linear species can also cyclize. The ODE system expresses rates of change of concentration for each species. This system is integrated numerically by R software (Version 2.8.1) using the ODE solver package deSolve. The theoretically predicted values of concentrations are used for nonlinear least-squares refinement of rate constants to fit the measured experimental data. R scripts and related files containing the code to solve the ODE system, perform the optimization, and generate plots of the results (e.g., Fig. 12.3C) are provided in Appendices A and C. From the fit, optimized values for the rates of intramolecular (k_{C_i}) and intermolecular (k_D) collisions are used to determine the desired *j*-factor, where

$$j_i = k_{C_i}/k_D \tag{12.10}$$

and i is taken to be 1 for monomer cyclization. Alternatively, using the substitution $k_D = k_{C_1} / j_1$, the j-factor of interest can be directly obtained from the curve fitting routine. If topoisomers are formed from a given linear molecule, each topoisomer has an independent j-factor.

Figure 12.3 shows an example of cyclization kinetics analysis. A method based on linear extrapolation of the ratio of monomer circle (C) to all forms of intermolecular products (D; Vologodskaia and Vologodskii, 2002; Vologodskii *et al.*, 2001) can be effective early in the reaction when DNA monomer is not significantly depleted (Fig. 12.3B). The more general fitting approach advocated here is shown in Fig. 12.3C.

Our fitting method is based on the standard assumptions about ligation reaction kinetics described above. The validity of these assumptions is subject to verification. Importantly, the experimental j-factor should not be dependent on the concentration of either ligase or DNA. Example analyses of these assumptions are shown in Fig. 12.5. In contrast to the analytical solutions or low-conversion approximation, the ODE method can be extended to consider different populations of linear substrates, such as mixtures of bound versus free DNA or populations with inactive DNA ends; where accurate determination of more rates or equilibrium constants will require correspondingly comprehensive experimental data.

3.6. Example data

Fig. 12.3D shows examples of data obtained by this method. The experimental j-factor was determined for labeled DNA probes of the indicated lengths, either as free DNA or in the presence of the indicated concentrations of sequence-nonspecific HMGB proteins known to enhance the apparent flexibility of DNA. Data from our laboratory for purified DNA (Fig. 12.3D, (fitted with WLC model)) are similar to those previously published (Vologodskaia and Vologodskii, 2002). The yeast (Nhp6A), mammalian (HMGB1), and bacterial (HU) architectural proteins dramatically increase apparent DNA flexibility as measured by the j-factor (Fig. 12.3D, as indicated). These results emphasize how the profound ability of HMGB and related architectural proteins to enhance the apparent flexibility of DNA can be quantitated by experimental measurement of the j-factor. The proteins add random kink sites, increasing the experimental j-factor by multiple orders of magnitude and serving to facilitate DNA transactions requiring small loops.

It should also be noted that there are a number of approaches to the quantitative prediction of j-factors from DNA sequence and structural models, but these are beyond the scope of this chapter.

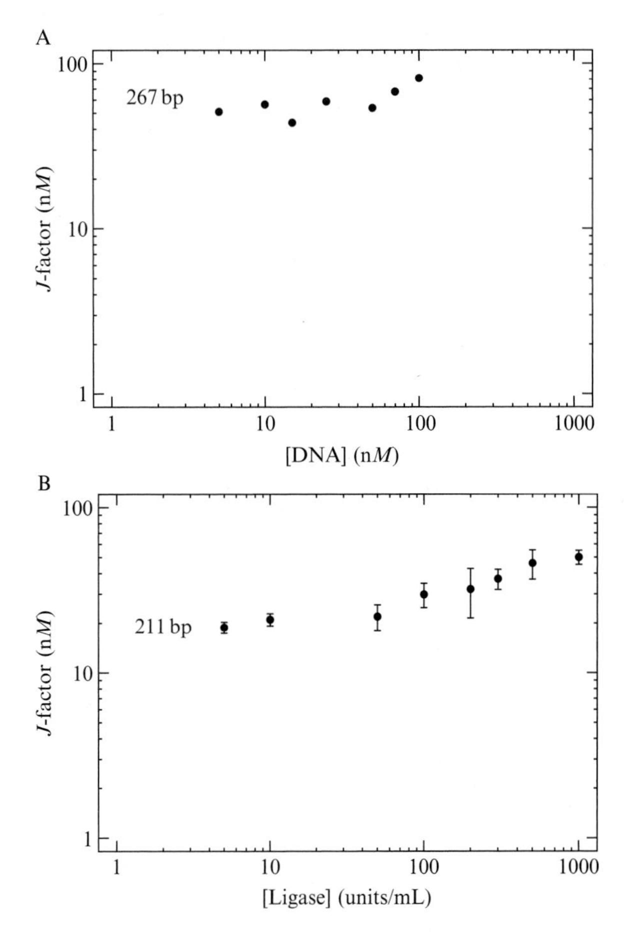

Figure 12.5 *Control experiments to ensure validity of experimental DNA cyclization kinetics.* The experimental *j*-factor should be estimated under conditions where the apparent *j*-factor is independent of DNA concentration (A) and ligase concentration (B).

4. *In Vivo* Analysis of *E. coli lac* Repression Loops

4.1. Concept

Certain *in vivo* assays provide the unusual opportunity to measure processes that depend upon DNA flexibility and looping in living cells. Assays based on components of the *E. coli lac* operon depend upon DNA flexibility through cooperativity at a distance (Fig. 12.1B; Becker *et al.*, 2005, 2007, 2008; Bellomy *et al.*, 1988; Kramer *et al.*, 1987; Law *et al.*, 1993; Mossing and Record, 1986; Muller *et al.*, 1996; Oehler *et al.*, 1990; Peters and Maher, 2010). Here, one operator (O_2) overlaps with the promoter of the *lacZ* reporter gene and serves as a very weak lac repressor binding site. The *lacZ* gene is only weakly repressed when this operator is present in isolation. Repression is strongly enhanced by placement of a high-affinity operator (O_{sym}) upstream of the promoter. The bidentate lac repressor tetramer bound at O_{sym} site has the potential to increase the local repressor concentration at O_2 by virtue of DNA looping (Mossing and Record, 1986; Oehler and Müller-Hill, 2009). Such looping requires different degrees of DNA (and repressor) distortion as a function of interoperator spacing and any intrinsic curvature or unusual flexibility in the loop DNA. The requirement for DNA twisting when the operators are out-of-phase with the DNA helical repeat (hr) makes looping less energetically favorable in these cases. Analysis of *lacZ* repression as a function of operator spacing has been shown to provide a sensitive assay of the bending and twisting flexibility of the DNA/lac repressor complex *in vivo* (Becker *et al.*, 2005, 2007, 2008; Bellomy *et al.*, 1988; Garcia *et al.*, 2007; Zhang *et al.*, 2006a,b).

The *E. coli lac* looping system permits quantitative analysis of DNA looping by measurement of β-galactosidase (the *lacZ* gene product) activity extracted from living cells. Analysis is performed in the absence (uninduced) or presence (induced) of IPTG, an allolactose analog that reduces lac repressor affinity for DNA. Thermodynamic modeling allows for identification of subtle trends in the data (Becker *et al.*, 2005, 2007, 2008; Bellomy *et al.*, 1988; Law *et al.*, 1993; Zhang *et al.*, 2006a,b). Although the present analysis is limited to *in vivo* DNA looping using components of the *E. coli lac* operon, other analyses have been performed in eukaryotic cells, as recently reviewed (Peters and Maher, 2010).

The goals of the *in vivo* work include assessing how the results differ from the *in vitro* WLC model. The bending and twisting flexibilities of DNA *in vivo* may be influenced by supercoiling, DNA-binding proteins, and ionic and osmotic conditions. Eventually, for quantitative modeling of gene regulation, it will be necessary to understand the *in vivo* physical chemistry of DNA in as much detail as we now have for DNA *in vitro*.

4.2. Experimental design

The thermodynamic model described here is based on the pioneering work of Record and coworkers (Bellomy *et al.*, 1988; Law *et al.*, 1993). Analysis is based on the premise that promoter repression is determined by the degree of occupancy of the proximal lac operator (O_2 in our system) at equilibrium; we do not consider the possible effects of looping *per se*. A variety of related models that differ in their descriptions of the effects of length variation have been developed (Bintu *et al.*, 2005; Zhang *et al.*, 2006a,b).

In our model, the extent of promoter repression is calculated by evaluating the distribution of possible states of the proximal *lac* operator. The promoter is repressed whenever the weak proximal O_2 operator is bound by lac repressor. This condition is most likely to arise from singly bound repressor at O_2 ("single bound") or from repressor bound to the proximal operator by virtue of DNA looping from the strong distal O_{sym} operator (a phase-dependent "specific loop"). Even the minimal observed repression looping (for out-of-phase operator separations) provides stronger repression than is observed when either of the operators is deleted. Other possible repressed states include repressor delivered to O_2 by non-phase-dependent transfer mechanisms (e.g., sliding or hopping) from the distal operator, or repressor delivered by looping to pseudooperator sites overlapping O_2. Such states are considered together in our model as nonspecific loops ("NS loop").

A partition function for the system expresses the sum of possible states of the O_2 operator:

$$[\mathrm{Free}] + [\mathrm{Specific\ Loop}] + [\mathrm{NS\ Loop}] + [\mathrm{Single\ bound}] = [\mathrm{O_2}] \quad (12.11)$$

This expression can be cast in terms of the equilibrium constants for the different states:

$$\begin{aligned} &[\mathrm{Free}](1 + K_{\mathrm{SL}} + K_{\mathrm{NSL}} + K_{\mathrm{O_2}}) = [\mathrm{O_2}] \\ &\text{where } K_{\mathrm{SL}} = \frac{[\mathrm{Specific\ Loop}]}{[\mathrm{Free}]},\ K_{\mathrm{NSL}} = \frac{[\mathrm{NS\ Loop}]}{[\mathrm{Free}]},\ \text{and} \\ &K_{\mathrm{O_2}} = \frac{[\mathrm{Single\ bound}]}{[\mathrm{Free}]} \end{aligned} \quad (12.12)$$

The constant cellular concentration of lac repressor has been absorbed into each of the equilibrium constants; we assume one or low copy number DNA template so there is no interaction between separate DNA molecules and no depletion of LacI.

The fraction of bound (repressed) O_2 operators is given by the sum of the statistical weights of the bound forms divided by the total partition function:

$$\begin{aligned} f_{\mathrm{bound}} &= ([\mathrm{Specific\ Loop}] + [\mathrm{NS\ Loop}] + [\mathrm{Single\ bound}])/[\mathrm{O_2}] = ([\mathrm{O_2}] - [\mathrm{Free}])/[\mathrm{O_2}] \\ &= (K_{\mathrm{SL}} + K_{\mathrm{NSL}} + K_{\mathrm{O_2}})/(1 + K_{\mathrm{SL}} + K_{\mathrm{NSL}} + K_{\mathrm{O_2}}) \end{aligned} \quad (12.13)$$

Experimentally, the fraction bound is given by

$$f_{\text{bound}} = \frac{\text{max induced activity} - \text{observed activity}}{\text{max induced activity}} \tag{12.14}$$

where the maximum induced activity is potentially different for each *E. coli* strain background. Control experiments (performed under both repressing and inducing conditions) with an isolated O_2 operator are used to determine K_{O_2} for each strain background. Based on the results of initial experiments, these values were fixed at 1 and 0 for repressing and inducing conditions, respectively.

Torsional flexibility of the repression loop is modeled by formulating a helical-phasing-dependent equilibrium constant for specific loop formation, K_{SL}. A sum of Gaussians expresses the total probability of twist deformations needed to bring the two operators into phase.

$$K_{SL} = \sum_{i=-5}^{5} K_{\max} e^{-\left(\text{sp}-\text{sp}_{\text{optimal}}+i.hr\right)^2/2\sigma_{\text{Tw}^2}} \tag{12.15}$$

The parameter sp gives the actual spacing (bp) between operator centers for a given construct; $\text{sp}_{\text{optimal}}$ is the spacing (bp) for optimal repression, $K_{\max}$ is the equilibrium constant for DNA loop formation when operators are perfectly phased, and σ_{Tw} is the standard deviation of the torsion angle between operators (given thermal fluctuations). Summation over the integer i captures all possible overtwisting or undertwisting needed to give the helical phasing required for optimal loop formation. Finally, σ_{Tw} is calculated from the torsional flexibility per base pair:

$$\sigma_{\text{TW}}{}^2 = \text{sp}\cdot\sigma_{\text{bp}}^2 \tag{12.16}$$

where σ_{bp} is the standard deviation of twist (per bp) given by

$$\sigma_{\text{bp}} = \sqrt{\frac{\ell kT}{C_{\text{app}}}}\text{radians}\cdot\frac{1\,\text{turn}}{2\pi\,\text{radians}}\cdot\frac{\text{hr bp}}{\text{turn}} = \frac{\text{hr}}{2\pi}\sqrt{\frac{\ell kT}{C_{\text{app}}}} \tag{12.17}$$

in units of bp twist increments. Here, ℓ is the average bp separation (3.4 Å), k Boltzmann's constant, T the absolute temperature, C_{app} the apparent torsional modulus for the DNA in the loop, and *hr* is the DNA helical repeat. Variability in the possible torsion angles permitted in a protein–DNA loop reduces the apparent torsional modulus, C_{app}, increasing the apparent σ_{Tw}.

In principle, additional K_{SL} terms could be included to consider different loop geometries: "open-form" versus V-shaped Lac repressor might well have different optimal spacing and K_{max} values. However, the data we have to date have not required consideration of multiple loop shapes for uninduced loops: complex repression efficiency curves derive instead from the relationship between induced and uninduced activities.

Surprisingly, our experimental data typically show no dependence of K_{SL} on operator spacing except through the effect of spacing on torsion: the apparent persistence length is effectively small enough that the loop free energy does not change markedly with distance. Since fitting the data did not require consideration of persistence length, our initial approach (Becker *et al.*, 2005, 2007) did not consider DNA length *per se*: f_{bound} was modeled as a function of DNA length (sp), with five adjustable parameters hr, C_{app}, K_{max}, K_{NSL}, and $sp_{optimal}$. Predicted absolute activities were calculated using f_{bound} and the measured maximal induced activity, and repression ratios were calculated from the ratios of these computed absolute activities.

A more general version of this model was subsequently developed for the analysis of data on the effects of eukaryotic HMGB proteins expressed in *E. coli*, which did exhibit twist-independent length dependence (Becker *et al.*, 2008). The experimental fraction of O_2 bound by repressor (f_{bound}) is now modeled as a function of DNA spacer length (sp) with six adjustable parameters. Three parameters reflect properties of the repressor–DNA complex. These are the optimal operator spacing in bp as above ($sp_{optimal}$), the equilibrium constant for specific O_{sym}–O_2 loop formation when operators are perfectly phased (K'_{max}, replacing the previous K_{max}, see below), and the equilibrium constant for all forms of O_{sym}-dependent enhanced binding to O_2 other than the specific loop as above (K_{NSL}). Three fitting parameters focus on properties of the looped DNA. The first two are the helical repeat and the apparent torsional modulus of the DNA loop (C_{app}) as above. The length dependence that is observed in the data is addressed by the third empirical fitting parameter, P_{app}, which reflects the expected decrease in DNA bending free energy as operator spacing (sp) increases:

$$\Delta G_{bend} = \frac{PRT}{2 \cdot sp}(\Delta\Theta)^2 = P_{app}/sp \tag{12.18}$$

The DNA persistence length P, the thermal energy RT, and the extent of bending ($\Delta\Theta$) have been included into a single constant P_{app} with units of bp $\times$ energy in increments of RT. Assuming a constant $\Delta\Theta$ is equivalent to a simplifying assumption of constant loop geometry for all sp; as above, alternative loops would have different P_{app} values. The bending energy above contributes an $e^{-P_{app}/sp}$ factor to the looping equilibrium constant K_{max}.

Therefore, normalization of the length dependence is accomplished by replacing K_{max} in the fitting routines with

$$K'_{max} = K_{max} \exp\left[P_{app}\left(\frac{1}{sp_{avg}} - \frac{1}{sp}\right)\right] \tag{12.19}$$

where sp_{avg} is the mean spacing over the data set. K_{max} (the value of K'_{max} at $sp = sp_{avg}$) can be compared directly to the K_{max} obtained using the prior fitting procedure (Becker *et al.*, 2005). An *in vivo* DNA persistence length can be recovered from P_{app} if $\Delta\Theta$ is known, but it is complicated because $\Delta\Theta$ may not be constant. The fact that the amount of actual DNA curvature in the loop likely changes with operator spacing is considered in P_{app}. We view the P_{app} parameter as a physically reasonable way to address cases where increased repression is observed with increasing operator spacing.

Finally, the latest implementation of the model takes into account the distinct values of K_{O_2} for each strain background, as determined from control experiments with an isolated O_2 operator and the maximal induced activity for the given strain. Consequently, comparison of the equilibrium constants requires normalization to K_{O_2}. Thus, we define $K_{max}^{\circ} = K_{max} / K_{O_2}$ and $K_{NSL}^{\circ} = K_{NSL} / K_{O_2}$; in our early work, K_{O_2} was found to be ~ 1 and the normalization was implicit, but in general, this is not the case.

4.3. Data handling and curve fitting

The experimental data comprise β-galactosidase activities (E) in the presence and absence of inducer and the conventional repression ratio (RR), defined as the reporter gene activity in the presence of IPTG (inducing conditions) divided by the activity in the absence of inducer (repressing conditions):

$$RR = \frac{E_{+IPTG}}{E_{-IPTG}} \tag{12.20}$$

Treating the quotients of activities as independent data points in addition to the activities themselves recognizes that changes in absolute E values due to small changes in promoter strength or day-to-day variations tend to compensate, so repression ratios show fewer outliers.

Curve fitting is performed in two steps. First, a global nonlinear least-squares refinement to each set of E values (repressed and induced) is performed with the six adjustable parameters for each data set. This first fitting is followed by a global nonlinear least-squares refinement to E (repressed and induced) and RR simultaneously, with K_{max}, K_{NSL}, and P_{app} estimates held fixed from the first fitting routine. Values and ranges for

hr, C_{app}, and $sp_{optimal}$ are reported from this second fitting. Matlab was used initially, but recently, the open source R environment (version 2.8.1) has been used for all data analysis and fitting. R scripts and related files containing this code appear in Appendices B and C.

Normalized reporter expression (E′) is plotted to allow for comparisons among experiments. E′ are obtained by dividing the observed β-galactosidase activity ($E_{O_{sym}O_2}$) by the activity observed in a bacterial strain lacking an upstream O_{sym} operator, (E_{O_2}), under either repressing or inducing conditions:

$$E' = \frac{E_{O_2 O_{sym}}}{E_{O_2}} \tag{12.21}$$

Additionally, as mentioned above, K_{max} and K_{NSL} are normalized to K_{O_2} to allow for comparisons among strains.

4.4. Reporter constructs for DNA looping in bacteria

Our DNA looping constructs are typically single-copy genes inserted into the F128 episome by way of the plasmid pJ992, which was created by modification of pFW11-null (Whipple, 1998). The wild-type (WT) *lacZ* O_2 at +401 (underlined below) was inactivated by site-directed mutagenesis (bold, QuikChange, Stratagene: La Jolla, CA). Upper and lower mutagenic primers are LJM-1921 5′-CG$_2$AGA$_2$TC$_2$GACGG**G**TG**C**TA **T**TC**AT**T**A**AC**T**TT**C**A$_2$TGT$_2$GATG and LJM-1922 5′-CATCA$_2$CAT$_2$-**G**AA**A**GT**T**A**AT**GA**A**TA**G**CA**C**CCCGTCG$_2$AT$_2$CTC$_2$G. Primers LJM-1930 (5′-CGTCGT$_4$ACA$_2$CGTCG) and LJM-1931 (5′-CAT$_2$GA$_3$G-T$_2$A$_2$T-GA$_2$TAGCAC) are used to monitor and confirm the *lacZ* mutations by PCR amplification. A 362-bp PCR product is amplified only when the mutagenized O_2 is present. A UV5 promoter was then installed between *Bam*HI and *Sal*I sites in pFW11-null. A new O_2 sequence (5′-A$_3$TGTGAGCGAGTA$_2$CA$_2$C$_2$) was cloned immediately downstream of the transcription start site between *Sal*I and *Pst*I sites. This step also installs a *Not*I site downstream of O_2. Additional spacing constructs are created by ligation of different operator inserts between *Sal*I and *Not*I sites of pJ992 with the upstream O_{sym} operator at spacings ranging from 49.5 to 90.5 bp (distances measured between operator centers). Constructs with a single upstream O_{sym} in the absence of a downstream O_2 are created by insertion of an inactive O_2 sequence (5′-GA$_3$GT$_2$A$_2$TGA$_2$TAGCAC$_3$) between the *Sal*I and *Not*I sites of pJ992.

LacZ looping constructs are moved from the plasmid onto the large single-copy F128 episome by homologous recombination. F128 encodes the *lacI* gene producing WT levels of lac repressor. Bacterial conjugation and selection for the desired recombinants (Fig. 12.6) is carried out as described (Whipple, 1998). Correct recombinants are confirmed by PCR amplification to detect the inactivated internal *lacZ* O_2 sequence using primers LJM-1930 and LJM-1931.

4.5. Disruption of hupA and hupB genes

For investigation of the effects of exogenous DNA-binding proteins, the endogenous *hupA* and *hupB* genes encoding HU protein (and genes encoding other nucleoid proteins) are disrupted in *E. coli* strain FW102 as described (Datsenko and Wanner, 2000). Gene-disruption reagents are obtained from the *E. coli* Genetic Stock Center (New Haven, CT). The entire *hupA* coding sequence is first replaced by recombination with a selectable marker (complementary sequence in bold) amplified with primer pair LJM 2201 5′ $GAT_3A_2CGC_2TGAT_3GTCGTAC_2TG_2AGTCT_2C_3T_3CGC_3$**$GTG$-$TAG_2CTG_2AGCTGC_2C$** and LJM-2202 5′-$G_3C_2AC_4T_2CGT_2A_4CTGT_2CACTGC_2ACGCA_2TCT_2AC$**$AT_2C_2G_4ATC_2GTC GAC_2$**. This is followed by the replacement of *hupB* with a selectable marker amplified with primer pair LJM-2203 5′-$CTGATATA_2CTGCT$

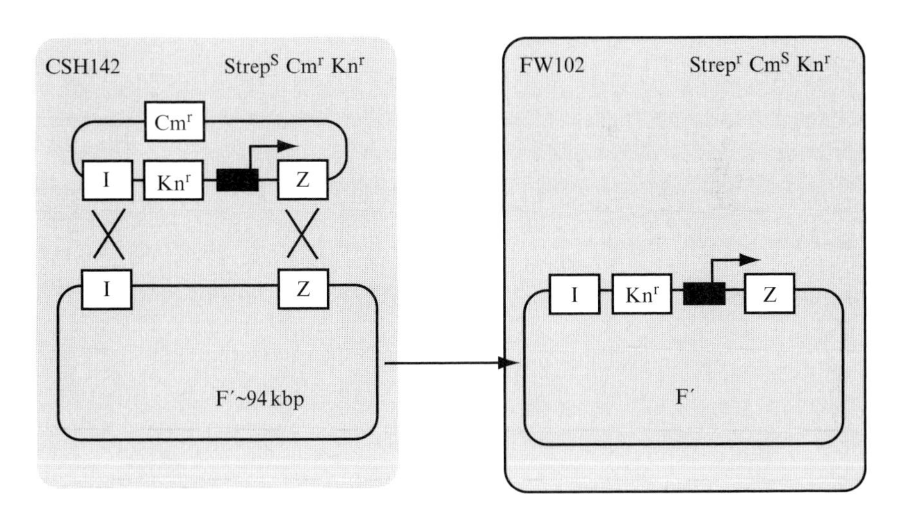

Figure 12.6 Bacterial recombination system (Whipple, 1998) used to move engineered lac promoter/operator constructs (solid box and broken arrow) from small recombinant plasmids to the large F′ episome of *E. coli* in strain CSH142 (wild-type levels of lac repressor), followed by conjugative transfer to FW102 cells. *LacI* and *lacZ* genes, sites of homologous recombination, and antibiotic resistance markers (Strep, streptomycin; Cm, chloramphenicol; Kn, kanamycin) are indicated.

$GCGCGT_2CGTAC_2T_2GA_2G_2AT_2CA_2GT$-$GCG\mathbf{GTGTAG_2CTG_2AG}$ $\mathbf{CTGCT_2C}$ and LJM-2204 5′-$GCAGA_3GTCAC_2G$-$A_2TACA_3TA_6GG$-$CACATCAGTAGATG\mathbf{AT_2C_2G_4ATC_2GTCGAC_2}$. In each case, the selectable marker is then removed in a second step (Datsenko and Wanner, 2000). Deletions are confirmed by PCR amplification. For example, primers flanking *hupA,* LJM-2243 (5′-$GCA_2TAT_2G_2TATA_2TTT$ TTC) and LJM-2244 (5′-$GA_2GTGA_2GAGT_2A$-$TGACTACAG$), amplify a 612-bp product from strain FW102 and a 297-bp product when *hupA* is deleted. Similarly, primers flanking *hupB,* LJM-2245 (5′-$C_2T_7GTCTCGC$ TA_2G) and LJM-2246 (5′-$G_2C_2TAT_2GTGACA_2GA_3C$), amplify a 550-bp product from strain FW102 and a 294-bp product when *hupB* is deleted. Genotypes of all ΔHU strains containing looping episomes are confirmed by PCR amplification with the same primer sets following conjugation and selection, and the presence of the looping episome is confirmed by PCR as above.

4.6. Protein expression

Sometimes, it is of interest to monitor repression looping *in vivo* after expression of *E. coli* proteins or heterologous proteins. Protein expression plasmids are created using a modified version of pXL20 (Whipple, 1998). The *Eco*RI to *Xba*I segment was removed. To create pJ1035, this region is replaced with the complementary oligonucleotide pair LJM-2085 (5′-$A_2T_2CTC_2TCT_2AG_2CAC_4AGGCT_3\underline{ACACT_3ATGCT_2C_2G_2CTCGTA}$ $\underline{TA_2T}GTGTG_2GA_2CTACATC_2TC_2GCTAG_2T_2CACACAG_2A_3T$) and LJM-2086 (5′-$CTAGAT_3C_2TGT$-$GTGA_2C_2TAGCG_2AG_2ATGTAGT_2$-$C_3ACAC\underline{AT_2ATACGAGC_2G_2A_2GCATA_3GTGTA}_3GC_2TG4TGC_2TA_2$ GAG_2AG), creating an *E. coli* promoter of moderate strength (underlined). As an example, protein expression plasmid pJ1043 is created by PCR amplification of rat HMGB1. Primers LJM-2092 (5′-$GCTCTAGA_2TG_3$-$CA_3G_2AGATC_2TA_2G$) and LJM-2093 (5′-$CG_3ATC_2T_2ACT_2CT_5CT_2G$ $CTCT_2CTC$) amplify the coding region from met-1 to lys-185, including both HMG boxes. Upper primer LJM-2092 installs a *Xba*I site and lower primer LJM-2093 installs both a TAA stop codon and *Bam*HI site for cloning into pJ1035. To recover HU expression in cells disrupted for HU expression, a *hupA*/*hupB* cocistronic gene-expression plasmid was created. Primers LJM-2361 (5′-$GCTCTAGA_2TG$-$A_2CA_2GACTCA_2CTGAT_2G$) and LJM-2362 (5′-$CGC_2ATG_2CGCAGT_2ACT_2A_2C$-$TGCGTCT_3CAG$) amplify *hupA* from FW102 genomic DNA. Primers LJM-2361 and LJM-2362 introduce terminal *Xba*I and *Nco*I sites. The *hupA* PCR product is cloned between *Xba*I and *Nco*I sites of pJ1035. To create the cocistronic *hupA*/*hupB* gene product, *hupB* is amplified from FW102 genomic DNA using primers LJM-2363 (5′-$GC_2ATG_2CAG_2A_2GAGA_2GA_2TGA_2$-$TA_3TCTCA_2T_2G$; introduces a *Nco*I restriction site and a Shine/Delgarno sequence prior to first ATG of *hupB*) and LJM-2364 (5′-$GCGGATC_2$-$T_2AGT_3ACCGCGTCT_3CAG$; introduces a *Bam*HI site for cloning). The *hupB* PCR product is then cloned between the *Nco*I and *Bam*HI sites of pJ1035 to produce a cocistronic *hupA*/*hupB* construct.

4.7. *E. coli* β-galactosidase reporter assays

Reagents are purchased from Sigma (St. Louis, MO). *LacZ* expression is measured using a liquid β-galactosidase colorimetric enzyme assay (Miller, 1992). Duplicate 2-mL subcultures of LB medium in 96-well boxes are inoculated with 100 μL of saturated overnight culture in the presence of either 0 or 2-m*M* IPTG. Results of preliminary experiments showed 2-m*M* IPTG to be saturating: higher levels of IPTG did not induce increased levels of β-galactosidase. Subcultures are grown at 37 °C until OD_{600} reaches ~0.3. For samples with low β-galactosidase activity, 1 mL of bacterial culture is assayed after centrifugation and resuspension in a 1-mL Z-buffer (60-m*M* Na_2HPO_4, 40-m*M* NaH_2PO_4, 10-m*M* KCl, 1-m*M* $MgSO_4$, 50-m*M* β-mercaptoethanol). For samples with high levels of β-galactosidase activity, 100 μL of bacterial culture is first diluted with a 900-μL Z-buffer. Cells are lysed by the addition of 50-μL chloroform and 25-μL 0.1% SDS, followed by vortex agitation for 10 s. Samples are then equilibrated at 30° C for 5 min, followed by the addition of 200-μL O-nitrophenylpyranogalactoside solution (ONPG; 4 mg/mL) in Z-buffer. Incubation is continued at 30° C with accurate timing until OD_{420} reaches ~ 0.5. Reactions are terminated with 500-μL 1 M Na_2CO_3 and the reaction time recorded. Cell debris is removed by centrifugation for 3 min at 15,000×*g*. Sample absorbance is measured on a Molecular Devices SpectraMax 340 microtiter plate reader. β-galactosidase activity (E) is calculated according to

$$E = 1000\frac{[OD_{420} - 1.75(OD_{550})]}{t \cdot v \cdot OD_{600}} \qquad (12.22)$$

where OD_x is the optical density at wavelength x, t indicates reaction time (min), and v indicates assay culture volume (mL). Assays are performed for a total of four colonies using two independent strains repeated on 2 days. Assays with ΔHU cells and different protein expression plasmids are typically performed with a total of six colonies repeated on 2 days. Repression due to specific DNA looping is expressed in terms of the normalized expression parameter E', as described in Eq. (12.21).

4.8. Example data and analysis

We have applied the experimental system (Fig. 12.7A) to study the properties of *lac* repression loops *in vivo* for *E. coli* cells altered in nucleoid proteins (Becker *et al.*, 2005, 2007, 2008; Sebastian *et al.*, 2009). Examples of data and fits are presented in Fig. 12.7B and Table 12.1. The data for *lac* repression looping in WT *E. coli* cells (Fig. 12.7B, left) are presented as

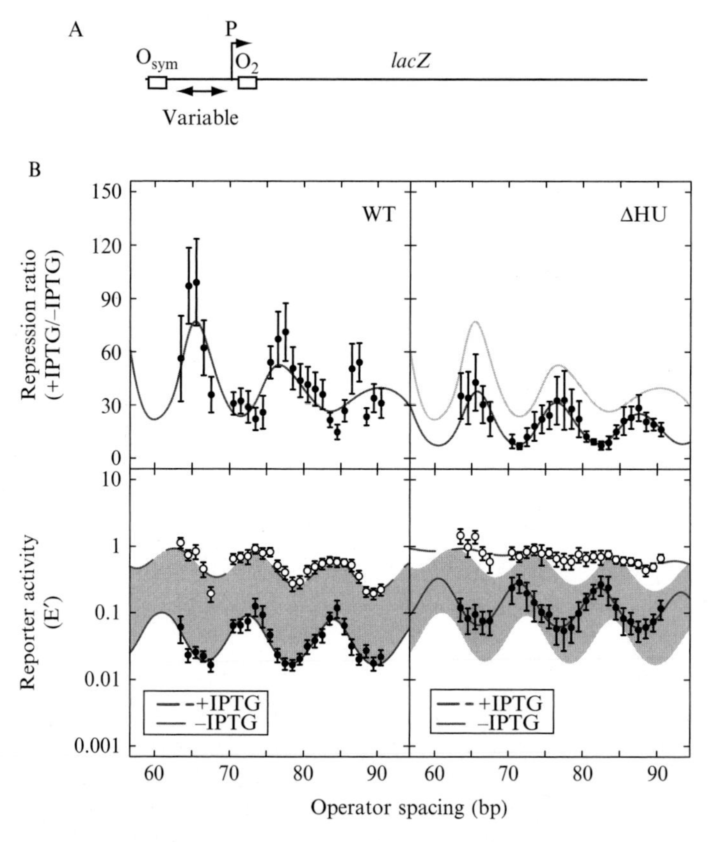

Figure 12.7 Example of quantitative *in vivo lac* repression analysis. (A) Experimental design with variable operator spacing to measure occupancy of O_2 as a function of increased local lac repressor concentration due to DNA looping-dependent collisions with bidentate lac repressor tetramer bound at the strong O_{sym} operator. (B) Experimental data (mean and standard deviation) with best least-squares fits to thermodynamic model for repression ratio (top) and normalized reporter activity (E′, bottom; lower and upper E′ curves are uninduced and induced, respectively). Left and right panels compare length-dependent DNA looping in wild-type *E. coli* cells (WT) versus cells deleted for both *hupA* and *hupB* genes encoding nucleoid heat unstable protein HU (ΔHU; Becker *et al.*, 2005).

Table 12.1 Example *lac* repressor looping fit parameters *in vivo*[a]

Parameter	WT fit value		ΔHU fit value	
	−IPTG	+IPTG	−IPTG	+IPTG
hr (bp)	11.1 ± 0.2	10.7 ± 0.7	11.3 ± 0.2	10.5 ± 2.5
C_{app} ($\times 10^{-19}$ erg-cm)	0.97 ± 0.03	1.65 ± 1.14	1.10 ± 0.13	0.35 ± 0.13
$sp_{optimal}$ (bp)	79.2 ± 0.1	80.2 ± 1.0	78.5 ± 0.3	79.6 ± 1.5
K_{O_2}	4.9	0.6	10.8	0.7
$K^{\circ}{}_{max}$	62 ± 15	0.9 ± 0.6	14 ± 3	0.1 ± 0.1
$K^{\circ}{}_{NSL}$	*0 ± 1*	0 ± 0.1	0 ± 1	0 ± 0.1
P_{app} (bp)	*0 ± 71*	*0 ± 390*	*0 ± 68*	*382 ± 500*

[a] Parameters not well determined by fitting are in italics.

the repression ratio (upper) and E′ reporter gene expression values in the absence (closed circles) and presence (open circles) of IPTG inducer (lower). Fits to the thermodynamic model are shown as smooth lines, with best fit values in Table 12.1. The right panels of Fig. 12.7B show corresponding data for *E. coli* cells lacking the HU protein. These data revealed several important insights that have been discussed (Becker *et al.*, 2005). WT cells show repression that oscillates with loop length (bottom). Surprisingly, an oscillating pattern (with slightly different phasing) is observed even in the presence of IPTG inducer, indicating residual weak DNA looping. The dephasing of these two traces gives rise to an irregular pattern in the repression ratio (Fig. 12.7B, top). Loss of HU protein disables *lac* repression looping (Fig. 12.7B, ΔHU).

Examination of best fit values in Table 12.1 is interesting. The helical repeat *in vivo* is generally higher than observed for relaxed DNA *in vitro*, consistent with the sites being in a plectonemic superhelix. The DNA twist constant is well below the accepted range of 2–4 $\times$ 10^{-19} erg-cm for pure DNA *in vitro*, suggesting that the flexibility of lac repressor (and/or other proteins) enhances the expected torsional inflexibility of DNA. The looping equilibrium constant, K_{max}, is dramatically weakened by IPTG induction, as expected, and is more than twofold lower in cells lacking HU protein. Finally, the fit value for the apparent persistence length of the lac repressor loop (including potential contributions from both protein and DNA) is less than 25% of the typical *in vitro* value for DNA. These fit values exemplify the complexity of understanding the mechanics of the protein–DNA loop when loop geometry is unknown. The work suggests a strong effect of HU on looping, but the thermodynamic model does not reveal the mechanism of this effect. Others have concluded that HU affects the overall apparent flexibility of the repression loop DNA (Zhang *et al.*, 2006a).

4.9. Concluding Comments

The methods presented here are firmly rooted in classic experiments in DNA biophysics. The approaches take advantage of two particularly tractable model systems: *in vitro* DNA cyclization kinetics and *in vivo* repression looping in bacteria. Though established through biophysical approaches more than 20 years ago, the latter gene regulatory system is still inadequately understood. Both systems offer the opportunity to quantitatively measure DNA physical properties and to assess how these properties are changed as the systems are perturbed by proteins. Here, we have shown adaptations that update the execution and interpretation of these experiments.

ACKNOWLEDGMENTS

Support was received from the Mayo Foundation and from NIH grants GM54411 and GM075965 (L. J. M.), NIH grant GM53620 (J. D. K.), and an NSF Career award to J. D. K.

Appendix A. R Code for *J*-Factor Experiments[1]

The following code can be copied into a text file and renamed "jfactor. Rnw". This script requires a space delimited input file (the example "jf-211mer-75nM-ligase100UmL.txt" is included in Appendix 3). Execution of this script requires running the R command: R CMD Sweave jfactor.Rnw, which performs the fitting and generates the output plots of the fits as ".eps" files, followed by running the latex command: latex jfactor.tex, which generates the final output "jfactor.dvi" which can be saved as a ".pdf" file.

```
\documentclass[11pt]{amsart}
\usepackage[margin=1in]{geometry}
\geometry{letterpaper,landscape}
\usepackage{graphicx,amssymb,epstopdf}
\usepackage{/Library/Frameworks/R.framework/Resources/share/
  texmf/Sweave}
\pagestyle{empty}
\begin{document}
\thispagestyle{empty}
<<results=hide,echo=FALSE>>=
###### to run script ::: Sweave("jfactor.Rnw") then will need to
run latex on jfactor.tex ##############
```

[1] Downloadable files for the appendices A, B and C are available at http://www.elsevierdirect.com/companions/9780123812681

```
library("deSolve")
library("xtable")
library("numDeriv")

###### labels and file names ###################################

label <- "jf-211mer-75nM-ligase100UmL"
file <- "jf-211mer-75nM-ligase100UmL.txt"

### "file.txt" contains the following columns: time, M, C, W, D,
T, Q, P (where the first row is time zero)

full <- paste(label,"-full.eps",sep="")
zoom <- paste(label,"-zoom.eps",sep="")

###### starting values for optimization ######################

start <- c(3.8e-03, 6.2e-02, 2.7e-02, 1.3e-02, 3.1e-09, 3.0, 75)

###### reading in data ########################################

dat <- NULL
dat <- read.table(file=file, header=FALSE)
dat <- as.data.frame(dat)

n <- dim(dat)[1]
time <- dat[,1]              ### timepoints in minutes
timesh <- dat[2:n,1]
conc <- dat[1,2:8]
M0 <- as.numeric(conc[1])    ### initial concentration in nM
M <- dat[2:n,2]
C <- dat[2:n,3]
W <- dat[2:n,4]
D <- dat[2:n,5]
T <- dat[2:n,6]
Q <- dat[2:n,7]
P <- dat[2:n,8]

###### FUNCTIONS ##############################################
###### ODE function ###########################################

odeFUN <- function(t,y,pars){
        with(as.list(c(y,pars)),{
                kc1=pars[1]
                kc2=pars[2]
                kc3=pars[3]
                kc4=pars[4]
                kc5=pars[5]
                j1=pars[6]
                kd <- kc1/j1
```

```
# rates of changes
dM  = -4*kd*M^2 -4*kd*M*D -4*kd*M*T -4*kd*M*Q -kc1*M -4*kd*M*P
dD  = 2*kd*M^2 -4*kd*M*D -4*kd*D^2 -4*kd*D*T -kc2*D -4*kd*D*Q -4*kd*D*P
dT  = 4*kd*M*D -4*kd*M*T -4*kd*D*T -kc3*T -4*kd*T^2 -4*kd*T*Q -4*kd*T*P
dQ  = 2*kd*D^2 +4*kd*M*T -4*kd*M*Q -kc4*Q -4*kd*D*Q -4*kd*T*Q -4*kd*Q^2
      -4*kd*Q*P
dP  = 4*kd*M*Q +4*kd*D*T -kc5*P -4*kd*M*P -4*kd*D*P -4*kd*T*P -4*kd*Q*P
      -4*kd*P^2
dCm = kc1*M
dCd = kc2*D
dCt = kc3*T
dCq = kc4*Q
dCp = kc5*P
        return(list(dy=c(dM,dD,dT,dQ,dP,dCm,dCd,dCt,dCq,dCp)))})}

###### optimization function #################################

optFUN <- function(names,start,time,n,M0,M,C,D,T,Q,P,W){

###### model function ########################################

      modelFUN<-function(s){
           parms <- s[1:6]

           M0par <- s[7]
           times <- time

           y <- c(M=M0par, D=0, T=0, Q=0, P=0, Cm=0, Cd=0, Ct=0, Cq=0,
           Cp=0)
           ODE  <-  as.data.frame(lsoda(y=y,times=times,func=ode-
           FUN,parms=parms))

           ### least squares
           lsse = sum((M-ODE$M[2:n])^2) +sum((C-ODE$Cm[2:n])^2) +sum
  ((D-2*ODE$D[2:n])^2) +sum((T-3*ODE$T[2:n])^2) +sum((Q-4*ODE$Q[2:
  n])^2) +sum((P-5*ODE$P[2:n])^2) +sum((W-(2*ODE$Cd[2:n]+3*ODE$Ct
  [2:n]+4*ODE$Cq[2:n]+5*ODE$Cp[2:n]))^2)
           if(is.na(lsse)){lsse <- 1e6}

           lsse}

###### output results into csv file and LaTeX table ##############

nlminbFit <- nlminb(start,modelFUN,lower=0,upper=Inf)
out <- NULL
out <- c(nlminbFit$par,nlminbFit$objective)
out <- as.data.frame(out)
hessian <- hessian(modelFUN,nlminbFit$par)
```

```
FIM <- solve(2*hessian)
se <- sqrt(abs(diag(FIM)))
outnames <- c("cost function","kc1","kc2","kc3","kc4","kc5","j1",
   "M0")
out <- cbind(outnames,out,c(NA,se))
names(out) <- c("Parameter","Estimate","Std.Error")
csv <- paste(label,"-summary.csv",sep="")
write.table(out,file=csv,sep=",",row.names=FALSE,col.names=TRUE,
   append=FALSE)
xtab <- xtable(out,digits=3)

###### generating best fit data for plots ########################

parmsfit <- nlminbFit$par
maxtime <- max(time)
times2 <- seq(0,maxtime,length=1000)
y2 <- c(M-parmsfit[7], D=0, T=0, Q=0, P=0, Cm-0, Cd-0, Ct-0, Cq-0, Cp-0)
ODEfit <- as.data.frame(lsoda(y=y2,times=times2,func=odeFUN,parm-
   s=parmsfit[1:6]))

################################################################
return <- NULL
return[[1]] <- xtab
return[[2]] <- ODEfit
return[[3]] <- out

return}
###### function for generating eps plots ########################

epsFUN <- function(full,zoom,timesh,M0,M,C,D,T,Q,P,W,ODEfit)
{

cexvar <- 1.75
mexvar <- 1.1
lwdv <- 3
lwdv2 <- 2

xmax <- max(timesh)
ymax1 <- round(M0*1.06)
ymax2 <- round(ymax1*0.25)

postscript(file=full, paper="letter", horizontal=TRUE, family="sans")

par(cex.axis=cexvar, cex.lab=cexvar, mex=mexvar, font.axis=2, font.
   lab=2, mar=c(5, 7, 4, 2) + 0.7, tck=0.02)

plot(timesh,M,xlim=c(0,xmax),ylim=c(0,ymax1),xlab="time (minutes)",
   ylab="concentration  (nM)",  pch=16,  col="tan3",  cex=cexvar,
   axes=FALSE)
```

```
box(lwd=lwdv)
axis(1,at=seq(0,xmax,by=round(xmax*0.25)),labels=TRUE,lwd=lwdv,
  las=1)
axis(2,at=seq(0,ymax1,by=round(ymax1*0.25)),labels=TRUE,
  lwd=lwdv,las=1)
axis(3,at=seq(0,xmax,by=round(xmax*0.25)),labels=FALSE,lwd=lwdv)
axis(4,at=seq(0,ymax1,by=round(ymax1*0.25)),labels=FALSE,
  lwd=lwdv)

points(ODEfit$time,   ODEfit$M,   pch=16,   type="l",col="tan3",
  lwd=lwdv2)
points(timesh,C,pch=16,cex=cexvar,col="orangered")
points(ODEfit$time, ODEfit$Cm, pch=16, col="orangered", type="l",
  lwd=lwdv2)
points(timesh,D,pch=16,cex=cexvar,col="darkblue")
points(ODEfit$time, 2*ODEfit$D, pch=16, col="darkblue", type="l",
  lwd=lwdv2)
points(timesh,W,pch=16,cex=cexvar,col="deeppink")
points(ODEfit$time,  2*ODEfit$Cd+3*ODEfit$Ct+4*ODEfit$Cq+5*ODEfit
  $Cp, pch=16, col="deeppink", type="l", lwd=lwdv2)

 legend("topright",c("linear monomer", "linear dimer", "linear
  trimer", "linear tetramer","linear pentamer", "circular mono-
  mer","circular multimers"), col=c("tan3","darkblue","spring-
  green","gold", "gray", "orangered","deeppink"), lty=1, bty="n",
  cex=cexvar, lwd=lwdv2)

dev.off()

#################################################################

postscript(file=zoom,     paper="letter",     horizontal=TRUE,
  family="sans")

par(cex.axis=cexvar, cex.lab=cexvar, mex=mexvar, font.axis=2, font.
  lab=2, mar=c(5, 7, 4, 2) + 0.7, tck=0.02)

plot(timesh,M,xlim=c(0,xmax),ylim=c(0,ymax2),xlab="time     (min-
  utes)",ylab="concentration (nM)", pch=16, col="tan3", cex=cexvar,
  axes=FALSE, type="n")

box(lwd=lwdv)
axis(1,at=seq(0,xmax,by=round(xmax*0.25)),labels=TRUE,lwd=lwdv,
  las=1)
axis(2,at=seq(0,ymax2,by=round(ymax2*0.25)),labels=TRUE,lwd=lwdv,
  las=1)
axis(3,at=seq(0,xmax,by=round(xmax*0.25)),labels=FALSE,lwd=lwdv)
axis(4,at=seq(0,ymax2,by=round(ymax2*0.25)),labels=FALSE,lwd=lwdv)
```

```
points(timesh,C,pch=16,cex=cexvar,col="orangered")
points(ODEfit$time, ODEfit$Cm, pch=16, col="orangered", type="l",
   lwd=lwdv2)
points(timesh,T,pch=16,cex=cexvar,col="springgreen")
points(ODEfit$time, 3*ODEfit$T, pch=16, col="springgreen", type=-
   "l", lwd=lwdv2)
points(timesh,Q,pch=16,cex=cexvar,col="gold")
points(ODEfit$time, 4*ODEfit$Q, pch=16, col="gold", type="l",
   lwd=lwdv2)
points(timesh,P,pch=16,cex=cexvar,col="gray")
points(ODEfit$time, 5*ODEfit$P, pch=16, col="gray", type="l",
   lwd=lwdv2)

legend("topright",c("linear monomer", "linear dimer", "linear
   trimer", "linear tetramer","linear pentamer", "circular mono-
   mer","circular multimers"), col=c("tan3","darkblue","spring-
   green","gold", "gray", "orangered","deeppink"), lty=1, bty="n",
   cex=cexvar, lwd=lwdv2)

dev.off()}

###### performing optimization and generating plots ###########

optimize <- optFUN(label,start,time,n,M0,M,C,D,T,Q,P,W)
tab <- optimize[[1]]
ODEfit <- optimize[[2]]
epsFUN(full,zoom,timesh,M0,M,C,D,T,Q,P,W,ODEfit)
graphics <- paste("%\r\r\\begin{center}\r\\includegraphics
   [angle=270,scale=0.45]{",full,"} \\includegraphics[angle=270,
   scale=0.45]{",zoom,"}\r\\end{center}", sep="")

@
<<results=tex,echo=FALSE>>=
cat("\\newpage")
cat(" \\ ",label,sep="")
print(tab,include.rownames=FALSE)
cat(graphics)
@

\end{document}
```

Appendix B. R Code for *lac* Looping Experiments

The following code can be copied into a text file and renamed "lac.Rnw". This script requires two comma delimited input files (the examples "E-IPTG.csv" and "E+IPTG.csv" are included in Appendix 3). Execution of this script requires running the R command: R CMD Sweave lac.Rnw, which performs the fitting and generates the output plots of the fits as ".eps" files, followed by running the latex command: latex lac.tex, which generates the final output "lac.dvi" which can be saved as a ".pdf" file.

```
\documentclass[11pt]{amsart}
\usepackage[margin=1in]{geometry}
\geometry{letterpaper}
\usepackage{graphicx,amssymb,epstopdf}
\usepackage{/Library/Frameworks/R.framework/Resources/share/
  texmf/Sweave}
\pagestyle{empty}
\begin{document}
\thispagestyle{empty}

<<results=hide,echo=FALSE>>=

###### to run script ::: Sweave("lac.Rnw") then will need to run
latex on lac.tex #################

library("numDeriv")
library("xtable")

###### reading in data #########################################

identifier <- "wild-type"

### data files contain 3 columns: sp, expression, and standard
deviation of expression

eu_dat <- read.table("E-IPTG.csv",header=TRUE,sep=",")
eu_dat <- as.data.frame(eu_dat)
ei_dat <- read.table("E+IPTG.csv",header=TRUE,sep=",")
ei_dat <- as.data.frame(ei_dat)

### last row is O2 alone strain
max <- dim(eu_dat)[1]
O2_U_exp <- eu_dat[max,2]
O2_U_exp_sd <- eu_dat[max,3]
O2_I_exp <- ei_dat[max,2]
O2_I_exp_sd <- ei_dat[max,3]

sp <- eu_dat[1:(max-1),1]
E_U_exp <- eu_dat[1:(max-1),2]
E_U_exp_sd <- eu_dat[1:(max-1),3]
E_I_exp <- ei_dat[1:(max-1),2]
E_I_exp_sd <- ei_dat[1:(max-1),3]

### calculating Eprime from E
Eprime_U_exp <- E_U_exp/O2_U_exp
Eprime_I_exp <- E_I_exp/O2_I_exp
```

```
### propagation of error in Eprime data

error_Eprime_U <- E_U_exp/O2_U_exp*sqrt((E_U_exp_sd/E_U_exp)^2 +
   (O2_U_exp_sd/O2_U_exp)^2)
error_Eprime_I <- E_I_exp/O2_I_exp*sqrt((E_I_exp_sd/E_I_exp)^2 +
   (O2_I_exp_sd/O2_I_exp)^2)

### experimental fbound (activity is normalized by the maximum
induced activity)
### fbound_exp <- (max induced activity - observed activity)/
(max induced activity)

max_U_exp <- max(E_U_exp,na.rm=TRUE)
max_I_exp <- max(E_I_exp,na.rm=TRUE)

fbound_r_exp <- (max_U_exp - E_U_exp)/(max_U_exp)
fbound_i_exp <- (max_I_exp - E_I_exp)/(max_I_exp)

### calculating repression ratio
### RR = E(+IPTG)/E(-IPTG)
RR_exp <- E_I_exp/E_U_exp
### propagation of error in RR data
error_RR    <-    E_I_exp/E_U_exp*sqrt((E_I_exp_sd/E_I_exp)^2    +
   (E_U_exp_sd/E_U_exp)^2)

### calculating KO2

K_O2_U <- (1-O2_U_exp/max_I_exp)/(1-(1-O2_U_exp/max_I_exp))
K_O2_I <- (1-O2_I_exp/max_I_exp)/(1-(1-O2_I_exp/max_I_exp))

###########################################################

start <- c(c(11.6,0.69,211,20.2,78.7,0),c(11,0.89,2.5,0,79.3,238))

###### global parameters ######################################

sp_U <- sp
sp_I <- sp
sp_mean <- mean(sp,na.rm=TRUE)
# l = separation between base pairs (units = cm)
l <- 3.4e-8
# k = Boltzmann's constant (units = erg/K)
k <- 1.3806503e-16
# T = absolute temperature (units = K)
T <- 293

###### FUNCTIONS ##########################################
###### adding error bars to plots ###########################

errAdd<-function(x,y,upper,lower,col){

       for(i in 1:length(x)){
```

```
            if(!is.na(y[i])){
                  if(lower[i]==Inf){
                        lower[i] <- 5
                        upper[i] <- 5}
            points(c(x[i],x[i]),c((y[i]-lower[i]),(y[i]+upper
            [i])),type="l",col=col)
            d<-0.01*(diff(range(x,na.rm=TRUE)))
            points(c((x[i]-d),(x[i]+d)),c((y[i]-lower[i]),(y
            [i]-lower[i])),type="l",col=col)
            points(c((x[i]-d),(x[i]+d)),c((y[i]+upper[i]),(y
            [i]+upper[i])),type="l",col=col)}}}

###### model function ##########################################

modelFUN<-function(s,sp){

### following the method described in Becker et al. 2005 and 2008
# DNA length - actual spacing between operators (center of
operator)
sp_U <- sp
sp_I <- sp

# DNA helical repeat
hr_U <- s[1]

# apparent torsional modulus for the DNA loop
C_app_U <- s[2]
### for comparision accepted torsional rigidity: C = 2.4e-19
(units = erg*cm)

# equilibrium constant for formation of a loop with perfect phasing
K_max_U <- s[3]

# equilibrium constant for non-specific loop formation
K_NSL_U <- s[4]

# optimal spacing between (center of) operators
sp_optimal_U <- s[5]

# P_app
P_app_U <- s[6]

# modifed K_max
K_maxp_U <- K_max_U*exp(P_app_U*(1/sp_mean - 1/sp_U))

# standard deviation of twist per base-pair
sigma_bp_U <- hr_U/(2*pi)*sqrt(l*k*T/(C_app_U*1e-19))

# standard deviation of torsion angle between sites (considering thermal
   fluctuations)
sigma_TW_U <- sqrt(sp_U*sigma_bp_U^2)
```

```
# phasing-dependent equilibrium constant for specific loop formation
K_SL_U <- 0
for(i in seq(-5,5,by=1)){
        K_SL_U <- K_SL_U + K_maxp_U*exp(-(sp_U-sp_optimal_U+i*hr_U)
        ^2/(2*sigma_TW_U^2))}

### fbound <- (K_SL + K_NSL + K_O2)/(1 + K_SL + K_NSL + K_O2)
fbound_U <- (K_SL_U + K_NSL_U + K_O2_U)/(K_SL_U + K_NSL_U + K_O2_U + 1)

hr_I <- s[7]
C_app_I <- s[8]
K_max_I <- s[9]
K_NSL_I <- s[10]
sp_optimal_I <- s[11]
P_app_I <- s[12]
K_maxp_I <- K_max_I*exp(P_app_I*(1/sp_mean - 1/sp_I))
sigma_bp_I <- hr_I/(2*pi)*sqrt(1*k*T/(C_app_I*1e 19))
sigma_TW_I <- sqrt(sp_I*sigma_bp_I^2)
K_SL_I <- 0
for(i in seq(-5,5,by=1)){
        K_SL_I <- K_SL_I + K_maxp_I*exp(-(sp_I-sp_optimal_I + i*hr_I)
        ^2/(2*sigma_TW_I^2))}
fbound_I <- (K_SL_I + K_NSL_I + K_O2_I)/(K_SL_I + K_NSL_I + K_O2_I + 1)

max_activity <- max(max_U_exp,max_I_exp)
E_U_theory <- (max_activity)*(1 - fbound_U)
E_I_theory <- (max_activity)*(1 - fbound_I)
Ep_U <- E_U_theory/O2_U_exp
Ep_I <- E_I_theory/O2_I_exp

### repression ratio
### RR = E(+IPTG)/E(-IPTG)
RR_theory <- E_I_theory/E_U_theory

out <- NULL

out[[1]] <- Ep_U
out[[2]] <- Ep_I
out[[3]] <- RR_theory

out}

###### model function 2 ########################################

modelFUN2<-function(hr_U,C_app_U,K_max_U,K_NSL_U,sp_optimal_U,
   P_app_U,hr_I,C_app_I,K_max_I,K_NSL_I,sp_optimal_I,P_app_I,sp){

sp_U <- sp
sp_I <- sp
```

```
K_maxp_U <- K_max_U*exp(P_app_U*(1/sp_mean - 1/sp_U))
sigma_bp_U <- hr_U/(2*pi)*sqrt(1*k*T/(C_app_U*1e-19))
sigma_TW_U <- sqrt(sp_U*sigma_bp_U^2)
K_SL_U <- 0
for(i in seq(-5,5,by=1)){
        K_SL_U <- K_SL_U + K_maxp_U*exp(-(sp_U-sp_optimal_U+i*hr_U)
        ^2/(2*sigma_TW_U^2))}
fbound_U <- (K_SL_U + K_NSL_U + K_O2_U)/(K_SL_U + K_NSL_U + K_O2_U + 1)

K_maxp_I <- K_max_I*exp(P_app_I*(1/sp_mean - 1/sp_I))
sigma_bp_I <- hr_I/(2*pi)*sqrt(1*k*T/(C_app_I*1e-19))
sigma_TW_I <- sqrt(sp_I*sigma_bp_I^2)
K_SL_I <- 0
for(i in seq(-5,5,by=1)){
        K_SL_I <- K_SL_I + K_maxp_I*exp(-(sp_I-sp_optimal_I + i*hr_I)
        ^2/(2*sigma_TW_I^2))}
fbound_I <- (K_SL_I + K_NSL_I + K_O2_I)/(K_SL_I + K_NSL_I + K_O2_I + 1)

max_activity <- max(max_U_exp,max_I_exp)
E_U_theory <- (max_activity)*(1 - fbound_U)
E_I_theory <- (max_activity)*(1 - fbound_I)
RR_theory <- E_I_theory/E_U_theory

out <- NULL

out[[1]] <- E_U_theory
out[[2]] <- E_I_theory
out[[3]] <- RR_theory

out}

###### fit function ###########################################

fitFUN <- function(s){

hr_U <- s[1]
C_app_U <- s[2]
K_max_U <- s[3]
K_NSL_U <- s[4]
sp_optimal_U <- s[5]
P_app_U <- s[6]

K_maxp_U <- K_max_U*exp(P_app_U*(1/sp_mean - 1/sp_U))
sigma_bp_U <- hr_U/(2*pi)*sqrt(1*k*T/(C_app_U*1e-19))
sigma_TW_U <- sqrt(sp_U*sigma_bp_U^2)
K_SL_U <- 0
for(i in seq(-5,5,by=1)){
        K_SL_U <- K_SL_U + K_maxp_U*exp(-(sp_U-sp_optimal_U+i*hr_U)
        ^2/(2*sigma_TW_U^2))}
```

```
fbound_U <- (K_SL_U + K_NSL_U + K_O2_U)/(K_SL_U + K_NSL_U + K_O2_U + 1)

hr_I <- s[7]
C_app_I <- s[8]
K_max_I <- s[9]
K_NSL_I <- s[10]
sp_optimal_I <- s[11]
P_app_I <- s[12]

K_maxp_I <- K_max_I*exp(P_app_I*(1/sp_mean - 1/sp_I))
sigma_bp_I <- hr_I/(2*pi)*sqrt(l*k*T/(C_app_I*1e-19))
sigma_TW_I <- sqrt(sp_I*sigma_bp_I^2)
K_SL_I <- 0
for(i in seq(-5,5,by=1)){
        K_SL_I <- K_SL_I + K_maxp_I*exp(-(sp_I-sp_optimal_I + i*hr_I)
        ^2/(2*sigma_TW_I^2))}
fbound_I <- (K_SL_I + K_NSL_I + K_O2_I)/(K_SL_I + K_NSL_I + K_O2_I + 1)

max_activity <- max(max_U_exp,max_I_exp)
E_U_theory <- (max_activity)*(1 - fbound_U)
E_I_theory <- (max_activity)*(1 - fbound_I)
RR_theory <- E_I_theory/E_U_theory

### non-linear least squares
lsse <- sum((E_U_theory - E_U_exp)^2/E_U_exp_sd^2,na.rm=TRUE) + sum
   ((E_I_theory - E_I_exp)^2/E_I_exp_sd^2,na.rm=TRUE)

lsse}
##########################################################

### initializing output
dat1 <- NULL
dat2 <- NULL
dat3 <- NULL
dat4 <- NULL
dat5 <- NULL

### first fitting
nlminbFit <- nlminb(start,fitFUN,lower=c(8,0.1,0,0,50,0),upper=c
   (15,3,300,100,90,500))
dat1$Parameter <- c("hr","C_app","K_max","K_NSL","sp_optimal","
   P_app")
fit1 <- nlminbFit$par
dat1$Estimate=nlminbFit$par
dat2$Cost.Function=nlminbFit$objective
hessian <- hessian(fitFUN,fit1)
FIM <- solve(2*hessian)
```

```
se <- sqrt(abs(diag(FIM)))
cl <- sqrt(qchisq(0.95,2*length(sp)-12))*se
cl1 <- cl
dat1$Confidence.Limits=cl

dat1 <- as.data.frame(dat1)
dat2 <- as.data.frame(dat2)
xtab1 <- xtable(dat1)
xtab2 <- xtable(dat2)
csv1 <- paste(identifier,"-parameters1.csv",sep="")
csv2 <- paste(identifier,"-costfuction1.csv",sep="")
write.table(dat1,file=csv1,sep=",",row.names=FALSE)
write.table(dat2,file=csv2,sep=",",row.names=FALSE)

### generating data points for plotting first fit
parfit <- nlminbFit$par
sp_seq <- seq(40,95,length=300)
test <- modelFUN(parfit,sp_seq)

#################################################################

parFUN <- function(s){

hr_U <- s[1]
C_app_U <- s[2]
K_max_U <- s[3]
K_NSL_U <- s[4]
sp_optimal_U <- s[5]
P_app_U <- s[6]
hr_I <- s[7]
C_app_I <- s[8]
K_max_I <- s[9]
K_NSL_I <- s[10]
sp_optimal_I <- s[11]
P_app_I <- s[12]

out <- NULL

out[[1]] <- c(hr_U,C_app_U,sp_optimal_U,hr_I,C_app_I,sp_optimal_I)
out[[2]] <- c(K_max_U,K_NSL_U,P_app_U,K_max_I,K_NSL_I,P_app_I)

out}

#################################################################

### second starting values
new_par <- parFUN(parfit)
start2 <- new_par[[1]]
fixed <- new_par[[2]]
```

```
###### fit function 2 ##########################################
fitFUN2 <- function(s){

hr_U <- s[1]
C_app_U <- s[2]
K_max_U <- fixed[1]
K_NSL_U <- fixed[2]
sp_optimal_U <- s[3]
P_app_U <- fixed[3]

hr_I <- s[4]
C_app_I <- s[5]
K_max_I <- fixed[4]
K_NSL_I <- fixed[5]
sp_optimal_I <- s[6]
P_app_I <- fixed[6]

K_maxp_U <- K_max_U*exp(P_app_U*(1/sp_mean - 1/sp_U))
sigma_bp_U <- hr_U/(2*pi)*sqrt(1*k*T/(C_app_U*1e-19))
sigma_TW_U <- sqrt(sp_U*sigma_bp_U^2)
K_SL_U <- 0
for(i in seq(-5,5,by=1)){
        K_SL_U <- K_SL_U + K_maxp_U*exp(-(sp_U-sp_optimal_U+i*hr_U)
        ^2/(2*sigma_TW_U^2))}
fbound_U <- (K_SL_U + K_NSL_U + K_O2_U)/(K_SL_U + K_NSL_U + K_O2_U + 1)

K_maxp_I <- K_max_I*exp(P_app_I*(1/sp_mean - 1/sp_I))
sigma_bp_I <- hr_I/(2*pi)*sqrt(1*k*T/(C_app_I*1e-19))
sigma_TW_I <- sqrt(sp_I*sigma_bp_I^2)
K_SL_I <- 0
for(i in seq(-5,5,by=1)){
        K_SL_I <- K_SL_I + K_maxp_I*exp(-(sp_I-sp_optimal_I + i*hr_I)
        ^2/(2*sigma_TW_I^2))}
fbound_I <- (K_SL_I + K_NSL_I + K_O2_I)/(K_SL_I + K_NSL_I + K_O2_I + 1)

max_activity <- max(max_U_exp,max_I_exp)
E_U_theory <- (max_activity)*(1 - fbound_U)
E_I_theory <- (max_activity)*(1 - fbound_I)
RR_theory <- E_I_theory/E_U_theory

### non-linear least squares
lsse <- sum((E_U_theory - E_U_exp)^2/E_U_exp_sd^2,na.rm=TRUE) + sum
   ((E_I_theory - E_I_exp)^2/E_I_exp_sd^2,na.rm=TRUE) + sum((RR_the-
   ory - RR_exp)^2/error_RR^2,na.rm=TRUE)

lsse}
```

```
#################################################################
### second fitting
nlminbFit <- nlminb(start2,fitFUN2,lower=0,upper=1000)
fit2 <- nlminbFit$par
dat3$Parameter <- c("hr","C_app","sp_optimal")
dat5$Parameter  <-  c("K_max","K_NSL","P_app")#,"K_max","K_NSL","
  P_app")
dat5$Estimate <- fixed
dat5$Confidence.Limits <- parFUN(cl)[[2]]
dat3$Estimate=nlminbFit$par
dat4$Cost.Function=nlminbFit$objective
hessian <- hessian(fitFUN2,fit2)
FIM <- solve(2*hessian)
se <- sqrt(abs(diag(FIM)))
cl <- sqrt(qchisq(0.95,3*length(sp)-6))*se
cl2 <- cl
dat3$Confidence.Limits=cl
dat3 <- as.data.frame(dat3)
dat4 <- as.data.frame(dat4)
dat5 <- as.data.frame(dat5)
dat5 <- cbind(dat3,dat5)
xtab3 <- xtable(dat5)
xtab4 <- xtable(dat4)
csv3 <- paste(identifier,"-parameters2.csv",sep="")
csv4 <- paste(identifier,"-costfunction2.csv",sep="")
write.table(dat3,file=csv3,sep=",",row.names=FALSE)
write.table(dat4,file=csv4,sep=",",row.names=FALSE)
#################################################################
parFUN2 <- function(s,s2){
out  <-  c(s[1],s[2],s2[1],s2[2],s[3],s2[3],s[4],s[5],s2[4],s2[5],
  s[6],s2[6])
out}
#################################################################
### generating data points for plotting second fit
parfit2 <- nlminbFit$par
parfit2full <- parFUN2(parfit2,fixed)
test2 <- modelFUN(parfit2full,sp_seq)
datfit <- parFUN2(dat5[[2]],dat5[[5]])
clfit <- parFUN2(dat5[[3]],dat5[[6]])
rindex <- rep(c(1,2,0,0,1,0),2)
```

```
datfitr <- NULL
clfitr <- NULL
comr <- NULL
for(k in 1:12){
        if(k<7){
                if(k==3|k==4){
                        datfit[k] <- datfit[k]/K_O2_U
                        clfit[k] <- clfit[k]/K_O2_U}
             datfitr[2*k-1] <- round(datfit[k],digits=rindex[k])
             clfitr[2*k-1] <- round(clfit[k],digits=rindex[k])
             if(clfit[k]/datfit[k] > 1){
                        datfitr[2*k-1]<-paste("(",datfitr
                        [2*k-1],")",sep="")}}
        if(k>6){
             if(k==9|k==10){
                        datfit[k] <- datfit[k]/K_O2_I
                        clfit[k] <- clfit[k]/K_O2_I}
             datfitr[2*(k-6)]<-round(datfit[k],digits=rindex[k])
             clfitr[2*(k-6)] <-round(clfit[k],digits=rindex[k])
               if(clfit[k]/datfit[k] > 1){
                        datfitr[2*(k-6)]   <-   paste("(",datfitr[2*
                        (k-6)],")",sep="")}}}
comr <- paste(datfitr," $\\pm$ ",clfitr,sep="")
comr  <-  rbind(cbind(comr[1:4]),round(K_O2_U,1),round(K_O2_I,1),
  cbind(comr[5:12]))
mat <- matrix(comr,nrow=2)
mat <- as.data.frame(mat)
row.names(mat) <- c("$-$IPTG","$+$IPTG")
names(mat)  <- c("hr (bp/turn)","C$_{app}$ ($\\times 10^{-19}$ erg-
  cm)","K$_{O2}$","K$^o_{max}$","K$^o_{NSL}$","sp$_{optimal}$
  (bp)","P$_{app}$ (bp)")
xtab5 <- xtable(mat)
names(mat) <- c("hr (bp/turn)","Capp (x10-19 erg-cm)","KO2","Kmax","
  KNSL","spoptimal (bp)","Papp (bp)")
csv5 <- paste(identifier,"-summary.csv",sep="")
write.table(mat,file=csv5,sep=",",append=FALSE)

##################################################################

plot1 <- paste(identifier,"-Eprime.eps",sep="")
plot2 <- paste(identifier,"-RR.eps",sep="")

lwdv <- 3
cexvar <- 2.25
mexvar <- 1.1
cexv <- 1.5
```

```
postscript(file=plot1, paper="letter", horizontal=TRUE,
   family="sans")

par(cex.axis=cexvar, cex.lab=cexvar,mex=mexvar,mar=c(5, 7, 4, 2) +
   0.7,font.axis=2, font.lab=2,tck=0.02)

plot(sp_U,Eprime_U_exp,pch=16,log="y",xlim=c(45,90),ylim=c
   (0.001,10),xlab="",ylab="",type="n",axes=FALSE)

mtext(expression(paste(bold(Operator)," ",bold(Spacing)," ",bold
   ((bp)),sep="")),side=1,line=4,cex=cexvar)
mtext(expression(paste(bold(Reporter),"          ",bold(Activity),
   sep="")),side=2,line=6,cex=cexvar)
mtext(expression(paste(bold((E*minute)),sep="")),side=2,line=4,
   cex=cexvar)

polygon(c(sp_seq,sp_seq[length(sp_seq):1]),c(test2[[1]],test2
   [[2]][length(sp_seq):1]),col="gray",border=NA)
points(sp_seq,test2[[1]],type="l",lty=1,col=2,lwd=lwdv,cex=cexv)
points(sp_seq,test2[[2]],type="l",lty=2,col=2,lwd=lwdv,cex=cexv)
points(sp_U,Eprime_U_exp,pch=16,cex=cexv)
errAdd(sp_U,Eprime_U_exp,error_Eprime_U,error_Eprime_U,col=1)
points(sp_I,Eprime_I_exp,cex=cexv)
errAdd(sp_I,Eprime_I_exp,error_Eprime_I,error_Eprime_I,col=1)
points(sp_I,Eprime_I_exp,pch=16,col="white",cex=cexv)
points(sp_I,Eprime_I_exp,cex=cexv)

box(lwd=lwdv)
axis(1,at=seq(45,90,by=5),labels=seq(45,90,by=5),las=1,lwd=lwdv)
axis(2,at=c(0.001,0.01,0.1,1,10),labels=c(0.001,0.01,0.1,1,10),
   las=1,lwd=lwdv)
axis(3,at=seq(45,90,by=5),labels=FALSE,lwd=lwdv)
axis(4,at=c(0.001,0.01,0.1,1,10),labels=FALSE,lwd=lwdv)

legend("topright",c(expression(bold(+IPTG)),expression(bold(-
   IPTG))), col=c("red","red"), lty=c(2,1), bty="n", inset=0.05,
   cex=cexv,lwd=lwdv)

dev.off()

##########################################################

postscript(file=plot2, paper="letter", horizontal=TRUE,
   family="sans")

par(cex.axis=cexvar, cex.lab=cexvar,mex=mexvar,mar=c(5, 7, 4, 2) +
   0.7,font.axis=2, font.lab=2,tck=0.02)

plot(sp_U,RR_exp,pch=16,xlim=c(45,90),ylim=c(0,180),xlab="",
   ylab="",type="n",axes=FALSE)
```

```
mtext(expression(paste(bold(Operator)," ",bold(Spacing)," ",bold
   ((bp)),sep="")),side=1,line=4,cex=cexvar)
mtext(expression(paste(bold(Repression)," ",bold(Ratio),sep="")),
   side=2,line=6,cex=cexvar)
mtext(expression(paste(bold((+IPTG/-IPTG)),sep="")),side=2,
   line=4,cex=cexvar)

points(sp_seq,test2[[3]],type="l",lwd=lwdv,col="gray",cex=cexv)
points(sp_seq,test2[[3]],type="l",lwd=lwdv,col=2,cex=cexv)
points(sp_U,RR_exp,pch=16,cex=cexv)
errAdd(sp_U,RR_exp,error_RR,error_RR,col=1)

box(lwd=lwdv)
axis(1,at=seq(45,90,by=5),las=1,lwd=lwdv)
axis(2,at=seq(0,180,30),las=1,lwd=lwdv)
axis(3,at=seq(45,90,by=5),labels=FALSE,lwd=lwdv)
axis(4,at=seq(0,180,20),labels=FALSE,lwd=lwdv)

dev.off()
@
<<results=tex,echo=FALSE>>=

        cat("%\n\n\\newpage")
        cat("%\n\n",identifier,"\n\n%",sep="")
        cat("%\n\n%")
        cat("%\n\n","*** uninduced (-IPTG) parameters are given
        first","\n\n%",sep="")
        cat("%\n\n%")
        cat("%\n\n","fit to E data (+/- IPTG)","\n\n%",sep="")
        print(xtab2,include.rownames=FALSE,floating=TRUE)
        print(xtab1,include.rownames=FALSE,floating=TRUE)
        cat("%\n\n%")
        cat("%\n\n","simultaneous fit to E data (+/- IPTG) and RR","
        \n\n%",sep="")
        print(xtab4,include.rownames=FALSE,floating=TRUE)
        print(xtab3,include.rownames=FALSE,floating=TRUE)
        cat("%\n\n%")
        cat("%\n\n","summary","\n\n%",sep="")
        cat("%\n\n","*** parentheses indicate parameters not well
        determined","\n\n%",sep="")
        print(xtab5,floating=FALSE,sanitize.text.function=
        function(x){x})
        cat("%\n\n\\newpage")
        cat("%\n\n",identifier,"\n\n%",sep="")
        graphics  <-  paste("%\n\n\\begin{center}\n\\include-
        graphics[angle=270,scale=0.75]{",plot1,"}
```

```
\\includegraphics[angle=270,scale=0.75]{",plot2,"}\n\\end{center}",
   sep="")
        cat(graphics)
@
\end{document}
```

Appendix C. Required Files for R scripts

jf-211mer-75nM-ligase100UmL.txt

```
0 75 0 0 0 0 0 0
2 45.080055326137 3.13754834457182 4.52876619762908 12.4953524807161
   5.85426352044715 2.57495147095558 1.32906265954329
4 33.2914310979581 4.72815611179549 6.90455253480098 16.70733890358
   82 7.94323178631252 3.5307552097175 1.89453435582711
6 27.8558990883505 5.51944962022432 8.23865362979177 17.03547326
   96023 9.44777091565846 4.59285289006137 2.30990058631128
8 25.0083855626908 6.14174562868144 9.30618026546597 16.3757667623
   754 10.1833197734334 5.28744752830441 2.69715447904854
```

E-IPTG.csv

```
E-IPTG,WT,WT-sd
63.5,16.113,6.525
64.5,6.118,1.026
65.5,6.776,0.81
66.5,5.911,0.577
67.5,4.344,0.641
70.5,16.895,2.489
71.5,17.092,3.022
72.5,19.596,4.24
73.5,32.972,8.475
74.5,24.414,7.842
75.5,12.227,1.621
76.5,6.179,1.054
77.5,4.586,0.579
78.5,4.385,0.452
79.5,5.395,0.574
80.5,8.299,1.528
81.5,10.145,1.603
82.5,12.257,2.145
83.5,21.979,3.04
84.5,31.112,7.855
85.5,16.982,3.687
```

```
86.5,8.355,1.82
87.5,5.304,0.476
88.5,7.154,1.058
89.5,4.62,0.776
90.5,5.787,1.207
,262.716,40.24
```

E+IPTG.csv

```
E+IPTGc,WT,WT-sd
63.5,905.081,130.881
64.5,593.962,85.541
65.5,671.263,144.889
66.5,367.261,84.026
67.5,155.75,37.047
70.5,522.824,74.554
71.5,553.266,72.633
72.5,568.988,124.858
73.5,734.275,86.648
74.5,635.396,97.21
75.5,660.628,67.778
76.5,414.782,65.594
77.5,326.426,62.478
78.5,222.558,46.877
79.5,236.466,43.836
80.5,345.038,53.205
81.5,396.226,72.239
82.5,441.869,65.91
83.5,476.839,62.281
84.5,464.427,55.665
85.5,456.611,29.975
86.5,421.753,73.024
87.5,286.891,51.174
88.5,168.242,24.018
89.5,157.051,25.593
90.5,180.356,29.883
,798.207,85.917
```

REFERENCES

Allemand, J. F., Cocco, S., Douarche, N., and Lia, G. (2006). Loops in DNA: An overview of experimental and theoretical approaches. *Eur. Phys. J. E* **19,** 293–302.

Becker, N. A., Kahn, J. D., and Maher, L. J., 3rd (2005). Bacterial repression loops require enhanced DNA flexibility. *J. Mol. Biol.* **349,** 716–730.

Becker, N. A., Kahn, J. D., and Maher, L. J., 3rd (2007). Effects of nucleoid proteins on DNA repression loop formation in *Escherichia coli*. *Nucleic Acids Res.* **35,** 3988–4000.

Becker, N. A., Kahn, J. D., and Maher, L. J., 3rd (2008). Eukaryotic HMGB proteins as replacements for HU in *E. coli* repression loop formation. *Nucleic Acids Res.* **36,** 4009–4021.

Bellomy, G., Mossing, M., and Record, M. (1988). Physical properties of DNA *in vivo* as probed by the length dependence of the *lac* operator looping process. *Biochemistry* **27,** 3900–3906.

Bintu, L., Buchler, N. E., Garcia, H. G., Gerland, U., Hwa, T., Kondev, J., and Phillips, R. (2005). Transcriptional regulation by the numbers: Models. *Curr. Opin. Genet. Dev.* **15,** 116–124.

Crothers, D. M., Drak, J., Kahn, J. D., and Levene, S. D. (1992). DNA bending, flexibility, and helical repeat by cyclization kinetics. *Meth. Enzymol.* **212,** 3–29.

Datsenko, K. A., and Wanner, B. L. (2000). One-step inactivation of chromosomal genes in *Escherichia coli* K-12 using PCR products. *Proc. Natl. Acad. Sci. USA* **97,** 6640–6645.

Davis, N. A., Majee, S. S., and Kahn, J. D. (1999). TATA box DNA deformation with and without the TATA box-binding protein. *J. Mol. Biol.* **291,** 249–265.

Du, Q., Smith, C., Shiffeldrim, N., Vologodskaia, M., and Vologodskii, A. (2005). Cyclization of short DNA fragments and bending fluctuations of the double helix. *Proc. Natl. Acad. Sci. USA* **102,** 5397–5402.

Garcia, H. G., Grayson, P., Han, L., Inamdar, M., Kondev, J., Nelson, P. C., Phillips, R., Widom, J., and Wiggins, P. A. (2007). Biological consequences of tightly bent DNA: The other life of a macromolecular celebrity. *Biopolymers* **85,** 115–130.

Hagerman, P. J., and Ramadevi, V. A. (1990). Application of the method of phage T4 DNA ligase-catalyzed ring-closure to the study of DNA structure. I. Computational analysis. *J. Mol. Biol.* **212,** 351–362.

Kahn, J. D., and Crothers, D. M. (1992). Protein-induced bending and DNA cyclization. *Proc. Natl. Acad. Sci. USA* **89,** 6343–6347.

Kramer, H., Niemoller, M., Amouyal, M., Revet, B., von Wilcken-Bergmann, B., and Müller-Hill, B. (1987). lac repressor forms loops with linear DNA carrying two suitably spaced lac operators. *EMBO J.* **6,** 1481–1491.

Kratky, O., and Porod, G. (1949). Röntgenuntersuchung gelöster fadenmoleküle. *Recl. Trav. Chim. Pays Bas* **68,** 1106–1123.

Law, S. M., Bellomy, G. R., Schlax, P. J., and Record, M. T., Jr. (1993). *In vivo* thermodynamic analysis of repression with and without looping in lac constructs. Estimates of free and local lac repressor concentrations and of physical properties of a region of supercoiled plasmid DNA *in vivo*. *J. Mol. Biol.* **230,** 161–173.

Levene, S. D., and Crothers, D. M. (1986). Ring closure probabilities for DNA fragments by Monte Carlo simulation. *J. Mol. Biol.* **189,** 61–72.

Lillian, T. D., Goyal, S., Kahn, J. D., Meyhofer, E., and Perkins, N. C. (2008). Computational analysis of looping of a large family of highly bent DNA by LacI. *Biophys. J.* **95,** 5832–5842.

Miller, J. (1992). A Short Course in Bacterial Genetics, Cold Spring Harbor Laboratory Press, New York.

Mossing, M. C., and Record, M. T., Jr. (1986). Upstream operators enhance repression of the *lac* promoter. *Science* **233,** 889–892.

Muller, J., Oehler, S., and Müller-Hill, B. (1996). Repression of lac promoter as a function of distance, phase and quality of an auxiliary lac operator. *J. Mol. Biol.* **257,** 21–29.

Oehler, S., and Müller-Hill, B. (2009). High local concentration: A fundamental strategy of life. *J. Mol. Biol.* **395,** 242–253.

Oehler, S., Eismann, E. R., Kramer, H., and Müller-Hill, B. (1990). The three operators of the lac operon cooperate in repression. *EMBO J.* **9,** 973–979.

Peters, J. P., and Maher, L. J., 3rd (2010). DNA curvature and flexibility *in vitro* and *in vivo*. *Q. Rev. Biophys.* **43,** 23–63.

Podtelezhnikov, A. A., Mao, C., Seeman, N. C., and Vologodskii, A. (2000). Multimerization-cyclization of DNA fragments as a method of conformational analysis. *Biophys. J.* **79,** 2692–2704.

Rippe, K. (2001). Making contacts on a nucleic acid polymer. *Trends Biochem. Sci.* **26,** 733–740.

Rippe, K., von Hippel, P. H., and Langowski, J. (1995). Action at a distance: DNA-looping and initiation of transcription. *Trends Biochem. Sci.* **20,** 500–506.

Sebastian, N. T., Bystry, E. M., Becker, N. A., and Maher, L. J., 3rd (2009). Enhancement of DNA flexibility *in vitro* and *in vivo* by HMGB box A proteins carrying box B residues. *Biochemistry* **48,** 2125–2134.

Shimada, J., and Yamakawa, H. (1984). Ring-closure probabilities for twisted wormlike chains. Application to DNA. *Macromolecules* **17,** 689–698.

Shore, D., and Baldwin, R. L. (1983). Energetics of DNA twisting. I. Relation between twist and cyclization probability. *J. Mol. Biol.* **170,** 957–981.

Shore, D., Langowski, J., and Baldwin, R. L. (1981). DNA flexibility studied by covalent closure of short fragments into circles. *Proc. Natl. Acad. Sci. USA* **78,** 4833–4837.

Swigon, D., Coleman, B. D., and Olson, W. K. (2006). Modeling the Lac repressor-operator assembly: The influence of DNA looping on Lac repressor conformation. *Proc. Natl. Acad. Sci. USA* **103,** 9879–9884.

Taylor, W. H., and Hagerman, P. J. (1990). Application of the method of phage T4 DNA ligase-catalyzed ring-closure to the study of DNA structure. II. NaCl-dependence of DNA flexibility and helical repeat. *J. Mol. Biol.* **212,** 363–376.

Vologodskaia, M., and Vologodskii, A. (2002). Contribution of the intrinsic curvature to measured DNA persistence length. *J. Mol. Biol.* **317,** 205–213.

Vologodskii, A. V., Zhang, W., Rybenkov, V. V., Podtelezhnikov, A. A., Subramanian, D., Griffith, J. D., and Cozzarelli, N. R. (2001). Mechanism of topology simplification by type II DNA topoisomerases. *Proc. Natl. Acad. Sci. USA* **98,** 3045–3049.

Whipple, F. W. (1998). Genetic analysis of prokaryotic and eukaryotic DNA-binding proteins in *Escherichia coli*. *Nucleic Acids Res.* **26,** 3700–3706.

Zhang, Y., and Crothers, D. M. (2003). High-throughput approach for detection of DNA bending and flexibility based on cyclization. *Proc. Natl. Acad. Sci. USA* **100,** 3161–3166.

Zhang, Y., McEwen, A. E., Crothers, D. M., and Levene, S. D. (2006a). Analysis of in-vivo LacR-mediated gene repression based on the mechanics of DNA looping. *PLoS ONE* **1,** e136.

Zhang, Y., McEwen, A. E., Crothers, D. M., and Levene, S. D. (2006b). Statistical-mechanical theory of DNA looping. *Biophys. J.* **90,** 1903–1912.

Author Index

E

F

G

N

O

W

X

Y

Z

Subject Index

M

N

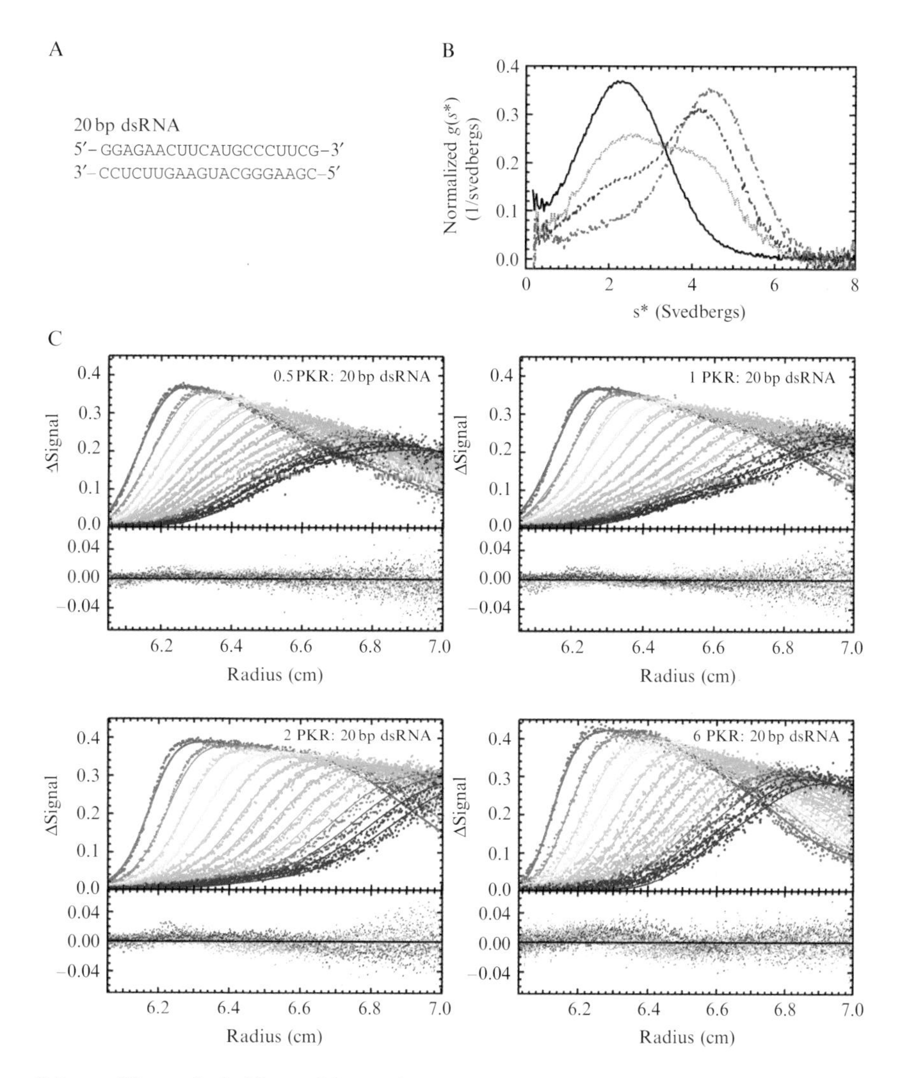

C. Jason Wong *et al.*, Figure 3.3 Sedimentation velocity analysis of PKR binding to a 20 bp dsRNA. (A) RNA structure. (B) Normalized $g(s^*)$ distributions of 1 μ*M* dsRNA (black, solid), dsRNA + 0.5 equiv. PKR (green, dot), dsRNA + 1 equiv. PKR (blue, dash), dsRNA + 2 equiv. PKR (red, dot–dash). The distributions are normalized by area. (C) Global analysis of sedimentation velocity difference curves. The data were subtracted in pairs and four data sets at the indicated ratios of PKR:dsRNA were fit to 1:1 binding stoichiometry model. The top panels show the data (points) and fit (solid lines) and the bottom panels show the residuals (points). For clarity, only every 2nd difference curve is shown. Conditions: rotor speed, 50,000 RPM; temperature, 20 °C; wavelength, 260 nm.

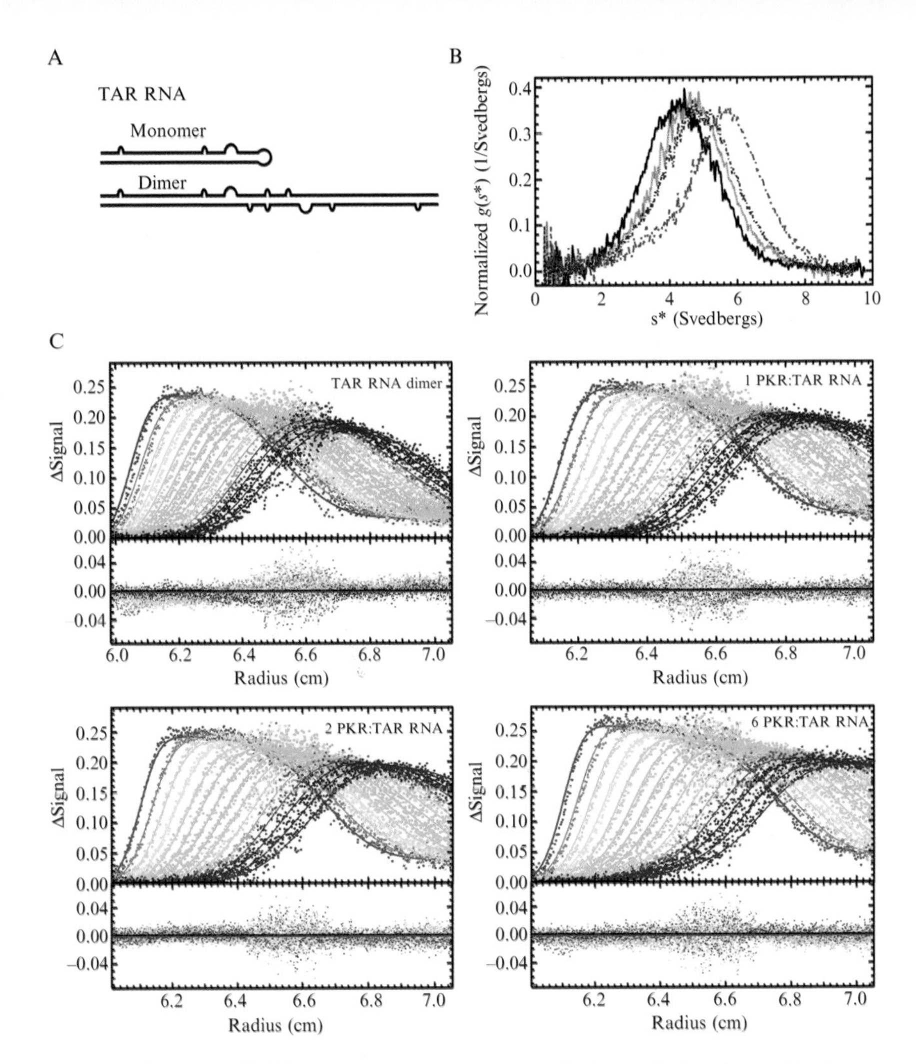

C. Jason Wong *et al.*, Figure 3.4 Sedimentation velocity analysis of PKR binding to HIV TAR RNA dimer. Measurements were performed in AU200 buffer at 20 °C and 40,000 RPM. (A) Structure of TAR monomer and dimer. (B) Plot of normalized $g(s^*)$ distributions for 0.5 μ*M* TAR dimer (black, solid) and 1 μ*M* TAR dimer plus 1 equiv. PKR (green, dot), 2 equiv. PKR (blue, dash), or 6 equiv. PKR (red, dot–dash). The distributions are normalized by area under the curve. (C) Global analysis of the sedimentation difference curves. Scans within each dataset were subtracted in pairs to remove time-invariant background and fit to a 1:2 binding model using SEDANAL. Top panels show data points and the solid lines represent fitting results using the parameters presented in Table 3.1. Residuals for each fit are shown in the bottom panels. Only every 2nd difference curve is shown for clarity. Measurements were performed in AU200 buffer at 20 °C and 40,000 RPM using absorbance detection at 260 nm.

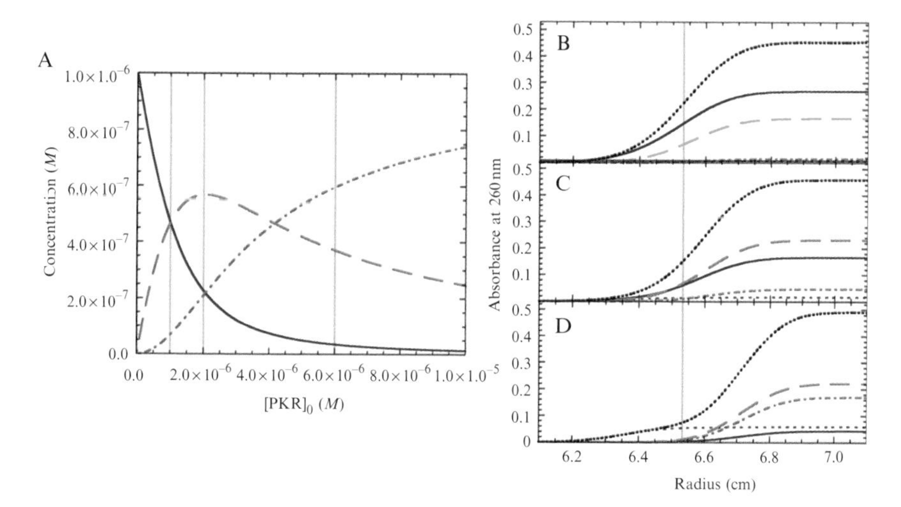

C. Jason Wong *et al.*, Figure 3.5 Species distributions for binding of PKR to TAR RNA dimer. Concentrations were calculated based on the best-fit parameters in Table 3.1 and the extinction coefficients of TAR and PKR at 260 nm. (A) Molar concentration distributions for TAR RNA dimer (blue, solid), RP complex (green, dash), and RP_2 complex (tan, dot–dash). The vertical gray lines indicate the three PKR concentrations used in the experiment depicted in Fig. 3.4. (B, C) Sedimentation velocity absorption profiles for TAR RNA dimer (blue, solid), PKR (red, dot), RP complex (green, dash), RP_2 complex (tan, dot–dash), and the total absorbance (black, dot) calculated for 1 equiv. PKR (B), 2 equiv. PKR (C), and 6 equiv. PKR (D). The profile was simulated at a time corresponding to the middle of the sedimentation run. The gray line indicates the midpoint of the total absorbance curve in (B).

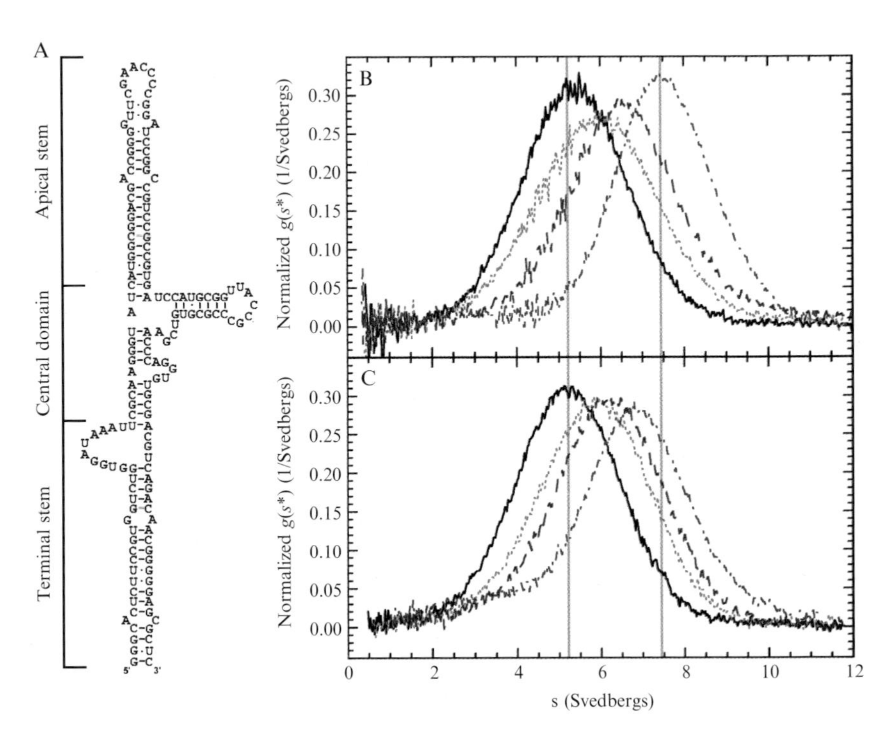

C. Jason Wong *et al.*, Figure 3.6 Sedimentation velocity analysis of PKR binding to VAI RNA: effects of divalent ion. (A) Structure of VAI. (B) Normalized $g(s^*)$ distributions obtained in the absence of Mg^{2+}: 0.4 μM VAI (black, solid), VAI + 1 equiv. PKR (green, dot), VAI + 2 equiv. PKR (blue, dash), and VAI + 6 equiv. PKR (red, dot–dash). The vertical gray line on the left corresponds to the peak of the distribution for the free VAI RNA and the line on the right corresponds to the peak in the presence of 6 equiv. of PKR. (C) Normalized $g(s^*)$ distributions obtained in the presence of 5 mM Mg^{2+}. The labeling is the same as in part B.

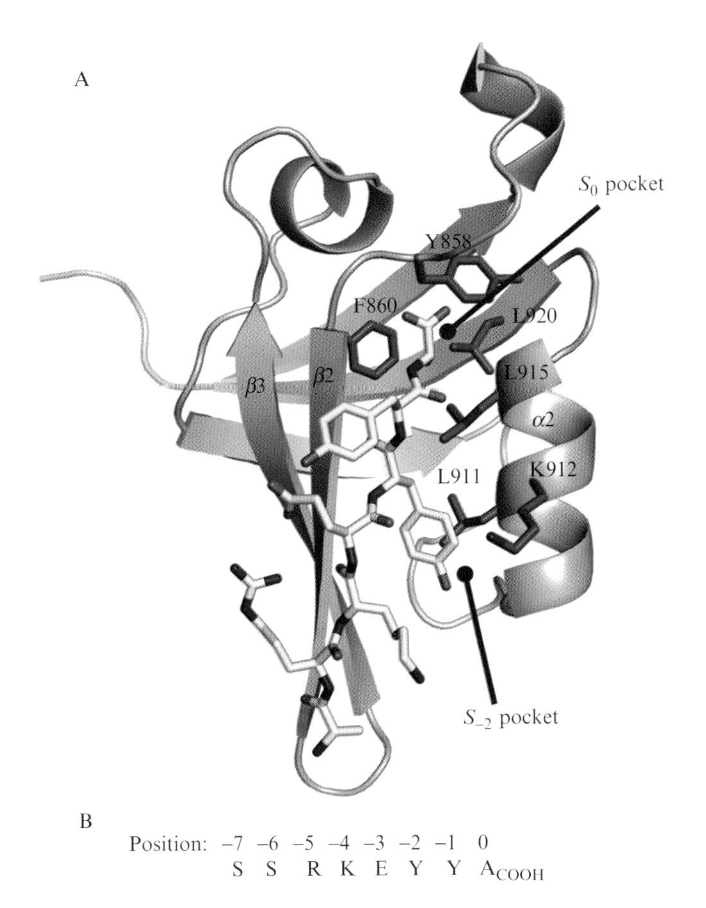

B

Position: −7 −6 −5 −4 −3 −2 −1 0
S S R K E Y Y A_{COOH}

Tyson R. Shepherd and Ernesto J. Fuentes, Figure 4.1 The structure of the Tiam1 PDZ domain bound to a model consensus peptide. (A) A cartoon representation of the PDZ/Model structure (chain B, PDB code 3KZE) is shown. The secondary structure of the PDZ domain is shown in gray and the Model peptide is colored in yellow. The specificity pockets of the PDZ domain (S_0 and S_{-2}) are labeled. Residues discussed in the text are colored red. (B) The sequence of the Model peptide used in the crystal structure determination is shown with the ligand position (P_x) shown above each residue (Shepherd *et al.*, 2010).

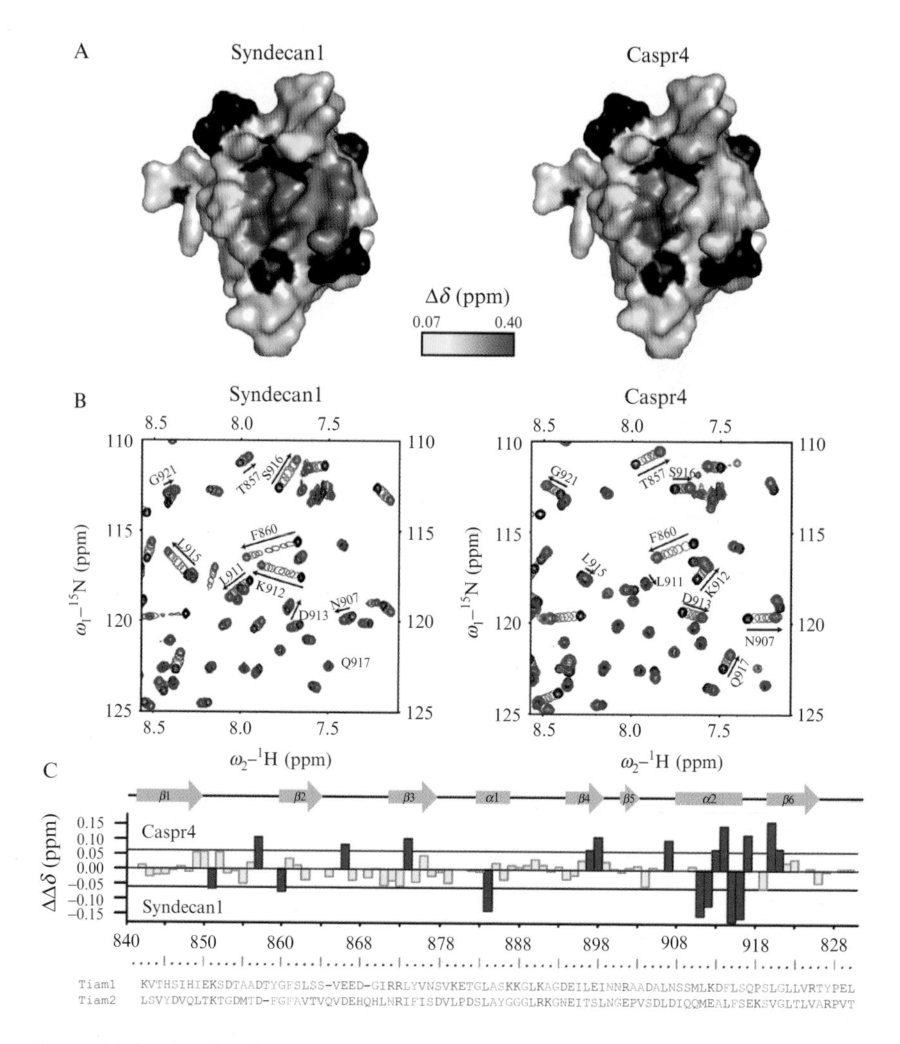

Tyson R. Shepherd and Ernesto J. Fuentes, Figure 4.2 Structural analysis of the Tiam1 PDZ domain identifies residues important for specificity. (A) The surface representation of the Tiam1 PDZ domain structure is color coded to indicate the extent of chemical shift changes $[\Delta\delta(^1\mathrm{H},^{15}\mathrm{N}) = (\Delta\delta(^1\mathrm{H})^2 + (0.4\Delta\delta(^{15}\mathrm{N}))^2)^{1/2}]$ upon titration with Syndecan1 or Caspr4 peptides. Residues shown in black could not be assigned and those in gray had a change in chemical shift that was less than the global average. Residues that had a significant change in chemical shift (greater than the global average) are colored continuously from yellow to red, where red indicates maximal changes. (B) An expanded view of ^{15}N–HSQC spectra obtained during a titration series for the Tiam1 PDZ domain with Syndecan1 and Caspr4 peptides. Labeled residues are implicated in Tiam1 PDZ specificity. (C) A histogram plot summarizing the chemical shift changes per residue in the Tiam1 PDZ domain upon titration with Syndecan1 (lower bars) and Caspr4 (upper bars). The value of each bar represents the absolute difference in chemical shift between the two complexes $[\Delta\Delta\delta = |\Delta\delta(\mathrm{Caspr4})| - |\Delta\delta(\mathrm{Syndecan1})|]$. A value of zero indicates that changes in chemical shift in the two complexes were identical in magnitude. Differences indicate unique chemical shifts changes in either complex. Residues in red underwent changes greater than 1σ from the average, while those in yellow underwent changes of less than 1σ from the average. Data taken from (Shepherd *et al.*, 2010).

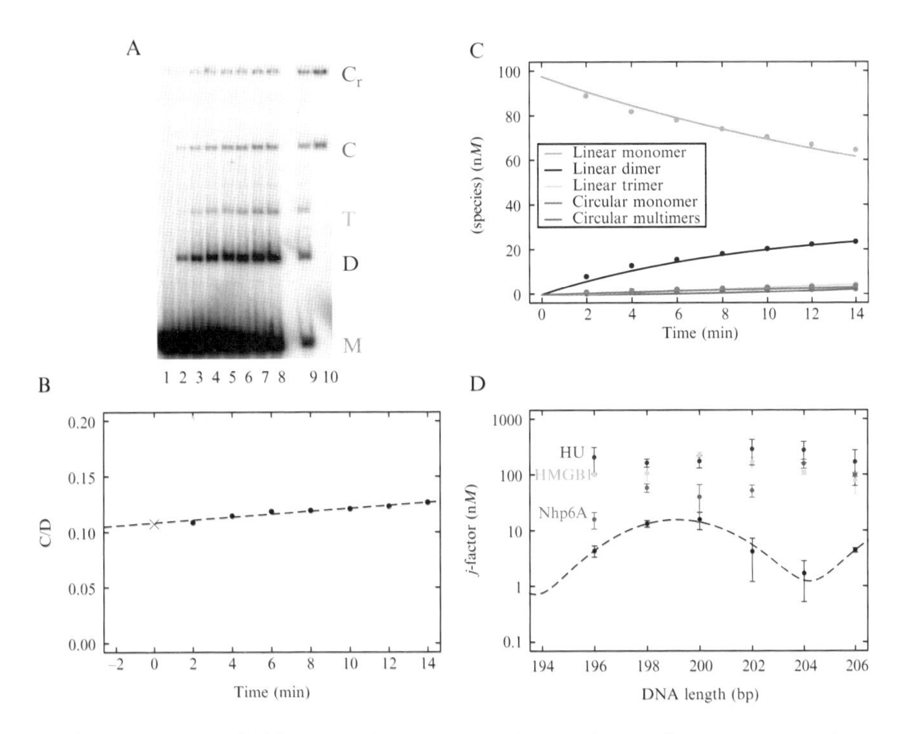

Justin P. Peters *et al.*, Figure 12.3 *Cyclization kinetics data analysis to estimate j-factor in vitro*. (A) Imaged native polyacrylamide gel. Lane 1 contains monomer (M) alone and lanes 2–8 are increasing 2 min time points showing the evolution of linear monomer (M), dimer (D), and trimer (T) as well as circular monomer (C) and circular dimer and trimer (collectively C_r). Lanes 9 and 10 are control treatment and Bal31 exonuclease treatment, respectively, of the final time point. Bal31 treatment (lane 10) provides verification of circular species. (B) Extrapolation method based on the ratio ($\mathbf{C}/\mathbf{D}$) of monomer circle products ($\mathbf{C} = C$) to the sum of all other ligation products ($\mathbf{D} = D + T + C_r$) under conditions that limit the extent of ligation reaction. The j-factor estimate is given by twice the initial probe DNA concentration multiplied by the y-intercept value (red cross). (C) Kinetic analysis applicable to full-time course of ligation reactions with multiple products and depletion of starting material. (D) Example of cyclization data published for intrinsically straight DNA (Vologodskaia and Vologodskii, 2002) and measured in our laboratory for either free DNA (fitted with WLC model) or in the presence of 40 nM Nhp6A (red), 40 nM HMGB1 (green), or 10 nM HU (blue).

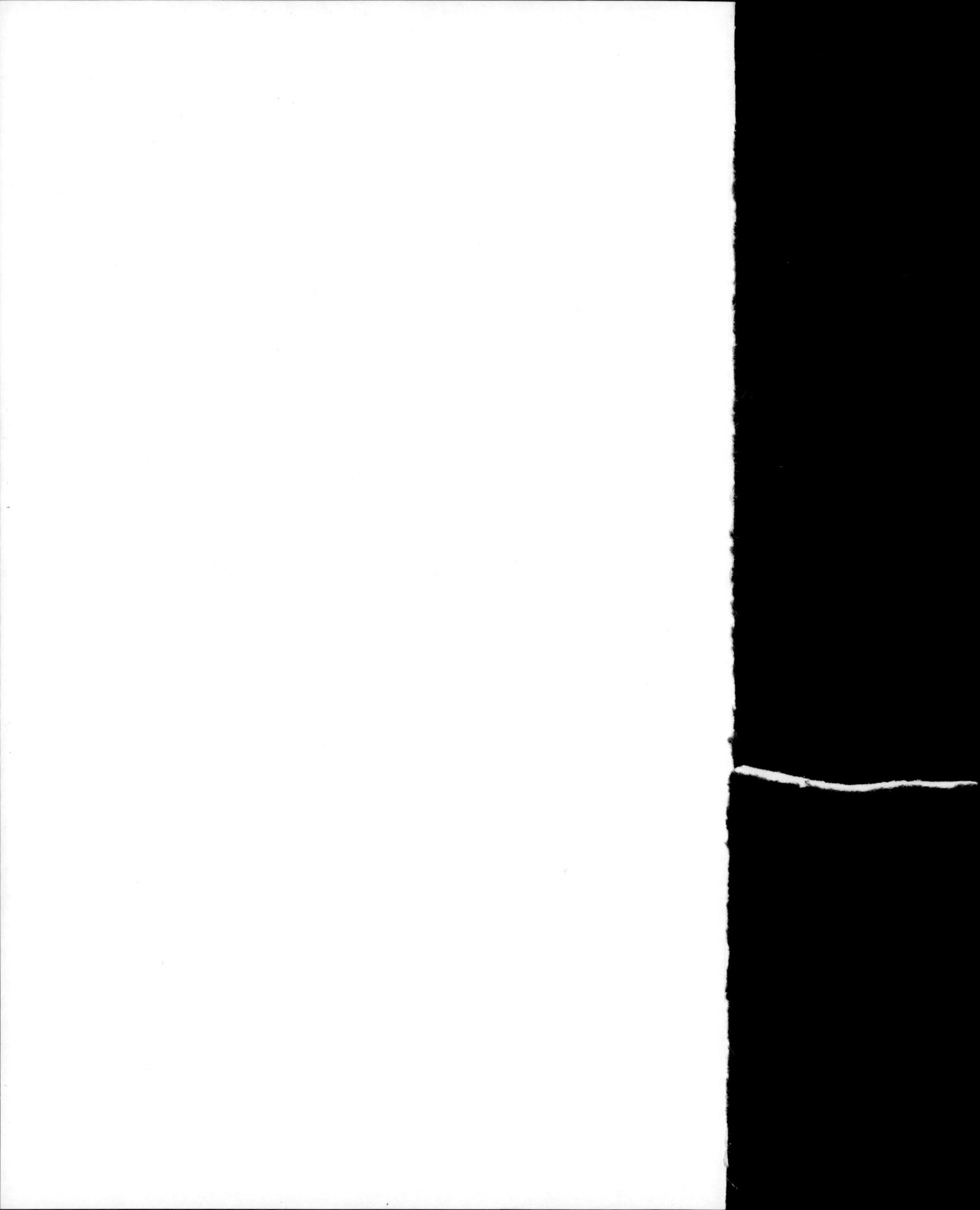